SELECTIONS

FROM

LIVES

OF

THE ENGINEERS

SAMUEL SMILES

From a Portrait by Sir George Reid

SELECTIONS

FROM

LIVES

OF

THE ENGINEERS

WITH

AN ACCOUNT OF THEIR PRINCIPAL WORKS

BY SAMUEL SMILES

edited and with an introduction

BY THOMAS PARKE HUGHES

"Bid Harbours open, Public Ways extend;
Bid Temples, worthier of God, ascend;
Bid the broad Arch the dang'rous flood contain,
The Mole projected, break the roaring main;
Back to his bounds their subject sea command,
And roll obedient rivers through the land.
These honours, Peace to happy Britain brings;
These are imperial works, and worthy kings."
POPE.

WITH PORTRAITS AND NUMEROUS ILLUSTRATIONS.

THE M.I.T. PRESS
MASSACHUSETTS INSTITUTE OF TECHNOLOGY
CAMBRIDGE, MASSACHUSETTS, AND LONDON, ENGLAND

Library of Congress Catalog Card Number: 66–19360
Printed in the United States of America

CONTENTS

v

Part II. — LIFE OF JOHN RENNIE

Part III. — LIFE OF THOMAS TELFORD

CHAPTER I.

CHAPTER II.

CHAPTER III.

CHAPTER IV.

CHAPTER V.

CHAPTER VI.

CHAPTER VII.

CHAPTER VIII.

CHAPTER IX.

INTRODUCTION

SAMUEL SMILES (1812–1904) was both a representative
and an eminent Victorian. His life spanned the reign of
Victoria, and his books, including the *Lives of the Engineers,* were immensely popular. He is best remembered for
Self-Help, published the same year, 1859, as Charles Dar-
win's *Origin of Species,* and John Stuart Mill's *Essay on
Liberty. Self-Help* sold twenty thousand copies in the
first year and almost a quarter of a million by 1900. Historian Asa Briggs, in an introduction to the seventy-second
impression (the "centenary edition"), wrote "there are
few books in history which have reflected the spirit of
their age more faithfully and successfully." [1]

Smiles reflected his age and also influenced it. He wrote
especially of engineers, inventors, and industrialists as they
transformed their environment — and society — through
rapid industrialization. His books appealed to those aspiring to achieve and who believed that self-help, thrift, duty,
and perseverance were the virtues of success. Ford Madox
Ford in *Parade's End* recalled him as an enormously popular novelist — he was in fact a reliable biographer and
historian of an age of achievement. His short character
sketches in books of the *Self-Help* genre (*Character,* 1871;
Duty, 1880; *Thrift,* 1875) are patently didactic, but Smiles
also wrote substantial biographies (Matthew Boulton, and
James Watt, 1865; Josiah Wedgwood, 1894; George Stephenson, 1857; and the *Lives of the Engineers*).

---

[1] Samuel Smiles, *Self-Help; With Illustrations of Conduct & Perseverance,* centenary edition introduced by Asa Briggs (London: John Murray,
1958).

Victorian Britain had developing men and developing areas, and Smiles struck this responsive chord. His *Lives of the Engineers,* published several years after *Self-Help,* also caught the spirit of an achieving society. A far more substantial work of historical research and literary craftsmanship, the *Lives* focused upon a field of endeavor and a kind of man epitomizing British character when Britain became the world's greatest industrial power. Prime Minister Gladstone wrote of the *Lives of the Engineers,* "it appears to me that you first have given practical expression to a weighty truth — namely, that the character of our engineers is a most signal and marked expression of British character . . ." [2] History may choose the British engineers of the Industrial Revolution as the heroic symbols of a great age whose deeds and role are comparable to the Elizabethan sea captains or the eighteenth-century landed gentry. In the meantime, however, their biographer's fame faded; his broadly conceived history of technology was neglected. Only within the past decade or so has a Smiles vogue among historians of the Victorian era developed; socially oriented history of technology remains an uncultivated field. [3]

[2] *The Autobiography of Samuel Smiles, L.L.D.,* edited by Thomas Mackay (London: John Murray, 1905), p. 256.

[3] Asa Briggs's *Victorian People: A Reassessment of Persons and Themes, 1851–67,* (Chicago: University of Chicago Press, 1955) has a chapter on Smiles that is based on Smiles's writings other than the *Lives.* Briggs comprehends the depths of Smiles's understanding which was masked by his common-sensical and sometimes popular approach to complex problems of society and psychology. The perceptive, balanced, and compassionate biography, *Samuel Smiles and his Surroundings* (London: Robert Hall, 1956) by his granddaughter, Aileen Smiles, should clarify misimpressions of Smiles. Still essential to understanding the man is his *Autobiography,* which has sections not as sharp and clear as the writings of the younger man, but it has been edited by Thomas Mackay (London: John Murray, 1905). Thomas Bowden Green's *Life and Work of Samuel Smiles* (1904) is a short compilation of selections from and about Smiles as well as publicity for the National Thrift Society of which Green was an officer. Sidney Lee wrote on "Samuel Smiles" in the *Dictionary of National Biography.*

The Life of Samuel Smiles

Smiles enjoyed fame and esteem during the decades between the Great Exhibition of 1851 and the onset of class conflict and economic recession in the seventies. Praising individualism, self-help, and demeaning generalized social programs, he won few friends among the socialist leaders and intellectuals rising to prominence as the century waned. Ignoring sex and deploring the irrational, Smiles appeared a quaint relic to those aware of Freud and the new psychology.

Self-discipline, thought by Smiles crucial for success, connoted repression within the context of the anti-Victorian reaction. Smiles found in his engineers a single-minded dedication to the cultivation of faculties necessary for professional fulfillment. For him professional fulfillment was the *sine qua non* of a rich life; the channeling of energies for purposeful work enriched the personality. Later generations, believing that the human spirit withered under the demands of organized business and technology, could interpret the work emphasis as a manifestation of guilt, and as an escape from freedom. After the burst of energy and material achievement ending about the last quarter of the nineteenth century, the sophisticated, sensitive, and affluent began to lament what had been lost on the way up — Smiles had been preoccupied by the way up.

The post-Victorian disillusionment survives today, and one of its minor manifestations is a reaction against Smiles's uncritical assertion of the obvious virtues. In reviewing the centennial edition of *Self-Help*,[4] the historian Geoffrey Barraclough observes that while Smiles tried to revitalize Victorian society with virtues he found in earlier generations, Matthew Arnold was criticizing all for which Smiles stood (his smugness, his Philistinism, and the maxim of every man for himself). The *Times Literary Supplement*, in a review of the biography of Smiles done

[4] *Spectator*, Feb. 6, 1959.

by his granddaughter, describes Smiles as a now (1956) shattered former idol of parents and schoolmasters.[5]

Before Smiles, the biographer of the engineers, can be appreciated again, the reaction against Smiles the popular moralist must be countered. The *Lives of the Engineers* was not as popular as *Self-Help,* but sales were brisk and the market broad from the publication of the first edition in 1861 to the last in 1904. This popularity has made him suspect among twentieth-century scholars. Cyril T. G. Boucher, in a recent painstaking and laudatory biography of John Rennie, a Smiles engineer, refers to the inevitable drawbacks of a study such as the *Lives* done for popular consumption. Boucher believes Smiles was on safe ground only in writing with a "popular" approach. From the technological point of view Boucher finds Smiles's biography sadly deficient.[6] L. T. C. Rolt, who has recently written a biography of Thomas Telford, another of the engineers of the *Lives,* appraises the pioneering Smiles's biographical study as brief and readable, but written in a style suggestive of a funeral oration. Because of the work's adulatory nature, Rolt does not believe the Smiles biography a true revelation of personality.[7]

The handsome and expensive[8] format given the first edition of the *Lives of the Engineers* by John Murray, its discriminating publisher, raises doubts whether the "popular consumption" envisaged by Smiles and his publisher would vulgarize the *Lives of the Engineers.* The reader may

[5] July 27, 1956, p. 444 (a review of Aileen Smiles, *Samuel Smiles and His Surroundings,* London, 1956). The reviewer would characterize Smiles — whose broad interests, activities, and complexities have been described in the biography — by an excerpt from a letter in which Smiles writes both of the funeral of an infant granddaughter and of the need of his house for repairs (this is intended to evidence banal materialism). It may manifest resignation in an age when few families did not suffer the loss of an infant child.

[6] *John Rennie, 1761–1821: The Life and Work of a Great Engineer* (Manchester: University Press, 1963), p. *ix.*

[7] *Thomas Telford* (London: Longmans, Green and Co., 1958), p. 197.

[8] Each volume was sold at a guinea (equivalent to at least $15 today).

find Smiles's knowledge of technology more than adequate for the intelligent nonspecialist. Contemporaries of Smiles with literary taste found his style that of a skilled literary craftsman — at least; few students of the art of biography believe there is a "true" revelation of personality. Smiles had a point of view and from it he saw clearly and evaluated honestly.

Smiles saw engineers achieving socially desirable goals by persevering in complex situations involving many factors. The evolution of his engineers and their works expressed the pervasive progress theme in the culture of the Victorians. He had the conceptual breadth to comprehend the many environmental forces influencing the engineers and their works. Smiles's environment was permeated by political, economic, social, and technological forces, and these forces selected for preeminence those aspiring engineers with the most favorable characteristics (thrift, duty, perseverance, etc.).

Samuel Smiles was suited by inclination and experience to write of the engineers and their environment during the Industrial Revolution — especially of civil engineers building facilities for transportation. His first intimate contact with the life of a great engineer resulted from his admiration for George Stephenson (1781–1848), the towering figure among the first generation of railway engineers. Having already found from his successful lectures to the Leeds Mutual Improvement Society that biography was philosophy teaching by example, Smiles inquired of Robert Stephenson — the engineer son of George — about writing a biography of the eminent railway engineer. Smiles envisaged George Stephenson as a suitable subject because of his striking character and the "wonderful impulse which he had given to civilization by the development of the railway locomotive," and because many men were still alive (1851) who could give first-hand information about him and his achievements.

Robert Stephenson gave his support to the project, but

doubted that many persons would be interested in a detailed biography. Stephenson thought that although they used the railway "they did not care how or by whom it is made." [9] Stephenson's support, despite his lack of enthusiasm for the endeavor, arose in part from the confidence Smiles inspired in those with whom he had contact, and also because of Smiles's outstanding qualifications for writing of railway affairs. Some years earlier Smiles had become administrative secretary of the Leeds and Thirsk Railway, and later in 1854 he became secretary to the larger South-Eastern Railway.

The enthusiastic response to the *Life of George Stephenson,* published by the prestigious John Murray of London in 1857 [10] determined the course of Smiles's writing career: He would become the earnest biographer of men who founded the industrial greatness of Britain. Smiles then turned to the research and writing of his *Lives of the Engineers.* He surveyed the past two centuries and chose those engineers about whom sufficient information existed for the writing of biography rich in personal details and the details of the environmental setting. As the footnotes in his *Lives* show, he relied upon printed sources — especially those contemporary to his subject — and drew as well upon the memories of those who had known the late eighteenth- and early nineteenth-century engineers. Smiles's methods were imaginative and energetic, for he used the transportation facilities provided by those about whom he wrote in order to collect the information he needed. On week ends free from his responsibilities as railway secretary he traveled the railways to visit libraries and the sources of oral history. He also employed researchers in distant libraries and commissioned the sketching of engineering works in their settings as a preliminary to the preparation of the cuts he deemed essential for his *Lives.* Furthermore, he persuaded relatives and business enterprises to give him

[9] *Autobiography of Smiles,* p. 163.
[10] Five editions in little more than one year.

access to the original source materials.[11] The excellence of his research has been evidenced by the heavy dependence of subsequent historians upon his biographies and the rare instances in which they have been able to justify their reworking of his material by citing his errors and setting the record straight.

Smiles was not only an indefatigable researcher but a strong organizer of his material. His organizing powers had been sharpened and tempered by responsibility for the affairs of a large railway company, and his broad conceptual powers had been developed by his varied experience. He had practiced medicine in his Scottish birthplace, Haddington, for several years after receiving his medical diploma from the University of Edinburgh in 1832. Finding his practice demanding but unrewarding and his quest for meaning and purpose unfulfilled, he applied himself in his free time to writing essays advocating social reform. A short book on *Physical Education* (a flourishing spirit is contingent upon a sound body), and a series of articles resulted: ("Health," "The Improvement and Education of Women," "Provide," "The Widows' and Orphans' Fund," and "Factory Women"). Allying himself with reform movements, he became especially active in anti-Corn Law agitation and in qualified support of the Manchester School of Economics.[12] His political sentiments, his writing ability, and a growing circle of influential friends in Eng-

[11] Smiles depended upon original sources for his life of James Brindley. Among these were the family papers and documents in the possession of Lord Ellesmere (the proprietor of the Bridgewater Canal when Smiles was writing). Sir John Rennie placed at Smiles's disposal materials for the life of his father, another of the *Lives*. Smiles drew upon letters written by Telford to Andrew Little and other friends in Eskdale to portray the early part of that engineer's career. Joseph Mitchell and George May, two of Telford's trainees, helped Smiles write of the events and achievements of later days.

[12] Smiles shared many views with Richard Cobden and the Manchester School, but he often disapproved of their tactics and even some of their principles. He drew his own distinction between economic laissez-faire, of which he approved, and social laissez-faire, which he did not. Asa Briggs has made this clear.

land secured him the editorship of the *Leeds Times,* a radical-reform newspaper. By 1842 he had had enough of newspaper editing ("I found it a rather unquiet life"), and resumed practice as a surgeon and as a part-time writer until his appointment as secretary of the Leeds and Thirsk Railway in 1845.

Smiles, then, as an experienced and able middle-aged man, had exceptional qualifications for the preparation of the *Lives of the Engineers* upon which he worked from 1858 to 1861. His medical education had exposed him to science (in Haddington he had given a series of well-attended adult-education lectures on chemistry and physics). His writing ability had been tested and proven by his editorship, his articles, and by his *Life of George Stephenson.* Through his sympathy for social reform and individual achievement he championed liberal causes and aspiring, humbly situated, young men. As a railway secretary he had learned how one maneuvered in the financial and political worlds to plan, build, and operate engineering facilities. His experience with the George Stephenson biography had convinced him that the engineer was not only his hero, but also a Victorian hero and a worthy subject. Through the *Lives of the Engineers,* Smiles believed he could provide a model for developing men and developing areas.

Smiles's enthusiasm for his subjects grew as he found them "strong-minded, resolute and ingenious men; impelled in their special pursuits by the force of their constructive instincts." [13] He found the sources rich enough for proceeding with the lives of Cornelius Vermuyden (1595–1655?) who undertook the drainage of the Fens; Hugh Myddelton (1559–1631), a goldsmith and member of Parliament who provided London with an adequate water supply; Captain John Perry (1669–1732), a naval and civil engineer who constructed canals for Peter the Great in Russia and who pioneered in land reclamation in

[13] *Autobiography of Smiles,* p. 249.

Britain; James Brindley (1716–1772), the first great British canal engineer; John Smeaton (1724–1792), who built the Eddystone lighthouse and developed the science of engineering; John Rennie (1761–1821), whose diversified achievements included mills of advanced design, fen drainage, harbor facilities, and London bridges; John Metcalf (1717–1810), who, despite blindness, laid out important roads in difficult and mountainous terrain; and Thomas Telford (1757–1834), whose notable bridges, canals, and roads merited his burial in Westminster Abbey.[14]

Smiles on Developing Men

Through these lives Smiles provided an account and philosophy of individual achievement. He wrote for those who aspired to achieve, and especially for those who wished to direct orderly achievement along socially desirable lines. (Perhaps his usefulness to those intent upon directing achievement explains the pejorative reference of a post-Victorian to Smiles as the idol of parents and schoolmasters.) The character and career of a Smiles engineer as portrayed in the three major biographies in the *Lives* can be summarized. When Telford, Rennie, and Brindley were children the rural setting was a major influence. Significantly, none was from an urban environment or from the working class. Telford's mother, the widow of a Scottish shepherd, was poverty-stricken, but she worked hard, contrived to live and "be cheerful." Brindley, too, was poor and depended upon the example of a steady and industrious mother, because his father, an agricultural laborer, "was fonder of sport than work." The young Scottish lad, Rennie, had the run of his father's substantial farm near East Lothian.

The economic status therefore varied, but there was a

---

[14] Later editions of *Lives of the Engineers* included the biography of George Stephenson as well as of Robert Stephenson, James Watt, and Matthew Boulton.

common factor more important than some others stressed
today in our sociological analyses — an exposure to the
natural environment. All these young men experienced the
untamed forces and disorder of nature; they also learned
of the patient and skilled work of millwrights and masons
who built the windmills, water wheels, bridges, and other
machines and structures that harnessed natural power
and overcame geographical barriers. If they had originated
from the industrial slums proliferating in Smiles's era,
these strong and resolute young men might have become
social reformers or labor-union leaders, but the young
heroes of Smiles were preoccupied by nature-reform.[15]

Formal training in engineering was not available for
Telford, Rennie, and Brindley. All of them apprenticed:
Brindley as a millwright; Telford as a stonemason; and
Rennie as a millwright. Rennie was unique in that he had
a university education, having studied for three years at
Edinburgh. There he had the rich experience of learning
from Professor John Robison of natural philosophy (physics) and from Professor Joseph Black of chemistry. From
them he acquired a scientific point of view and an understanding that would manifest itself in his engineering
works. The mind of young and poor Telford was cultivated
by an elderly lady of his district who gave him access to
her small library of classics. This taste, once acquired, was
never lost. His professional enthusiasm led him later to
engineering, scientific, and architectural studies (for example, as a young mason he borrowed a manuscript copy
of Professor Black's lecture notes and bought his "Experiments upon Magnesia Alba, Quicklime . . ." so that —

---

[15] Smiles not only advocated improvement of the material environment
but also social reform. The concern about social reform does not emerge
so clearly from the *Lives* as it does from others of his works. Smiles
assumed, as did many Victorians, that social progress depended upon
material improvement, but he also thought that constructive social work
had to be done. Asa Briggs in *Victorian People* and in his introduction
to *Self-Help* stresses Smiles's desire for social reform and his broad
sympathies.

among other things — he could inquire into the nature of lime). Brindley, by contrast, was barely literate, and his engineering problems were formulated and solved by complex mental processes conceivable only to thinking men unable to organize their thoughts on paper.

The remarkable careers of these engineers help explain the strong bias against formal training in engineering that persisted in Britain into the twentieth century. These were the giants of a heroic age, and succeeding generations found it unpopular to challenge the precedent. The *École des Ponts et Chausées*, the *École Mines*, the *École des Arts et Métiers*, and the *École Polytechnique* had not graduated a comparable breed in France — so at least the popular British argument ran as the nineteenth century (and the heroic age of British engineering) drew to a close.

When Telford, Rennie, and Brindley reached maturity and the heights of the engineering profession, they manifested common characteristics that can be associated with the Smiles model. The successful engineer was orderly, regular in his habits, disciplined, predictable, methodical in his problem-solving, even-tempered, and law-abiding. He had brought order out of the chaos of his natural instincts; sensuousness, self-indulgence, recklessness, untidiness and emotional outbursts were foreign to him.

The engineer was anti-nature, if by nature is meant the uncultivated, whether in a man's personality or in the world about him. The natural world connoted for the engineers — and Smiles — cold, drought, storm, flood, and famine. From the engineers' point of view the uncultivated man in his idleness, drunkenness, disorder, improvidence, and violence could not cope with this hostile world. Many descriptive passages in Smiles associate undeveloped men with uncultivated, or undeveloped, areas before the engineer appeared to drain the land, restrain the flood, bridge the river, shelter the harbor, open the channels of commerce, and harness wind and water.

The Smiles engineer, then, had cultivated first the world

within and then externalized the force of his character. An existentialist philosophy calling for asymmetry, spontaneity, violence, and disorder would have mortally threatened the temper of the men and their times. Britain and her engineers needed the psychological, sociological, and political order within which painstaking material progress could take place. Smiles instinctively knew that the agitator, demagogue, and the revolutionary flourished when passions ran strong and the structure of society crumbled, but not the patient constructor whose creativity and works depended upon constraint, balance, and attention to detail. The engineer was a pragmatist made conservative by the palpable failures of structures and machines hastily contrived.

An unusual instance of Smiles finding fault with an engineer is his reaction to Telford's espousing for a time the revolutionary doctrine of France. Telford found Thomas Paine's *Rights of Man* stimulating and persuasive and wrote to a friend in 1791 that nothing short of revolution could prevent Britain from sinking into bankruptcy, slavery, and insignificance. Smiles resorts to sarcasm (a rare expedient) to hold up this lapse for opprobrium: "Had his [Telford's] advice been asked about the foundations of a bridge, or the security of an arch, he would have read and studied much before giving it; he would have carefully inquired into the chemical qualities of different kinds of lime — into the mechanical principles of weight and resistance, and such like; but he had no such hesitation in giving an opinion about the foundations of a constitution more than a thousand years old." [16] "Fortunately," Telford's radicalism was short-lived.

Smiles believed that the political situation in Britain was conducive to evolutionary change and orderly progress. Since in his opinion there were no reactionary immovable obstacles to technological innovation and economic change, he held up for emulation the endeavors of his engineers to

[16] See p. 327.

work within the constitution and with the liberal-minded authorities. To the errant Telford, he had ironically apostrophized: "those rosy-cheeked old country gentlemen who came riding . . . to quarter sessions, and were so fond of their young Scotch surveyor — occupying themselves in building bridges, maintaining infirmaries, making roads, and regulating gaols — these county magistrates and members of Parliament, aristocrats all, were the very men who, according to Paine, were carrying the country headlong to ruin."

The Smiles engineers, then, cooperated constructively with Parliament and local authority. Since they were engineers in private practice concerned with what would be called public works projects today, there were frequent encounters with government. Rennie was a founder in 1793 of a society whose purpose was to bring together the civil (nonmilitary) engineers who gathered in London when Parliament met. They discussed their projects needing parliamentary approval and their testimony before sharp-tongued committees. Telford took permanent lodgings near Parliament in the Salopian Coffee House, and it became a focal point for the domestic and foreign engineers, establishing a close relationship between engineering and politics.

Rennie was extensively engaged in government projects, and Telford was deeply involved in one of the greatest government-subsidized regional development projects of the era — the providing of Scotland with a modern transportation system. According to Smiles, they set an example for those who must negotiate with the wily politician or the crafty contractor. (In contrast to the "glib and unscrupulous style which has since become the fashion," the engineers spoke simply, directly, honestly, and courageously.)

The frequency with which Telford and Rennie testified before or negotiated with local authorities and Parliamentary committees suggests that the estrangement be-

tween science and government, or engineering and politics,
is a product of the more recent era of organized technology.
These men were professionals in private practice who
charged fees for their services as consultants or overseers
of construction. There were no engineering departments of
great corporations, research laboratories, or research-and-
development teams to organize them and speak for them.
They were charismatic leaders who, dealing directly with
society, imposed their styles upon their immediate followers and succeeding generations.

In his indifference to organized endeavor and his emphasis upon individual achievement, Smiles assumed that
the engineers in particular and society in general could
depend upon self-regulation and self-discipline, in combination with the overarching challenge of work to structure the
relations between individuals. Because the problem of
ordering nature and supplying its deficiences by technology
seemed so obvious and so rational a means to achieve the
desirable end of improving the material conditions of
society, Smiles saw no need for elaborate social systems or
redefinition of social values. He believed, like Disraeli, that
society put too much faith in social systems and too little
in men; he thought, like William James later, that society
needed the moral equivalent of war. Smiles was convinced
that his engineers had found it in constructive work in
developing areas.

Though a proselytizer for the gospel of work, Smiles
did not begrudge his engineers their relaxation, the quiet
amenities of home, the satisfaction of warm friendships,
or the pleasures of the finer arts. He was of the opinion
that Telford occasionally overindulged his taste for writing
poetry, yet he nonetheless reprinted lines of Telford's
Eskdale ("the poem," Smiles comments, "describes very
pleasantly the fine pastoral scenery of the district").[17] He
thought that Rennie's genius was not impaired by his
flute and violin playing nor his sensible economies violated

[17] See footnote, p. 304.

by collecting rare books. Incidentally, Smiles himself was not the thorough Philistine that his critics have portrayed him. Not only did he approve of the limited cultural activities of his engineers but, after a stroke severely curtailed his other activities, he began frequent visits to London's National Gallery to copy his favorite paintings.[18]

So little is written of the wives of the engineers by Smiles that no analysis of this relationship is possible. The wives were helpmates and homemakers whose children respected and admired the head of the household. In his autobiography Smiles portrays his own home and family in the same bland colors as he did those of the engineers. Cyril T. G. Boucher, in his recent biography of Rennie, prints a letter in which that engineer appears to have longed inordinately for his intended bride's fortune, but he married another who was sensible and friendly and who bore him nine children.[19] We can imagine well-upholstered furnishings and well-repressed emotions in these homes, but from them came admirable children. John Rennie's oldest son George (1791–1866) had a distinguished career as a civil engineer and machine builder, and his second son Sir John (1794–1874) also built great bridges and railroads.

Smiles approved of his engineers' — and his own — associations with enlightened members of the aristocracy, the progressive political establishment, and the more conservative of the literary elite.[20] He deplored class conflict (conceivably his engineers of humble birth could have viewed the privileged with some resentment), but he did not object to class status.

[18] Henrietta Keddie (Sarah Tytler), *Three Generations: The Story of a Middle-Class Scottish Family* (London: John Murray, 1911), pp. 307–310.

[19] C. T. G. Boucher, *John Rennie* (Manchester: University Press, 1963), p. 14.

[20] When Telford joined Robert Southey for a tour of Scotland, the poet was the laureate and not unaware of his status as a court poet. Introduction by C. H. Herford to Robert Southey, *Journal of a Tour in Scotland* (London: John Murray, 1929), pp. vii-viii.

His relating of the distinguished and leisurely company which the elderly Telford kept — and especially his warm friendship with the poet Southey — may seem to some a betrayal by both Smiles and Telford of the plain origins and Spartan vocation of the engineers. Yet this was the twilight of life with no longer so much work to be done.

Smiles on Developing Areas

Smiles wrote of developing areas as well as of developing men. He was more than a biographer of engineers — his *Lives of the Engineers* is a history of technology, economic development, and social change. Smiles adroitly related the engineers and their works to the economic, political, social, and psychological trends that together have been called the Industrial Revolution. Smiles did not use the expression "industrial revolution." He realized, however, that a remarkable transformation "in all that relates to skilled industry" had taken place in the century after 1760 — the year when Brindley began the canal for the Duke of Bridgewater.[21]

Smiles described this remarkable transformation in terms of technological changes — especially the technology of transportation and the effect that these changes had upon the British people. His emphasis upon the technology of transportation follows from his choice of civil engineers as the subjects for the *Lives*. This choice of civil engineers, however, resulted from his awareness that canals, roads, improved harbors (and subsequently railroads) [22] were the channels along which the forces of industrialization flowed. Smiles's accounts of the effect of a new canal or

[21] In the preface to the 1861 edition Smiles defines his concept of the remarkable transformation we call the Industrial Revolution. In the revised preface of the 1874 edition the definition is fuller (London: John Murray) I, pp. 3–23.

[22] Smiles's *Life of George Stephenson* (London: John Murray, 1857) is a history of railway transportation as well as a biography. Later it became the third volume of the *Lives*.

road system upon an undeveloped area suggests that of an irrigation system upon an area long arid. In his lifetime he witnessed the dramatic impact of transportation technology upon regions denied navigable rivers, level terrain, and sheltered harbors. The steam engine, the machine tool, the coke-fired blast furnace, and the factory system were all agents of change with which Smiles was thoroughly familiar, but his *Lives* leaves the impression that these were derivative effects. Until transportation technology made possible the flow of goods, men, and ideas the effects of the new prime mover, the more complex machines and the more productive processes, were limited.

Even though the *Lives* — subtitled "The History of Inland Communication" — cannot be accepted as a balanced account of the Industrial Revolution in Britain, it should be welcomed as a counterweight to the emphasis that many industrial histories have placed upon the engine and machine components of the developing industrial system. Transportation networks systematized the development involving those components. Canals and improved roads linked blast furnace to mine and machine shop; these arteries connected the wharf laden with imported cotton with the textile mill tied to the upland water-power site; and they provided as well the means to bring fertilizer to exhausted soil whose products could be exchanged for urban goods, services, and ideas. We take the technology of transportation for granted; two centuries ago only natural transportation was commonplace.

Smiles, concerned as he was with social progress, took especial interest in the effect of the new transportation system upon backward regions in Britain. His concept of a backward or undeveloped region emphasized its natural condition. If a region were favored by nature with arable land and a salubrious clime, then the population was sustained as was the vegetation; if a region had mineral resources and natural transportation then an exchange economy might develop; or if a region were favored by a combination of rich land, and also by deep rivers or safe harbors, then urbanization was probable; otherwise a region remained depressed. Not until the artificial river, the man-made harbor, and the engineer's road appeared could a change take place. The depressed regions, then, were in a natural state and subject to cultivation by technology. Smiles's account of the transformation of many of these regions after 1760 leads to a simple but powerful conception of technology — especially transportation technology — as man's improvement upon nature.

To dramatize his account of technology, economic development, and social change, Smiles described the state of undeveloped Britain before the transportation revolution. Many of these descriptions were included in the first four parts of the *Lives* which have not been included in this new edition.[23] The accounts, however, of the regions of Manchester,[24] the Potteries,[25] and the Highlands of Scotland,[26] before the penetration of a road and canal system, are characteristic.

Smiles's knowledge of the hardships in these undeveloped regions was too deep to permit him to generalize about the bucolic satisfactions of the simple rustic or to imply a political solution. He knew that the poor inhabitants of isolated northerly regions could suffer acutely from cold while coal lay only miles away at the pit head. Famine could strike a region and there was little recourse to the plentiful food only hundreds of miles distant. He realized that mineral deposits and water-power sites remained largely unexploited in regions denied transportation facilities by nature. Inland iron manufacturers suffered from a fuel famine while not too distant coal went unmined and timber rotted. There were no transportation networks to

[23] See especially "Early Roads and Modes of Travelling" and "Bridges, Harbours, and Ferries" in the edition of 1861 (I, pp. 155–304).

[24] See p. 54.

[25] See pp. 158–159.

[26] See pp. 336–338.

rectify for these regions the random dispersion of resources by nature. Many men may have been created equal — few regions have.

The population in these naturally deprived areas bore the mark of the environment. In the eighteenth century people of remote and isolated Dartmoor in Devonshire remained "a piece of England of the Middle Ages left behind on the march." Old manners, customs, language, and superstitions remained, including the belief in three kinds of witches and the use of agricultural implements long forgotten in advanced areas.[27] The Potteries district before the construction of the Grand Trunk Canal and the rise of dependent industries such as that of Josiah Wedgwood was populated by people as "rough as their roads," with coarse manners and brutal amusements. The landscape was broken by clay hovels housing half-naked males and dirty children layered with rags. These were not scenes from which Smiles turned in puritanical righteousness; these were the settings within which he anticipated the drama of creative technology. This was the challenge that he believed justified the self-help and self-discipline of his engineers. He gives a synopsis of the drama: [28]

> [about the middle of the last century] we found a country [in Scotland] without roads, fields lying uncultivated, mines unexplored, and all branches of industry languishing, in the midst of an idle, miserable, and haggard population. Fifty years passed, and the state of the Lowlands had become completely changed. Roads had been made, canals dug, coal-mines opened up, iron works established; manufactures were extended in all directions; and Scotch agriculture, instead of being the worst, was admitted to be the best in the island.

From Smiles's point of view, the British Industrial Revolution becomes a set of regional revolutions; some in sequence, some contemporaneous. This rapid industrialization of Britain between 1760 and 1860 becomes the spread of an ocean and river — or nature-based — economy into

[27] *Lives* (1861), I, p. 192.
[28] See p. 332.

the hinterland by means of artificial transportation and the creation there of a technology-based economy. The penetration of the remoter regions with their energy and material resources, and the linking of them with the more ancient centers of culture created an industrial system of hitherto unknown power to create and control the material environment. Smiles summarized the situation: [29]

> It mattered not that England was provided with convenient natural havens situated on the margin of the world's great highway, the ocean, and with fine tidal rivers capable of accommodating ships of the largest burden. Unless the country inland could be effectually connected with these ports and tidal rivers, the general extension of commerce and its civilizing influences upon the community must necessarily be in a great measure prevented.

London can be taken to represent the earlier nature-based society; Manchester, the technology-based industrial system. Located at the navigable head of the Thames in a fertile interior region, London thrived with a natural system of inland communication. Thames tides carried sailing ships up the narrow channel to the inland port. Like so many European cities before the transportation revolution, London's economic and cultural ties were closer to some continental port cities than to the remote interior of her own nation. Smiles points out that in 1700 goods came cheaper by sea to London than by land from Manchester.

He attributes the extraordinary growth of Manchester, in the decades after the Liverpool-Manchester canal system was built, to the linking of this community with its nearby coal mines, its neighboring water-power sites, and its tradition of small-scale textile activity, to the world market through Liverpool and to the interior through the connecting Grand Trunk Canal. In his, *Life of George Stephenson,* Smiles describes the acceleration of this growth after the completion of the Liverpool-Manchester Railway in 1830.

[29] *Lives* (1861), I, p. 299.

The extension of the transportation system went beyond the termini of canals and roads. The dynamism of the continuous flow of goods along the canals and roads characterized the engineering style of Brindley and Rennie, and their style carried over from transportation systems to their materials-handling systems. Brindley did more than plan and construct a canal for the Duke of Bridgewater from his mines at Worsley to the market at Manchester — he constructed a dynamic system. He provided that the drainage water from the Worsley mines supply the canal at the summit level. He planned a materials-handling system involving the extension of the canal into the Worsley mine so that boats could be loaded by gravity feed from the face of the coal seams. He used a water-wheel hoist to raise the coal from canal terminus to street level in Manchester. He systematically organized the construction work on his canals so that tools and materials could be transported along the completed sections to those under construction. He designed a floating blacksmith shop and a floating carpenter shop that moved with the work gangs. Engineering of the eighteenth century, when described and analyzed by Smiles, loses the connotation of quaintness found in some histories.

John Rennie also exploited the possibilities of labor-saving machines for materials handling. He recommended steam-driven cranes and a system of railways for his new London docks. Rennie's recommendations were not accepted because (in the opinion of Smiles) "labour-saving processes were not then valued as they now are." [30] At the London docks of the East India Company, however, Rennie installed machinery for transporting immense blocks of mahogany by a system of railways and locomotive cranes. The system paid for itself in six months (the resistance of the dockworkers to these innovations must have been bitter).

Rennie's philosophy of labor-saving by investment in

[30] See p. 260.

systematic materials handling also manifested itself in his contribution to the design of the Albion Mills at London. Rennie, recently begun on his engineering career, was employed in 1784 by Boulton and Watt to design and install the machinery of the mills. Powered by two 75-hp steam engines, the mill was probably the largest concentration of power in British history. He located the steam-driven machine components (for grinding, fanning, sifting, dressing, hoisting, lifting, loading, unloading) to minimize energy expenditure. His philosophy of design strikingly paralleled that of Oliver Evans, who was building his highly mechanized mill at Redclay Creek in the United States about the same time.

The common characteristic of the transportation and materials-handling designs of Smiles's engineers offers a clue to understanding a great change wrought by the Industrial Revolution — the transition from a relatively static economy to a dynamic one. The movement of heavy goods over substantial distances at a regular velocity on canals and later on railroads, impossible with pack horses over rough terrain or wagons over poor roads, was literally dynamic. If it were possible to measure the mass of goods by the distance moved over the interior of Britain during the year 1760 and to compare this figure with the comparable one for 1850, a quantitative measure of the dynamic transformation of the British economy would result. This dynamism was in turn in accord with the increased capital investment and the imperative demand for rapid turn-over resulting from an awareness of the continuity of interest payments.[31]

Smiles understood the importance of the economic factor in technological change. Writing of the consulting done

[31] While the engineers' awareness of the need for constant flow of raw materials and commodities is reflected in their work and while the *Lives* shows their concern for obvious labor-saving devices, there is little in the *Lives* that would help understand the attitude of the engineers toward the nuances of labor cost. The *Lives* leaves no doubt, however, that economic factors greatly influenced the engineers' style.

by his engineers in the eighteenth century, he observed that their excellent plans often remained sterile because local governments or private entrepreneurs could not raise the requisite capital. An undrained district, a dangerous and inaccessible harbor, or a decaying bridge remained, for "the country was too poor or too spiritless to undertake their improvements on any comprehensive scale . . . much excellent and carefully-considered advice" brought no action until the country had become richer and "a new race of capitalists, engineers, and contractors" appeared.[32] The closing decades of the century brought the riches and the "new race," as the Smiles narrative demonstrates.

James Brindley worked with capital raised by landed aristocracy (Duke of Bridgewater) with entrepreneurial drive or by industrialists whose vision extended beyond their own enterprise to related regional factors (Josiah Wedgwood). They could draw upon capital accumulated during an earlier commercial revolution and from colonial trade. There was in addition a feedback relationship which Smiles's history lucidly illustrates. The Duke of Bridgewater could repay the advances from his own tenantry, and the loans from his London bankers from the income of coal mines, after the coal could be transported by canal.

Josiah Wedgwood (realizing that reduced transportation cost of raw materials would repay his efforts) helped raise the capital for the Grand Trunk Canal. Confident that the connection provided by the Grand Trunk from the Potteries district to the northwestern port of Liverpool, the southwestern port of Bristol, and the eastern port of Hull would dramatically benefit pottery manufacture, he built between 1769 and 1771 the "first manufactory of the kind" on the banks of the yet-to-be-completed Grand Trunk. His judgment was good, for gypsum from Northwich, clay and flints from the seaports could be brought in, and the finished pottery could be moved out at about one fourth the previous rates. Furthermore, the raw materials could

[32] *Lives* (1861), II, pp. 49–50.

be transported in quantities inconceivable by wagon and packhorse.

Smiles illustrated other ways in which capital investments in technology proved self-amortizing. Usually his case history involved transportation technology, but he also referred on occasion to this characteristic of land-reclamation projects.[33] In the case of the land drainage by John Rennie, Smiles cites figures showing that the investment of about £580,000 by local landowners in drainage of the Lincoln Fens increased the land value by not less than £81,000 per annum (allowing for five percent interest).

A substantial number of Rennie's plans for improvements upon natural harbors and construction of unloading docks were solicited and executed by private investors anticipating that the increase in trade would repay them or their creditors. A company formed by London merchants in 1800 appointed him engineer for the construction of the London Docks, completed in 1805. In 1803 another company persuaded of the economy of floating docks and capitalized at £660,000, named him co-engineer for construction of docks to accommodate vessels of the East India Company.

As Smiles observes, "it would occupy much space to mention in detail the various harbours in the United Kingdom which Mr. Rennie was employed to examine, report upon, and improve," but the capacity of local and private capital to support some of the greatest public works of the engineers is demonstrated in concrete detail by Smiles's history. Smiles did not proselytize for the effectiveness of private enterprise — he assumed it. The prominent role of local initiative in carrying out engineering projects as described by Smiles helps make understandable the way in which an industrial system could evolve rationally without central planning, (the Liverpool

[33] Land reclamation was the major activity of the Smiles engineer that cannot be subsumed under the transportation-technology rubric.

merchants supported the Liverpool-Manchester Canal when cotton destined for the interior accumulated on their docks). The localities felt the pressure of the dynamic economic forces as canals, harbors, and roads tied them into the evolving production system. The response of the locality was determined in part by the obvious bottlenecks and potentials resulting from the new relationship with the expanding economy.

Smiles's Influence

There is no denying that Smiles, determinedly cheerful as he was, looked away from the slums that evolved around the nodal points in the growing industrial system — also without central planning. This does not show, however, that he was disinterested in social effects. After recounting the drama of transformation of the material environment by technology he seldom failed to describe in glowing terms the transformation of the society within that environment. He assumed — as did many Victorians — that material welfare would nourish the human spirit and prove the salvation of society. The *Lives* reveals that he had to look primarily to the rural regions for evidence of the progress he so earnestly desired. The transformation of rural Scotland, the Potteries countryside, the Lincoln Fens, and the Manchester-Liverpool district (compare for startling contrast with Friedrich Engels' *The Condition of the Working Class in England* [urban Manchester] *in 1844*) was edifying and inspiring.

His account of technology, economic development, and social change in the Scottish highlands, resulting from the new road system, was as likely to encourage aspiring regions as the lives of his eminently successful engineers were to encourage aspiring young men. With the introduction, over a period of eighteen years, of the 920 miles of capital roads,

. . . manure was no longer carried to the field on women's backs.
Sloth and idleness disappeared before the energy, activity, and industry which were called into life by the improved communications;
better built cottages took the place of the old mud biggins with a
hole in the roof to let out the smoke; the pigs and cattle were treated
to a separate table; the dunghill was turned to the outside of the
house; tartan tatters gave place to the produce of Manchester and
Glasgow looms; and very soon few young persons were to be found
who could not both read and write English.[34]

From Telford's report Smiles added: "About two hundred thousand pounds had been granted in fifteen years . . .
the means of advancing the country at least a century."

It was no accident therefore that Smiles and the philosophy he expounded in *Self-Help* and the history he re-
counted in the *Lives of the Engineers* enjoyed remarkable
popularity in the developing regions of the late nineteenth
century.[35] Smiles was diffusing the ethos of achievement
pervading an earlier Britain to the nations with rising
expectations. In Italy he was especially popular and in 1879
had private audiences with both Garibaldi and Queen
Margherita. It was symbolic of Smiles's influence that
the Italian Queen received him when Victoria had not paid
him the honor. By 1880 the undeveloped regions were
abroad; Britain's new heroes of social progress were the
organizers of labor and the reformers of the man-made
environment, not the organizers of energy and matter. In
Italy, by contrast, a deputy of the Italian Parliament and
the president of the press association wrote that "amongst
contemporary English authors there is no one better known
or more heartily admired in Italy than yourself." [36] The
Italians also informed him that "his story of the triumphs
and heroisms of English industry was educating the rising
generation of Italians in honesty, courage, and perseverance." [37]

---

[34] See pp. 346–347.

[35] In addition to American editions there were many translations.

[36] *Autobiography of Smiles,* p. 410.

[37] *Ibid.,* p. 411.

There are many other instances of Smiles's effectiveness
as a diffuser of the achievement ethic to undeveloped
regions. The King of Serbia bestowed upon him the Cross
of Knight Commander of the Order of St. Sava; the
Khedive of Egypt placed texts from Smiles's *Self-Help*
(not the Koran) upon the walls of one of his palaces.
Whether the King and the Khedive read the more substantial version of Smiles's interpretation in the *Lives,* we
do not know. Curious juxtapositions arise among those
influenced by him. Samuel Insull, creator of an American
electric-utilities empire, was inspired by the *Lives* and
Self-Help, absorbing the message without reservation and
making it work.[38] Mao Tse-tung also came under the influence of Smiles, for Mao's highly regarded professor of
philosophy at the Changsha Normal School and future
father-in-law had been deeply persuaded by Smiles's
ethic when studying in England.[39]

Mao and other leaders in the developing regions of the
world are now far more interested in history's revolutionary
political reformers than in the lives of her engineers.
Perhaps Mao may not have encountered "rosy-cheeked
old country gentlemen . . . occupying themselves in building bridges, maintaining infirmaries, making roads, and
regulating gaols [the exception, perhaps]." These gentlemen, according to Smiles, brought Telford back to his
political senses and his engineering works; perhaps the
establishment of a new political order will bring the leaders
in the developing areas back to fundamental problems
faced by Smiles's engineers.

[38] Forrest McDonald, *Insull* (Chicago: Univ. of Chicago, 1962), p. 10.

[39] Stuart Schram, ed., *The Political Thought of Mao Tse-tung* (New
York: Praeger, 1963), pp. 11–12. An early article of Mao's was on physical
culture — the same subject as Smiles's first book.

This edition of Samuel Smiles's *Lives of the Engineers* is taken from an 1862 revision of the first edition published by John Murray in 1861. The first edition had two volumes including the lives of Brindley, Smeaton,[40] Rennie, and Telford, and shorter biographies of Cornelius Vermuyden, Captain John Perry, Sir Hugh Myddelton, John Metcalf, William Edwards, and Andrew Meikle. In conjunction with the shorter biographies, Smiles gave a richly illustrated social history of inland transportation, land reclamation, and water supply. These engineers, the subjects of the shorter biographies, flourished before the beginning of the canal phase of the British Industrial Revolution (around 1760).

The shorter biographies, the biography of John Smeaton, and early social history have been removed from this new edition. The selections included are from the lives of Brindley, Rennie, and Telford. Where deletions have been made, summaries of the deleted portion have been supplied by the editor. The cuts and engravings from the original have been reproduced. Percival Skelton, R. P. Leitch, and E. M. Wimperis did many of the drawings; James Cooper, the cuts. The result is a one-volume edition of Smiles that includes much of what he wrote in the *Lives* on three distinguished engineers who flourished during the canal era of the British Industrial Revolution. After 1861 new editions of the *Lives* were published. First the volume on George and Robert Stephenson was added (1862) and then a volume on James Watt and Matthew Boulton. The Stephenson volume carried the Smiles narrative on inland

[40] Because no full biography of John Smeaton exists, the editor reluctantly removed Smiles's eighty-five-page biography. The decision accorded with two criteria governing the creation of a one-volume edition: to include those sections that give Smiles's concept of the role of inland transportation technology in the Industrial Revolution and to present his social history of technology. The Smeaton biography—especially the lengthy section on the construction of the Eddystone Lighthouse—does not fulfill these objectives as well as do the biographies of Brindley, Rennie, and Telford. The Smeaton biography has a different emphasis and an integrity meriting separate publication.

transportation into the railway phase of the Industrial Revolution, but with the addition of Boulton and Watt Smiles shifted from the history of inland navigation to that of the steam engine.[41]

In preparing this new edition of selections from the *Lives,* the editor has used asterisks in the text to refer to his own notes, which will be found beginning on p. 418. Smiles's footnotes should prove helpful to those who wish to explore the earlier literature; the editor's bibliography lists some of the more recent studies of the engineers. In his notes, the editor has indicated some of the errors in fact and questionable generalizations of Smiles that have been brought to light by the recent studies.

The editor has placed brackets about the summaries he has prepared to replace deleted sections. Quotations within the summaries are from Smiles, unless otherwise indicated.

[41] See the Selected Bibliography, pp. 430–435 for a fuller account of Smiles's works.

JAMES BRINDLEY

Engraved by W. Holl, after the portrait by F. Parsons

LIFE OF JAMES BRINDLEY.

CHAPTER I.

THE WHEELWRIGHT'S APPRENTICE.

In the third year of the reign of George I., whilst the British Government were occupied in extinguishing the embers of the Jacobite rebellion which had occurred in the preceding year, the first English canal engineer was born in a remote hamlet in the High Peak of Derby, in the midst of a rough country, then inhabited by quite as rough a people.

The nearest town of any importance was Macclesfield, where a considerable number of persons were employed, about the middle of last century, in making wrought buttons in silk, mohair, and twist—such being then the staple trade of the place. Those articles were sold throughout the country by pedestrian hawkers, most of whom lived in the wild country called " The Flash," from a hamlet of that name situated between Buxton, Leek, and Macclesfield. They squatted on the waste lands and commons in the district, and were notorious for their wild, half-barbarous manners, and brutal pastimes. Travelling about from fair to fair, and using a cant or slang dialect, they became generally known as " Flash men," and the name still survives. Their numbers so grew, and their encroachments on the land became so great, that it was at length found necessary to root them out ; but for some time no bailiff was found sufficiently bold to attempt to serve a writ in the district. At length an officer was found who undertook to arrest several of

them, and other landowners, taking courage, followed
the example; when those who refused to become tenants
left, to squat elsewhere; and the others then consented
to settle down to the cultivation of their farms.   Another set of travelling rogues belonging to the same

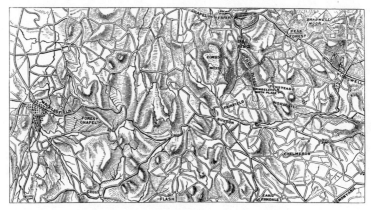

BRINDLEY'S NATIVE DISTRICT.   [Ordnance Survey.]

neighbourhood was called the "Broken Cross Gang,"
from a place called Broken Cross, situated to the south-
east of Macclesfield.   Those fellows consorted a good deal
with the Flash men, frequenting markets and travelling
from fair to fair, practising the pea-and-thimble trick,
and enticing honest country people into the temptation
of gambling.   They proceeded to more open thieving
and pocket-picking, until at length the magistrates of
the district took active measures to root them out of
Broken Cross, and the gang became broken up.   Such
was the district and such the population in the neighbourhood of which our hero was born.

James Brindley first saw the light in a humble cottage
standing about midway between the hamlet of Great
Rocks and that of Tunstead, in the liberty of Thornsett,
some three miles to the north-east of Buxton.   The
house in which he was born, in the year 1716, has long
since fallen to ruins—the Brindley family having been
its last occupants.   The walls stood long after the roof

had fallen in, and at length the materials were removed
to build cowhouses; but in the middle of the ruin there
grew up a young ash tree, forcing up one of the flags
of the cottage-floor.   It looked so healthy and thriving
a plant, that the labourer employed to remove the stones
for the purpose of forming the pathway to the neighbouring farm-house, spared the seedling, and it grew
up into the large and flourishing tree, six feet nine
inches in girth, standing in the middle of the Croft,
and now known as " Brindley's Tree."   This ash tree is
nature's own memorial of the birth-place of the engineer,
and it is the only one as yet raised to the genius of
Brindley.

BRINDLEY'S CROFT.[1]

[By Percival Skelton, after a Sketch by Mrs. Fleming.]

Although the enclosure is called Brindley's Croft, this
name was only given to it of late years by its tenant, in
memory of the engineer who was born there.   The state-

---

[1] The site of the Croft is very elevated, and commands an extensive
view as far as Topley Pike, between
Bakewell and Buxton, at the top of
what is called the Long Hill.   Topley
Pike is behind the spectator in looking at the Croft in the above aspect.
The rising ground behind the ash
tree is called Wormhill Common,
though now enclosed.   The old road
from Buxton to Tideswell skirts the
front of the rising ground.

ment made in Mr. Henshall's memoir of Brindley,[1] to the effect that Brindley's father was the freehold owner of his croft, does not appear to have any foundation; as the present owner of the property, Dr. Fleming, informs us that it was purchased, about the beginning of the present century, from the heirs of the last of the Heywards, who became its owners in 1688. No such name as Brindley occurs in any of the title-deeds belonging to the property; and it is probable that the engineer's father was an under-tenant, and merely rented the old cottage in which our hero was born. There is no record of his birth, nor does the name of Brindley occur in the register of the parish of Wormhill, in which the cottage was situated; but registers in those days were very imperfectly kept, and part of that of Wormhill has been lost.

It is probable that Brindley's father maintained his family by the proceeds of his little croft, and that he was not much, if at all, above the rank of a cottier. It is indeed recorded of him that he was by no means a steady man, and was fonder of sport than of work. He went shooting and hunting, when he should have been labouring; and if there was a bull-running within twenty miles, he was sure to be there. The Bull Ring of the district lay less than three miles off, at the north end of Long Ridge Lane, which almost passed his door; and of that place of popular resort Brindley's father was a regular frequenter. These associations led him into bad company, and very soon reduced him to poverty. He neglected his children, not only setting before them a bad example, but permitting them to grow up without education. Fortunately, Brindley's mother in a great measure supplied the father's shortcomings: she did what she could to teach them what she knew, though that was but small; but, perhaps more important still, she encou-

---

[1] Kippis's 'Biographia Britannica,' Art. Brindley.

raged them in the formation of good habits by her own steady industry.[1]

The different members of the family, of whom James was the eldest, were thus under the necessity of going out to work at a very early age to provide for the family wants. James worked at any ordinary labourer's employment which offered until he was about seventeen years old. His mechanical bias had, however, early displayed itself, and he was especially clever with his knife, making models of mills, which he set to work in little mill-streams of his contrivance. It is said that one of the things in which he took most delight when a boy, was to visit a neighbouring grist-mill and examine the water-wheels, cog-wheels, drum-wheels, and other attached machinery, until he could carry away the details in his head; afterwards imitating the arrangements by means of his knife and such little bits of wood as he could obtain for the purpose. We can thus readily understand how he should have turned his thoughts in the direction in which we afterwards find him employed, and that, encouraged by his mother, he should have determined to bind himself, on the first opportunity that offered, to the business of a millwright.

The demands of trade were so small at the time, that Brindley had no great choice of masters; but at the village of Sutton, near Macclesfield, there lived one Abraham Bennett, a wheelwright and millwright, to whom young Brindley offered himself as apprentice; and in the year 1733, after a few weeks' trial, he became bound to that master for the term of seven years. Although the employment of millwrights was then of a very limited character, a great deal of valuable practical

---

[1] Brindley's father seems afterwards to have somewhat recovered himself; for we find him, in 1729, purchasing an undivided share of a small estate at Lowe Hill, within a mile of Leek, in Staffordshire, where he had before gone to settle; and he contrived to realise the remaining portion before his death, and to leave it to his son James. None of the Brindley family remained at Wormhill, and the name has disappeared in the district.

information was obtained whilst carrying on their business. The millwrights were as yet the only engineers.
In the course of their trade they worked at the foot-lathe,
the carpenter's bench, and the anvil, by turns; thus
cultivating the faculties of observation and comparison,
acquiring practical knowledge of the strength and
qualities of materials, and dexterity in the handling of
tools of many different kinds. In country places, where
division of labour could not be carried so far as in the
larger towns, millwrights were compelled to draw largely
upon their own resources, and to devise expedients to
meet pressing emergencies as they arose. Necessity
thus made them dexterous, expert, and skilful in mechanical arrangements, more particularly those connected
with mill-work, steam-engines, pumps, cranes, and such
like. Hence millwrights in those early days were looked
upon as a very important class of workmen. The nature
of their business tended to render them self-reliant, and
they prided themselves on the importance of their
calling. On occasions of difficulty the millwright was
invariably resorted to for help; and as the demand for
mechanical skill arose, in the course of the progress of
manufacturing and agricultural industry, the men trained
in millwrights' shops, such as Brindley, Meikle, Rennie,
and Fairbairn, were borne up by the force of their practical skill and constructive genius into the highest rank
of skilled and scientific engineering.*

Brindley, however, only acquired his skill by slow
degrees. Indeed, his master thought him slower than
most lads, and even stupid. Bennett, like many well-
paid master mechanics at that time, was of intemperate
habits, and gave very little attention to his apprentice,
leaving him to the tender mercies of his journeymen,
who were for the most part a rough and drunken set.
Much of the lad's time was occupied in running for beer,
and when he sought for information he was often met
with a rebuff. Skilled workmen were then very jealous

of new hands, and those who were in any lucrative employment usually put their shoulders together to exclude those who were out. Brindley had thus to find out nearly everything for himself, and he only worked his way to dexterity through a succession of blunders. He was frequently left in sole charge of the wheelwrights' shop—the men being absent at jobs in the country, and the master at the public-house, from which he could not easily be drawn. Hence, when customers called at the shop to get any urgent repairs done, the apprentice was under the necessity of doing them in the best way he could, and that often very badly. When the men came home and found tools blunted and timber spoiled, they abused Brindley and complained to the master of his bungling apprentice's handiwork, declaring him to be a mere "spoiler of wood." On one occasion, while Bennett and the journeymen were absent, he had to fit in the spokes of a cart-wheel, and was so intent on completing his job that he did not find out that he had fitted them all in the wrong way until he had applied the gauge-stick. Not long after this occurrence, Brindley was left by himself in the shop for an entire week, working at a piece of common enough wheelwright's work, without any directions; and he made such a "mess" of it, that on the master's return he was so enraged that he threatened, there and then, to cancel the indentures and send the young man back to farm-labourer's work, which Bennett declared was the only thing for which he was fit.

Brindley had now been two years at the business, and in his master's opinion had learnt next to nothing; though it shortly turned out that, notwithstanding the apprentice's many blunders, he had really groped his way to much valuable practical information on matters relating to his trade. Bennett's shop would have been a bad school for an ordinary youth, but it proved a prolific one for Brindley, who was anxious to learn, and

determined to make a way for himself if he could not
find one. He must have had a brave spirit to withstand
the many difficulties he had to contend against, to learn
dexterity through blunders, and success through defeats.
But this is necessarily the case with all self-taught workmen ; and Brindley was mainly self-taught, as we have
seen, even in the details of the business to which he had
bound himself apprentice.

In the autumn of 1735 a small silk-mill at Macclesfield, the property of Mr. Michael Daintry, sustained
considerable injury from a fire at one of the gudgeons
inside the mill, and Bennett was called upon to execute
the necessary repairs. Whilst the men were employed
at the shop in executing the new work, Brindley was
sent to the mill to remove the damaged machinery,
under the directions of Mr. James Milner, the superintendent of the factory. Milner had thus frequent occasion to enter into conversation with the young man, and
was struck with the pertinence of his remarks as to the
causes of the recent fire and the best means of avoiding
similar accidents in future. He even applied to Bennett,
his master, to permit the apprentice to assist in executing
the repairs of certain parts of the work, which was
reluctantly assented to. Bennett closely watched his
" bungling apprentice," as he called him ; but Brindley,
encouraged by the superintendent of the mill, succeeded
in satisfactorily executing his allotted portion of the
repairs, not less to the surprise of his master than to the
mortification of his men. Many years after, Brindley,
in describing this first successful piece of mill-work which
he had executed, observed, " I can yet remember the
delight which I felt when my work was fixed and fitted
complete ; and I could not understand why my master
and the other workmen, instead of being pleased, seemed
to be dissatisfied with the insertion of every fresh part
in its proper place."

The completion of the job was followed by the usual

supper and drink at the only tavern in the town, then on Parsonage Green. Brindley's share in the work was a good deal ridiculed by the men when the drink began to operate; on which Mr. Milner, to whose intercession his participation in the work had been entirely attributable, interposed and said, " I will wager a gallon of the best ale in the house, that before the lad's apprenticeship is out he will be a cleverer workman than any here, whether master or man." We have not been informed whether the wager was accepted; but it was long remembered, and Brindley was so often taunted with it by the workmen, that he was not himself allowed to forget that it had been offered. Indeed, from that time forward, he zealously endeavoured so to apply himself as to justify the prediction, for it was nothing less, of his kind friend Mr. Milner; and before the end of his third year's apprenticeship his master was himself constrained to admit that Brindley was not the " fool " and the " blundering blockhead " which he and his men had so often called him. Very much to the chagrin of the latter, and to the surprise of Bennett himself, the neighbouring millers, when sending for a workman to execute repairs in their machinery, would specially request that " the young man Brindley " should be sent them in preference to any other of the workmen. Some of them would even have the apprentice in preference to the master himself. At this Bennett was greatly surprised, and, quite unable to understand the mystery, he even went so far as to inquire of Brindley where he had obtained his knowledge of mill-work! Brindley could not tell; it " came natural-like;" but the whole secret consisted in Brindley working with his head as well as with his hands. The apprentice had already been found peculiarly expert in executing mill repairs, in the course of which he would frequently suggest alterations and improvements, more especially in the application of the water-power, which no one had before thought of, but

which proved to be founded on correct principles, and
worked to the millers' entire satisfaction. Bennett, on
afterwards inspecting the gearing of one of the mills
repaired by Brindley, found it so securely and substantially fitted, that he even complained to him of his
style of work. " Jem," said he, " if thou persist in this
foolish way of working, there will be very little trade
left to be done when thou comes out of thy time : thou
knaws firmness of wark's th' ruin o' trade." Brindley,
however, gave no heed whatever to the unprincipled
suggestion, and considered it the duty and the pride of
the mechanic always to execute the best possible work.

Among the other jobs which Brindley's master was
employed to execute about this time, was the machinery
of a new paper-mill proposed to be erected on the river
Dane. The arrangements were to be the same as those
adopted in the Smedley paper-mill on the Irk, and at
Throstle-Nest on the Irwell, near Manchester; and
Bennett went over to inspect the machinery at those
places. But Brindley was afterwards of opinion that he
must have inspected the taverns in Manchester much
more closely than the paper-mills in the neighbourhood;
for when he returned, the practical information he
brought with him proved almost a blank. Nevertheless,
Bennett could not let slip the opportunity of undertaking
so lucrative a piece of employment in his special line,
and, ill-informed though he was, he set his men to work
upon the machinery of the proposed paper-mill.

It very soon appeared that Bennett was altogether
unfitted for the performance of the contract which he
had undertaken. The machinery, when made, would
not fit; it would not work; and, what with drink and
what with perplexity, Bennett soon got completely be-
wildered. Yet to give up the job altogether would be
to admit his own incompetency as a mechanic, and must
necessarily affect his future employment as a millwright.
He and his men, therefore, continued distractedly to

persevere in their operations, but without the slightest appearance of satisfactory progress. About this time an old hand, who happened to be passing the place at which the men were at work, looked in upon them and examined what they were about, as a mere matter of curiosity. When he had done so, he went on to the nearest public-house and uttered his sentiments on the subject very freely. He declared that the job was a farce, and that Abraham Bennett was only throwing his employer's money away. The statement of what the " experienced hand " had said, was repeated until it came to the ears of young Brindley. Concerned for the honour of his shop as well as for the credit of his master —though he probably owed him no great obligation on the score either of treatment or instruction—Brindley formed the immediate resolution of attempting to master the difficulty so as to enable the work to be brought to a satisfactory completion.

At the end of the week's work Brindley left the mill without saying a word of his intention to any one, and instead of returning to his master's house, where he lodged, he took the road for Manchester. Bennett was in a state of great alarm lest he should have run away ; for Brindley, now in the fourth year of his apprentice-ship, had reached the age of twenty-one, and the master feared that, taking advantage of his legal majority, he had left his service never to return. A messenger was despatched in the course of the evening to his mother's house ; but he was not there. Sunday came and passed —still no word of young Brindley : he must have run away ! On Monday morning Bennett went to the paper-mill to proceed with his fruitless work ; and lo ! the first person he saw was Brindley, with his coat off, working away with greater energy than ever. His disappear-ance was soon explained. He had been to Smedley Mill to inspect the machinery there with his own eyes, and clear up his master's difficulty. He had walked the twenty-five miles thither on the Saturday night, and on the following Sunday morning he had waited on Mr. Appleton, the proprietor of the mill, and requested permission to inspect the machinery. With an unusual degree of liberality Mr. Appleton gave the required consent, and Brindley spent the whole of that Sunday in the most minute inspection of the entire arrangements of the mill. He could not make notes, but he stored up the particulars carefully in his head; and believing that he had now thoroughly mastered the difficulty, he set out upon his return journey, and walked the twenty-five miles back to Macclesfield again.

Having given this proof of his determination, as he had already given of his skill in mechanics, Bennett was only too glad to give up the whole conduct of the contract thenceforth to his apprentice; Brindley assuring him that he should now have no difficulty in completing it to his satisfaction. No time was lost in revising the whole design; many parts of the work already fixed were rejected by Brindley, and removed; others, after his own design, were substituted; several entirely new improvements were added; and in the course of a few weeks the work was brought to a conclusion, within the stipulated time, to the satisfaction of the proprietors of the mill.

There was now no longer any question as to the extraordinary mechanical skill of Bennett's apprentice. The old man felt that he had been in a measure saved by young Brindley, and thenceforth, during the remainder of his apprenticeship, he left him in principal charge of the shop. Thus for several years Brindley maintained his old master and his family in respectability and comfort; and when Bennett died, Brindley carried on the concern until the work in hand had been completed and the accounts wound up; after which he removed from Macclesfield to begin business on his own account at the town of Leek, in Staffordshire.

[After seven years as an apprentice and two as journeyman, Brindley established himself as a wheelwright and millwright at the small market town of Leek near the Staffordshire Potteries district. He achieved a reputation locally as an erector of flint mills for the pottery makers and for work on a silk mill at Congleton. His attention attracted to the use of steam power in manufacturing operations, he contrived in 1756 an engine which he believed would prove more economical than other Newcomen engines. Smiles's account of Brindley's Fenton Vivian engine reads:

With this idea in his head, he proceeded to contrive an improved engine, the main object of which was to ensure greater economy in fuel. In 1756 we find him erecting a steam-engine for one Mr. Broade, at Fenton Vivian, in Staffordshire, in which he adopted the expedient, afterwards tried by James Watt, of wooden cylinders made in the manner of coopers' ware, instead of cylinders of iron. He also substituted wood for iron in the chains which worked at the end of the beam. Like Watt, however, he was under the necessity of abandoning the wooden cylinders; but he surrounded his metal cylinders with a wooden case, filling the intermediate space with wood-ashes; and by this means, and using no more injection of cold water than was necessary for the purpose of condensation, he succeeded in reducing the waste of steam by almost one-half. Whilst busy with Mr. Broade's engine, we find from the entries in his pocket-book that Brindley occasionally spent several days together at Coalbrookdale, to superintend the making of the boiler-plates, the pipes, and other iron-work. Returned to Fenton Vivian, he proceeded with the erection of his engine-house and the fitting of the machinery, whilst, during five days more, he appears to have been occupied in making the hoops for the cylinders. It takes him five days to get the "great leavor fixed," thirty-nine days to put the boiler together, and thirteen days to get the pit prepared; and as he charges only workman's wages for those days, we infer that the greater part of the work was done by his own hands. He even seems to have himself felled the requisite timber for the work, as we infer from the entry in his pocket-book of "falling big tree 3½ days."

The engine was at length ready after about a year's work, and was set a-going in November, 1757, after which we find these significant entries: "Bad louk [luck] five days;" then, again, "Bad louk" for three days more; and, after that, "Midlin louk;" and so on with "Midlin louk" until the entries under that head come to an end. In the spring of the following year we find him again striving to get his "engon at woork," and it seems at length to have been fairly started on the 19th of March, when we have the entry "Engon at woork 3 days." There is then a stoppage of four days, and again the engine works for seven days more, with a sort of "loud cheer" in the words added to the entry, of "driv a-Heyd!" Other intervals occur, until, on the 16th of April, we have the words "at woor good ordor 3 days," when the entries come to a sudden close. The engine must certainly have given Brindley a great deal of trouble, and almost driven him to despair, as we now know how very imperfect an engine with wooden hooped cylinders must have been; and we are not therefore surprised at the entry which he honestly makes in his pocket-book on the 21st of April, immediately after the one last mentioned, when the engine had, doubtless, a second time broken down, "to Run about a Drinking, 0:1:6." Perhaps he intended the entry to stand there as a warning against giving way to future despair; for he underlined the words, as if to mark them with unusual emphasis.*

Another engine he erected in **1763** for the Walker Colliery at Newcastle was pronounced a "complete and noble piece of ironwork." Brindley's interests, however, then turned in another direction.]

CHAPTER II.

VERY little had as yet been done to open up the inland navigation of England, beyond dredging and clearing out in a very imperfect manner the channels of several of the larger rivers, so as to admit of the passage of small barges. Several attempts had been made in Lancashire and Cheshire, as we have already shown, to open up the navigation of the Mersey and the Irwell from Liverpool to Manchester. There were similar projects for improving the Weaver from Frodsham, where it joins the Mersey, to Winford Bridge above Northwich; and the Douglas, from the Ribble to Wigan. About the same time like schemes were started in Yorkshire, with the object of opening up the navigation of the Aire and Calder to Leeds and Wakefield, and of the Don from Doncaster to near Sheffield. One of the Acts passed by Parliament in 1737 is worthy of notice, as probably the beginning of the Bridgewater Canal enterprise : we allude to the Act for making navigable the Worsley Brook to its junction with the river Irwell, near Manchester. A similar Act was obtained in 1755, for making navigable the Sankey Brook from the Mersey, about two miles below Warrington, to St. Helens, Gerrard Bridge, and Penny Bridge. In this case the canal was constructed separate from the brook, but alongside of it; and at several points locks were provided to adapt the canal to the level of the lands passed through.*

The same year in which application was made to Parliament for powers to construct the Sankey Canal,

the Corporation of Liverpool had under their consideration a much larger scheme—no less than a canal to unite
the Trent and the Mersey, and thus open a water-communication between the ports of Liverpool and Hull.
It was proposed that the line should proceed by Chester,
Stafford, Derby, and Nottingham. A survey was made,
principally at the instance of Mr. Hardman, a public
spirited merchant of Liverpool, and for many years one
of its representatives in Parliament. Another survey
was made at the instance of Earl Gower, afterwards
Marquis of Stafford, and it was in making this survey
that Brindley's attention was first directed to the business
of canal engineering. We find his first entry relating to
the subject was on the 5th of February, 1758—" novo-
cion [navigation] 5 days ;" the second, a little better
spelt, on the 19th of the same month—" a bout the novogation 3 days ;" and afterwards—" surveing the novogation from Long brigg to Kinges Milles 12 days ½." It
does not, however, appear that the scheme made much
progress, or that steps were taken at that time to bring
the measure before Parliament ; and Brindley continued to pursue his other employments, more especially
the erection of " fire-engines " after his new patent. This
continued until the following year, when we find him in
close consultation with the Duke of Bridgewater relative
to the construction of his proposed canal from Worsley
to Manchester.

The early career of this distinguished nobleman was of
a somewhat remarkable character. He was born in 1736,
the fifth and youngest son of Scroop, third Earl and first
Duke of Bridgewater, by Lady Rachel Russell. He lost
his father when only five years old, and all his brothers
died by the time that he had reached his twelfth year, at
which early age he succeeded to the title of Duke of
Bridgewater. He was a weak and sickly child, and his
mental capacity was thought so defective, that steps were
even in contemplation to set him aside in favour of the

next heir to the title and estates.   His mother seems
almost entirely to have neglected him.   In the first year
of her widowhood she married Sir Richard Lyttleton,
and from that time forward took the least possible notice
of her boy.   He did not give much promise of surviving
his consumptive brothers, and his mind was considered
so incapable of improvement, that he was left in a great
measure without either domestic guidance or intellectual
discipline and culture.   Horace Walpole writes to Mann
in 1761 : " You will be happy in Sir Richard Lyttleton
and his Duchess; they are the best-humoured people in
the world."   But the good humour of this handsome
couple was mostly displayed in the world of gay life,
very little of it being reserved for home use.   Possibly,
however, it may have been even fortunate for the young
Duke that he was left so much to himself, and to profit
by the wholesome neglect of special nurses and tutors,
who are not always the most judicious in their bringing
up of delicate children.

At seventeen, the young Duke's guardians, the Duke
of Bedford and Lord Trentham, finding him still alive
and likely to live, determined to send him abroad on his
travels—the wisest thing they could have done.   They
selected for his tutor the celebrated traveller, Robert
Wood, author of the well-known work on Troy, Baal-
bec, and Palmyra; afterwards made Under-Secretary of
State by the Earl of Chatham.   Wood was an accomplished scholar, a persevering traveller, and withal a man
of good business qualities.   His habits of intelligent observation could not fail to be of service to his pupil, and
it is not unnatural to suppose that the great artificial
watercourses and canals which they saw in the course
of their travels had some effect in afterwards determining
the latter to undertake the important works of a similar
character by which his name became so famous.* During
their residence in Italy the Duke and his tutor visited all
the galleries, and Mr. Wood sat to Mengs for his portrait,

which still forms part of the Bridgewater collection. The
Duke also purchased works of sculpture at Rome; but
that he himself entertained no great enthusiasm for art is
evident from the fact related by the late Earl of Ellesmere, that these works remained in their original packing-
cases until after his death.[1]

Returned to England, he seems to have led the usual
life of a gay young nobleman of the time, with plenty of
money at his command. In 1756, when he was only
twenty years of age, he appears from the 'Racing Calen-
dar' to have kept race-horses; and he occasionally rode
them himself. Though in after life a very bulky man,
he was so light as a youth, that on one occasion, Lord
Ellesmere says a bet was jokingly offered that he would
be blown off his horse. Dressed in a livery of blue
silk and silver, with a jockey cap, he once rode a race
against His Royal Highness the Duke of Cumberland,
on the long terrace at the back of the wood in Trentham
Park, the seat of his relative, Earl Gower. During His
Royal Highness's visit, the large old green-house, since
taken down, was hastily run up for the playing of skittles;
and prison-bars and other village games were instituted
for the recreation of the guests. Those occupations of
the Duke were varied by an occasional visit to his racing-
stud at Newmarket, where he had a house for some time,
and by the usual round of London gaieties during the
season.

A young nobleman of tender age, moving freely in
circles where were to be seen some of the finest specimens of female beauty in the world, could scarcely be
expected to pass heart-whole; and hence the occurrence
of the event in his London life which, singularly enough,
is said to have driven him in a great measure from
society, and induced him to devote himself to the con-

---

[1] 'Essays in History, Biography, | the late Earl of Ellesmere. London,
Geography, Engineering,' &c. By | 1858. P. 226.

struction of canals! We find various allusions in the
letters of the time to the rumoured marriage of the
young Duke of Bridgewater. One rumour pointed to
the only daughter and heiress of Mr. Thomas Revell,
formerly M.P. for Dover, as the object of his choice.[1]
But it appears that the lady to whom he became the
most strongly attached was one of the Gunnings—the
comparatively portionless daughters of an Irish gentleman, who were then the reigning beauties at court.
The object of the Duke's affection was Elizabeth, the
youngest daughter, and perhaps the most beautiful of
the three. She had been married to the fourth Duke of
Hamilton, in Keith's Chapel, Mayfair, in 1752, "with a
ring of the bed-curtain, half-an-hour after twelve at
night," [2] but the Duke dying shortly after, she was now
a gay and beautiful widow, with many lovers in her
train. In the same year in which she had been clandes-
tinely married to the Duke of Hamilton, her eldest
sister was married to the sixth Earl of Coventry.

The Duke of Bridgewater paid his court to the young
widow, proposed, and was accepted. The arrangements
for the marriage were in progress, when certain rumours
reached his ear reflecting seriously upon the character of
Lady Coventry, his intended bride's elder sister, who
was certainly more fair than she was wise. Believing
the reports, he required the Duchess to desist from
further intimacy with her sister, a condition which her
high spirit would not brook, and, the Duke remaining

---

[1] Thomas and Maria. Revell were
both servants in the family of Mr.
Nightingale, of Knibsworth. They
afterwards married, and took a farm
at Shingay, under my Lord Orford,
who, taking a liking to their two
eldest sons, Thomas and Russell, gave
them an English education, and got
them both places in the Victualling
Office. The eldest, Thomas, was
M.P. for Dover, and, dying in 1752
at Bath, was buried, as I think, at or
near Leatherhead, Surrey, leaving an
only daughter behind him, to whom
he left about 120,000*l.* or 130,000*l.* It
is thought she is to be married to the
present Duke of Bridgewater, her
cousin.—'The Cole MSS.' (British
Museum), vol. ix., 113.

[2] 'Walpole to Mann,' Feb. 27th,
1752.

PORTRAIT OF THE YOUNG DUKE.

[By T. D. Scott.]

firm, the match was broken off. From that time forward he is said never to have addressed another woman in the language of gallantry.[1] The Duchess of Hamilton, however, did not remain long a widow. In the course of a few months she was engaged to, and afterwards married, John Campbell, subsequently Duke of Argyll. Horace Walpole, writing of the affair to Marshal Conway, January 28th, 1759, says : " You and M. de Bareil do not exchange prisoners with half as much alacrity as Jack Campbell and the Duchess of Hamilton have exchanged hearts. . . . It is the prettiest match in the world since yours, and everybody likes it but the Duke of Bridgewater and Lord Conway. What an extraordinary fate is attached to these two women ! Who could have believed that a Gunning would unite the two great houses of Campbell and Hamilton ? For my part, I expect to see my Lady Coventry Queen of Prussia. I would not venture to marry either of them these thirty years, for fear of being shuffled out of the world prematurely to make room for the rest of their adventures."

The Duke of Bridgewater, like a wise man, seems to have taken refuge from his disappointment in active and useful occupation. Instead of retiring to his beautiful seat at Ashridge, we find him straightway proceeding to his estate at Worsley, on the borders of Chat Moss, in Lancashire, and conferring with John Gilbert, his land-steward, as to the practicability of cutting a canal by which the coals found upon his Worsley estate might be readily conveyed to market at Manchester.

Manchester and Liverpool at that time were improving towns, gradually rising in importance and increasing in population. The former place had long been noted for

---

[1] Chalmers, in his 'Biographical Dictionary,' vol. xiii., 94, gives another account of the rumoured cause of the Duke's subsequent antipathy to women ; but the above statement of the late Earl of Ellesmere, confirmed as it is by certain passages in Walpole's Letters, is more likely to be the correct one.

its manufacture of coarse cottons or coatings made of wool, in imitation of the goods known on the Continent by that name. The Manchester people also made fustians, mixed stuffs, and small wares, amongst which leather-laces for women's bodice, shoe-ties, and points were the more important. But the operations of manufacture were still carried on in a clumsy way, entirely by hand. The wool was spun into yarn by means of the common spinning-wheel, for the spinning-jenny had not yet been contrived, and the yarn was woven into cloth by the common hand-loom. There was no whirr of engine-wheels then to be heard; for Watt's steam-engine had not yet been invented. The air was free from smoke, except what arose from household fires, and there was not a single factory-chimney in Manchester. In 1724 Dr. Stukeley says Manchester contained no fewer than 2400 families, and that their trade

VIEW OF MANCHESTER IN 1740.

[Fac-simile of an Engraving of the period by J. Harris, published by Robert Whitworth.]

was "incredibly large" in tapes, ticking, girth-webb, and fustians. In 1757 the united population of Manchester and Salford was only 20,000 ;[1] it is now, after the lapse of a century, 460,000 ! The Manchester

---

[1] Aikin's 'Description of the Country from Thirty to Forty Miles round Manchester.' London, 1795.

manufacturer was then a very humble personage compared with his modern representative. He was part
chapman, part weaver, and part merchant—working
hard, living frugally, principally on oatmeal,[1] and contriving to save a little money. As trade increased, its
operations became more subdivided, and special classes
and ranks began to spring into importance. The manufacturers sent out riders to take orders, and chapmen
with gangs of pack-horses to distribute the goods and
bring back wool in exchange. The chapmen used pack-
horses because the roads were as yet mostly impracticable for waggons, and it was more difficult then to
reach a village twenty miles out of Manchester than
it is to make the journey from thence to London now.
Indeed, the only coach to London plied but every second
day, and it was four days and a-half in making the
journey, there being a post only three times a week.[2]
The roads in most districts of Lancashire were what
were called " mill roads," along which a horse with a
load of oats upon its back might proceed towards the
mill where they were to be ground. There was no private carriage kept by any person in business in Manchester until the year 1758, when the first was set up
by some specially luxurious individual. But wealth led
to increase of expenditure, and Aikin mentions that
there was " an evening club of the most opulent manufacturers, at which the expenses of each person were fixed

---

[1] Dr. Aikin, in 1795, gave the following description of the Manchester
manufacturer in the first half of the
eighteenth century : " An eminent
manufacturer in that age," said he,
" used to be in his warehouse before
six in the morning, accompanied by
his children and apprentices. At
seven they all came in to breakfast,
which consisted of one large dish of
water-pottage, made of oatmeal, water,
and a little salt, boiled thick, and
poured into a dish. At the side was
a pan or basin of milk, and the master
and apprentices, each with a wooden

spoon in his hand, without loss of
time, dipped into the same dish, and
thence into the milk-pan, and as soon
as it was finished they all returned to
their work." What a contrast with
the " eminent manufacturer" of our
own day !

[2] March 3rd, 1760, the Flying
Machine was started, and advertised
to perform the journey, " if God permit," in three days, by John Han-
forth, Matthew Howe, Samuel Granville, and William Richardson. Fare
inside, 2l. 5s.; outside, half-price.

at fourpence-halfpenny—fourpence for ale, and a half-penny for tobacco." The progress of luxury was further aided by the holding of a dancing assembly once a-week in a room situated at the middle of the now fashionable street called King Street, the charge for which was half-a-crown the quarter; the ladies having their maids to come with lanterns and pattens to conduct them home; "nor," adds Aikin, "was it unusual for their partners also to attend them." [1]

The imperfect state of the communications leading to and from Manchester rendered it a matter of some difficulty at certain seasons to provide food for so large a population. In winter, when the roads were closed, the place was in the condition of a beleaguered town; and even in summer, the land about Manchester itself being comparatively sterile, the place was badly supplied with fruit, vegetables, and potatoes, which, being brought from considerable distances slung across horses' backs, were so dear as to be beyond the reach of the mass of the population. The distress caused by this frequent dearth of provisions was not effectually remedied until the canal navigation became completely opened up. Thus a great scarcity of food occurred in Manchester and the neighbourhood in 1757, which the common people attributed to the millers and corndealers; and unfortunately the notion was not confined to the poor who were starving, but was equally entertained by the well-to-do classes who had enough to eat. An epigram by Dr. Byrom, the town clergyman, written in 1737, on two millers (tenants of the School corn-mills), who, from their spare habits, had been nicknamed "Skin" and "Bone," was now revived, and tended to fan the popular fury. It ran thus:—

> "Bone and Skin, two millers thin,
>   Would starve the town, or near it;
> But be it known to Skin and Bone,
>   That Flesh and Blood can't bear it."

---

[1] Aikin, p. 187.

The result of the popular hunger was, that a great commotion occurred, which at length broke out in open outrage, and a riot took place in 1758, long after remembered in Manchester as the " Shude Hill fight," in which several lives were unhappily lost.[1]

For the same reason the supply of coals was scanty in winter ; and though abundance of the article lay underground, within a few miles of Manchester, in nearly every direction, those few miles of transport, in the then state of the roads, were an almost insurmountable difficulty. The coals were sold at the pit mouth at so much the horse-load, weighing 280 lbs., and measuring two baskets, each thirty inches by twenty, and ten inches deep ; that is, as much as an average horse could carry on its back.[2]   The price of the coals at the pit mouth was 10d. the horse-load ; but by the time the article reached the door of the consumer in Manchester, the price was usually more than doubled, in consequence of the difficulty and cost of conveyance.   The carriage alone amounted to about nine or ten shillings the ton.   There was as yet no connection of the navigation of the Mersey and Irwell with any of the collieries situated to the eastward of Manchester, by which a supply could reach the town in boats ; and although the Duke's collieries were only a comparatively short distance from the Irwell, the coals had to be carried on horses' backs or in carts from the pits to the river to be loaded, and after reaching Manchester they had again

---

[1] In 1715 the first London baker settled in Manchester, Mr. Thomas Hatfield, known by his styptic. His apprentices took the mills in the vicinity, and in time reduced the inhabitants to the necessity of buying flour of them. Monopolies at length took place in consequence of these changes, which, at different times, produced riots ; one of which, occasioned by a large party of country people coming to Manchester in order to destroy the mills, ended in the loss of several lives, at a fray known by the name of Shude Hill fight, in the year 1758. Since that time until the present [1795] the demand for corn and flour has been increasing to a vast amount, and new sources of supply have been opened from distant parts by the navigations, so that monopoly or scarcity cannot be apprehended.—Aikin's ' Manchester.'

[2] This "load " is still used as a measure of weight, though the practice of carrying all sorts of commodities on horses' backs, in which it originated, has long since ceased.

to be carried to the doors of the consumers,—so that there was little if any saving to be effected by that route. Besides, the minimum charge insisted on by the Mersey Navigation Company of 3s. 4d. a ton for even the shortest distance, proved an effectual barrier against any coal reaching Manchester by the river.

The same difficulty stood in the way of the transit of goods between Manchester and Liverpool. By road the charge was 40s. a ton, and by river 12s. a ton; that between Warrington and Manchester being 10s. a ton : besides, there was great risk of delay, loss, and damage by the way. Some idea of the tediousness of the river-navigation may be formed from the fact, that the boats were dragged up and down stream exclusively by the labour of men, and that horses and mules were not employed for this purpose until after the Duke's canal had been made. It was, indeed, obvious that unless some means could be devised for facilitating and cheapening the cost of transport between the seaport and the manufacturing towns, there was little prospect of any considerable further development being effected in the industry of the district.

Such was the state of things when the Duke of Bridgewater turned his attention to the making of a water-road for the passage of his coal from Worsley to Manchester. The Old Mersey Company would give him no facilities for sending his coals by their navigation, but levied the full charge of 3s. 4d. for every ton he might send to Manchester by river even in his own boats. He therefore perceived that to obtain a vend for his article, it was necessary he should make a way for himself; and it became obvious to him that if he could but form a water-road or canal between the two points, he would at once be enabled to secure a ready sale for all the coals that he could raise from his Worsley pits.

CHAPTER III.

WE have already stated that, as early as 1737, an Act had been obtained by the Duke's father, to enable the Worsley Brook to be made navigable to the point at which it entered the Irwell. But the enterprise seemed to be too difficult, and its cost too great; so the powers of the Act were allowed to expire without anything being done to carry them out. The young Duke now determined to revive the Act in another form, and in the early part of 1759 he applied to Parliament for the requisite powers to enable him to cut a navigable canal from Worsley Mill eastward to Salford, and to carry the same westward to a point on the river Mersey, called Hollin Ferry. He introduced into the bill several important concessions to the inhabitants of Manchester. He bound himself not to exceed the freight of 2s. 6d. per ton on all coals brought from Worsley to Manchester, and not to sell the coal so brought from the mines to that town at more than 4d. per hundred, which was less than half the then average price. It was clear that, could such a canal be made and the navigation opened up as proposed, it would prove a great public boon to the inhabitants of Manchester, and it was hailed by them as such accordingly. The bill was well supported, and it passed the legislature without opposition, receiving the Royal assent in March, 1759.

The Duke gave further indications of his promptitude and energy, in the steps which he adopted to have the works carried out without loss of time. He had no intention of allowing the powers of this Act to remain a dead

letter, as the former had done.   Accordingly, no sooner
had it passed than he set out for his seat at Worsley to
take the requisite measures for constructing the canal.   The
Duke was fortunate in having for his land-agent a very
shrewd, practical, and enterprising person, in John Gilbert, whom he consulted on all occasions of difficulty.*
Mr. Gilbert was the brother of Thomas Gilbert, the
originator of the Gilbert Unions, then agent to the Duke's
brother-in-law, Lord Gower.   That nobleman had for
some time been promoting the survey of a canal to unite
the Mersey and the Trent, on which Brindley had been
employed, who was thus well known to Gilbert as well
as to his brother.   We find from an entry in his pocket-
book, that the millwright had sundry interviews with
Thomas Gilbert on matters of business previous to the
passing of the first Bridgewater Canal Bill, though there
is no evidence that Brindley was employed in making
the survey.   Indeed, it is questionable whether any survey was made of the first scheme,—engineering projects
being then submitted to Parliamentary Committees in
a very rough state ; levels being guessed at rather than
surveyed and calculated ; and merely general powers
taken enabling such property to be purchased as might
by possibility be required for the execution of the works
—the prices of land and compensation for damage being
assessed by a local committee appointed by the Act for
the purpose.

When the Duke proceeded to consider with Gilbert
the best mode of carrying out the proposed canal, it
very shortly appeared that the plan originally contemplated was faulty in many respects, and that an application must be made to Parliament for further powers.
By the original Act it was intended to descend from
the level of the coal-mines at Worsley by a series of
locks into the river Irwell.   This, it was found, would
necessarily involve both a heavy cost in the construction
and working of the canal, as well as considerable delay

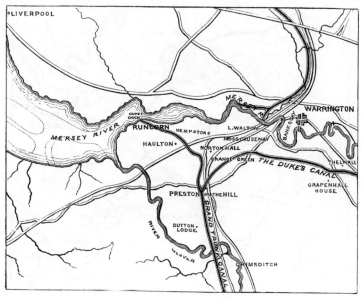

MAP OF THE DUKE'S CANAL.

[Western Part.]

in the conduct of the traffic, which it was most desirable
to avoid. Neither the Duke nor Gilbert had any practical knowledge of engineering ; nor, indeed, were there
many men in the country at that time who knew
much of the subject. For it must be remembered that
this canal of the Duke's was the very first project in
England for cutting a navigable trench through the dry
land, and carrying merchandise across the country in it
independent of the course of the existing streams.

It was in this emergency that Gilbert advised the
Duke to call to his aid James Brindley, whose fertility
of resources and skill in overcoming mechanical difficulties had long been the theme of general admiration
in the district. Doubtless the Duke was as much impressed by the native vigour and originality of the un-
lettered genius thus introduced to him, as were all with
whom he was brought in contact. Certain it was that
the Duke showed his confidence in Brindley by en-

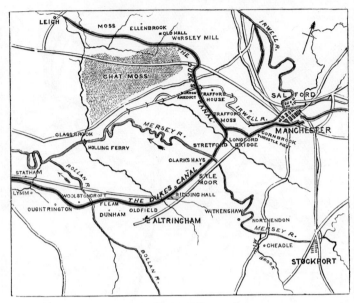

MAP OF THE DUKE'S CANAL.

[Eastern Part.]

trusting him with the conduct of the proposed work ;
and, as the first step, he was desired to go over the
ground at once, and give his opinion as to the best
plan to be adopted for carrying it out with dispatch.
Brindley, accordingly, after making what he termed an
" ochilor [ocular] servey or a ricconitoring," speedily
formed his conclusion, and came back to the Duke with
his advice. It was that, instead of carrying the canal
down into the Irwell by a flight of locks, and so up
again on the other side to the proposed level, it should be
carried right over the river, and constructed on one
entire level throughout. But this, it was clear, would
involve a series of formidable works, the like of which
had never before been attempted in England. In the first
place, the low ground on the north side of the Irwell
would have to be filled up by a formidable embankment,
and united with the land on the other bank by means of
a large aqueduct of stone. Would it be practicable or

possible to execute works of such magnitude ? Brindley expressed so strong and decided an opinion of their practicability, that the Duke became won over to his views, and determined again to go to Parliament for the requisite powers to enable the design to be carried out.

Many were the deliberations which took place about this time between the Duke, Gilbert, and Brindley, in the Old Hall at Worsley, where the Duke had now taken up his abode. We find from Brindley's pocket-book memoranda, that in the month of July, 1759, he had taken up his temporary quarters at the Old Hall; and from time to time, in the course of the same year, while the details of the plan were being prepared with a view to the intended application to Parliament, he occasionally stayed with the Duke for several weeks together. He made a detailed survey of the new line, and at the same time, in order to facilitate the completion of the undertaking when the new powers had been obtained, he proceeded with the construction of the sough or level at Worsley Mill, and such other portions of the work as could be executed under the original powers.

During the same period Brindley travelled backwards and forwards a great deal, on matters connected with his various business in the Pottery district. We find, from his record, that he was occupied at intervals in carrying forward his survey of the proposed canal through Staffordshire, visiting with this object the neighbourhood of Newcastle-under-Lyme, Lichfield, and Tamworth. He also continued to give his attention to mills, water-wheels, cranes, and fire-engines, which he had erected, or required repairs, in various parts of the same district. In short, he seems at this time to have been fully employed as a millwright; and although, as we have seen, the remuneration which he received for his skill was comparatively small, being a man of frugal habits he had saved a little money; for about this time we find him able to raise a sum of 543*l.* 6*s.* 8*d.*, being his fourth share of the purchase-money of the Turnhurst estate, situated near Golden Hill, in the county of Stafford.   It appears, however, from his own record that he borrowed the principal part of this amount from his friend Mr. Launcelot, of Leek ; showing that, amongst his townsmen and neighbours, who knew him best, he stood in good credit and repute.   His other partners in the purchase were Mr. Thomas Gilbert (Earl Gower's agent), Mr. Henshall (afterwards his brother-in-law), and his brother John Brindley.   The estate was understood to be full of minerals, the knowledge of which had most probably been obtained by Brindley in the course of his surveying of the proposed Staffordshire canal ; and we shall afterwards find that the purchase proved a good investment.

At length the new plans of the canal from Worsley to Manchester were completed and ready for deposit ; and on the 23rd of January, after a visit to the Duke and Gilbert at the Hall, we find the entry in Brindley's pocket-book of " Sot out for London."   On the occasion of his visits to London, Brindley adopted the then most convenient method of travelling on horseback, the journey usually occupying five days.   We find him varying his route according to the state of the weather and of the roads.   In summer he was accustomed to go by Coventry, but in winter he made for the Great North Road by Northampton, which was usually in better condition for winter travelling.

The second Act passed without opposition, like the first, early in the session of 1760.   It enabled the Duke to carry his proposed canal *over* the river Irwell, near Barton Bridge, some five miles westward of Manchester, by means of a series of arches, and to vary its course accordingly ; whilst it further authorised him to extend a short branch to Longford Bridge, near Stretford,—that to Hollin Ferry, authorised by the original Act, being abandoned.   In the mean time the works near

Worsley had been actively pushed forward, and considerable progress had been made by the time the additional powers had been obtained. That part of the canal which lay between Worsley Mill and the public highway leading from Manchester to Warrington had been made; the sough or level between Worsley Mill and Middlewood, for the purpose of supplying water to the canal, was considerably advanced; and operations had also been begun in the neighbourhood of Salford and on the south of the river Irwell.

The most difficult part of the undertaking, however, was that authorised by the new Act; and the Duke looked forward to its execution with the greatest possible anxiety. Although aqueducts of a far more formidable description had been executed abroad, nothing of the kind had until then been projected in this country; and many regarded the plan of Brindley as altogether wild and impracticable. The proposal to confine and carry a body of water within a water-tight trunk of earth upon the top of an embankment across the low grounds on either side of the Irwell, was considered foolish and impossible enough; but to propose to carry ships upon a lofty bridge, over the head of other ships navigating the Irwell which flowed underneath, was laughed at as the dream of a madman. Brindley, by leaving the beaten path, thus found himself exposed to the usual penalties which befall originality and genius.

The Duke was expostulated with by his friends, and strongly advised not to throw away his money upon so desperate an undertaking. Who ever heard of so large a body of water being carried over another in the manner proposed? Brindley was himself appealed to; but he could only repeat his conviction as to the entire practicability of his design. At length, by his own desire and to allay the Duke's apprehensions, another engineer was called in and consulted as to the scheme. To Brindley's surprise and dismay, the person consulted concurred in the view so strongly expressed by the public. He characterised the plan of the Barton aqueduct and embankment as instinct with recklessness and folly; and after expressing his unqualified opinion as to the impracticability of executing the design, he concluded his report to the Duke thus: "I have often heard of castles in the air; but never before saw where any of them were to be erected." [1]

It is to the credit of the Duke that, notwithstanding these strongly adverse opinions, he continued to give his confidence to the engineer whom he had selected to carry out the work. Brindley's common-sense explanations, though they might not remove his doubts, nevertheless determined the Duke to give him the full opportunity of carrying out his design; and he was accordingly authorised to proceed with the erection of his "castle in the air." Its progress was watched with great interest, and people flocked from all parts to see it.

The Barton aqueduct is about two hundred yards in length and twelve yards wide, the centre part being sustained by a bridge of three semicircular arches, the middle one being of sixty-three feet span. It carries the canal over the Irwell at a height of thirty-nine feet above the river—this head-room being sufficient to enable the largest barges to pass underneath without lowering their masts. The bridge is entirely of stone blocks, those on the faces being dressed on the front, beds, and joints, and cramped with iron. The canal, in passing over the arches, is confined within a puddled [2] channel to prevent leakage, and is in as good

---

[1] We have heard the name of Smeaton mentioned as that of the engineer consulted on the occasion, but we are unable to speak with certainty on the point. Excepting Smeaton, however, there was then no other engineer in the country of recognised eminence in the profession.

[2] The process of puddling is of considerable importance in canal engineering. Puddle is formed by a mixture of well-tempered clay and sand re-

duced to a semi-fluid state, and rendered impervious to water by manual labour, as by working and chopping it about with spades. It is usually applied in three or more strata to a depth or thickness of about three feet; and care is taken at each operation so to work the new layer of puddling stuff as to unite it with the stratum immediately beneath. Over the top course a layer of common soil is usually laid. It is only by the

a state now as on the day on which it was completed.
Although the Barton aqueduct has since been thrown
into the shade by the vastly greater works of modern

BARTON AQUEDUCT.

[By Percival Skelton, after his original Drawing.]

engineers, it was unquestionably a very bold and ingenious enterprise, if we take into account the time at which
it was erected. Humble though it now appears, it was
the parent of the magnificent aqueducts of Rennie and
Telford, and of the viaducts of Stephenson and Brunel,
which rival the greatest works of any age or country.

The embankments formed across the low grounds on

careful employment of puddling that
the filtration of the water of canals
into the neighbouring lower lands
through which they pass can be effectually prevented.

either side of the Barton viaduct were also considered very formidable works at that day. A contemporary writer speaks of the embankment across Stretford Meadows as an amazing bank of earth 900 yards long, 112 feet in breadth across the base, 24 feet at top, and 17 feet high. The greatest difficulty anticipated, was the holding of so large a body of water within a hollow channel formed of soft materials. It was supposed at first that the water would soak through the bank, which its weight would soon burst, and wash away all before it. But Brindley, in the course of his experience, had learnt something of the powers of clay-puddle to resist the passage of water. He had already succeeded in stopping the breaches of rivers flowing through low grounds by this means ; and the thorough manner in which he finished the bed of this canal, and made it impervious to water, may be cited as not the least remarkable illustration of the engineer's practical skill, taking into account the early period at which this work was executed. Another very difficult part of the undertaking was the formation of the canal across Trafford Moss, where the weight of the embankment pressed down and "blew up" the soft oozy stuff on either side ; but the difficulty was again overcome by the engineer's specific of clay-puddle, which proved completely successful. Indeed, the execution of these embankments by Brindley was regarded at the time as something quite as extraordinary in their way as the erection of the Barton aqueduct itself.

The rest of the canal between Longford and Man-chester, being mostly on sidelong ground, was cut down on the upper side and embanked up on the other by means of the excavated earth. This was comparatively easy work ; but a matter of greater difficulty was to accommodate the streams which flowed across the course of the canal, and which were provided for in a highly ingenious manner. For instance, a stream called Corn-brook was found too high to pass under the canal at its

natural level.   Accordingly, Brindley contrived a weir,
over which the stream fell into a large basin, from
whence it flowed into a smaller one open at the bottom.
From this point a culvert, constructed under the bed of
the canal, carried the waters across to a well situated on its
further side, where the waters, rising up to their natural
level, again flowed away in their proper channel.   A
similar expedient was adopted at the Manchester terminus
of the canal, at the point at which it joined the waters
of the Medlock.   It was a principle of Brindley's never
to permit the waters of any river or brook to intermix
with those of the canal except for the purpose of supply ;
as it was clear that in a time of flood such intermingling
would be a source of great danger to the navigation.   In
order, therefore, to provide for the free passage of the
Medlock without causing a rush into the canal, a weir
was contrived, 366 yards in circumference, over which
its waters flowed into a lower level, and from thence into
a well several yards in depth, down which the whole
river fell.   It was received at the bottom in a subter-
ranean passage, by which it passed into the river Irwell,
near at hand.   The weir was very ingeniously contrived,
though it was afterwards found necessary to make considerable alterations and improvements in it, as experience suggested, in order effectually to accommodate the
flood-waters of the Medlock.   Arthur Young, when
visiting the canal, shortly after it was opened up to
Manchester, says, " The whole plan of these works shows
a capacity and extent of mind which foresees difficulties,
and invents remedies in anticipation of possible evils.
The connection and dependence of the parts upon each
other are happily imagined ; and all are exerted in
concert, to command by every means the wished-for
success." [1]

Brindley's labours, however, were not confined to the

---

[1] 'Six Months' Tour through the North of England,' vol. iii., p. 258.
Ed. 1770.

WORSLEY BASIN.

[By Percival Skelton, after his original Drawing.]

construction of the canal, but his attention seems to have been equally directed to the contrivance of the whole arrangements and machinery by which it was worked. The open navigation between Worsley Mill and Manchester was 10¼ miles in length. At Worsley, where a large basin was excavated of sufficient capacity to contain a great many boats, and to serve as a head for the navigation, the canal did not stop, but entered the bottom of the hill by a subterraneous channel which extended for a great distance,—connecting the different workings of the mine, and enabling the coals readily to be transported in boats to their place of sale. In Brindley's time, this subterraneous canal, hewn out of the rock, was only about a mile in length, but it now extends to nearly forty miles in all directions underground.[1] Where the tunnel passed through earth or coal, the arching was of brick-work; but where it passed through rock, it was simply hewn out. This tunnel acts not only as a drain and water-feeder for the canal itself, but as a means of carrying the facilities of the navigation through the very heart of the collieries; and it will readily be seen of how great a value it must have proved in the economical working of the navigation, as well as of the mines, so far as the traffic in coals was concerned.

---

[1] Worsley-basin lies at the base of a cliff of sandstone, some hundred feet in height. [See the illustration.] Luxuriant foliage overhangs its precipitous side, and beyond is seen the graceful spire of Worsley church. In contrast to this bright nature above, lies the almost stagnant pool beneath. The barges are deeply laden with their black freight, which they have brought from the mine through the two low, semi-circular arches opening at the base of the rock, such being the entrances to the underground canals. The smaller aperture is the mouth of a canal of only half a mile in length, serving to prevent the obstruction which would be caused by the entrance and egress of so many barges through a single passage. The other archway is the entrance of a wider channel, extending nearly six miles in the direction of Bolton, and from which various other canals diverge in different directions. The barges are narrow and long, each conveying about ten tons of coal. They are drawn along the tunnels by means of staples fixed along the sides. When they are empty, and consequently higher in the water, they are so near the roof that the bargemen, lying on their backs, can propel them with their feet.

At every point Brindley's originality and skill were at work. He invented the cranes for the purpose of more readily loading the boats with the boxes filled with the Duke's " black diamonds." He also contrived and laid down within the mines a system of underground railways, all leading from the face of the coal (where the miners were at work) to the wells which he had made at different points in the tunnels, through which the coals were shot into the boats waiting below to receive them. At Manchester, where they were unloaded for sale, the contrivances which he employed were equally ingenious. It was at first intended that the canal should terminate at the foot of Castle Hill, up which the coals were dragged by their purchasers from the boats in wheelbarrows or carts. But the toil of dragging the loads up the hill was found very great; and, to remedy the inconvenience, Brindley contrived to extend the canal for some way into the hill, opening a shaft from the surface of the ground down to the level of the water. The barges having made their way to the foot of this shaft, the boxes of coal were hoisted to the surface by a crane, worked by a box waterwheel of 30 feet diameter and 4 feet 4 inches wide, driven by the waterfall of the river Medlock. In this contrivance Brindley was only adopting a modification of the losing and gaining bucket, moved on a vertical pillar, which he had before successfully employed in drawing water out of coal-mines. By these means the coals were rapidly raised to the higher ground, where they were sold and distributed, greatly to the convenience of those who came to purchase them.

Brindley's practical ability was equally displayed in planning and building a viaduct and in fitting up a crane —in carrying out an embankment or in contriving a coal-barge. The range and fertility of his constructive genius were extraordinary. For the Duke, he invented water-weights at Rough Close, riddles to wash coal for the forges, rising dams, and numerous other contrivances of well-adapted mechanism. At Worsley he erected a steam-engine for draining those parts of the mine which were beneath the level of the canal, and consequently could not be drained into it; and he is said to have erected, at a cost of only 150*l*., an engine which until that time no one had known how to construct for less than 500*l*. At the mouth of one of the mines he erected a water-bellows for the purpose of forcing fresh air into the interior, and thus ventilating the workings.[1] At the entrance of the underground canal he designed and built an over-shot mill of a new construction, driven by a wheel twenty-four feet in diameter, which worked three pair of stones for grinding corn, besides a dressing or boulting mill, and a machine for sifting sand and mixing mortar.[2] Brindley's quickness of observation and readiness in turning circumstances to account, were equally displayed in the mode by which he contrived to obtain an ample supply of lime for building purposes during the progress of the works. We give the account as related by Arthur Young :—" In carrying on the navigation," he observes, "a vast quantity of masonry was necessary for building aqueducts, bridges, warehouses, wharves, &c., and the want of lime was felt severely. The search that was made for matters that would burn into lime was for a long time fruitless. At last Mr. Brindley met with a substance of a chalky kind, which, like the rest, he tried; but found (though it was of a limestone nature—lime-

---

[1] A writer in the ' St. James's Chronicle,' under date the 30th of September, 1763, gives the following account of this apparatus, long since removed :—" At the mouth of the cavern is erected a water-bellows, being the body of a tree, forming a hollow cylinder, standing upright. Upon this a wooden bason is fixed, in the form of a funnel, which receives a current of water from the higher ground. This water falls into the cylinder, and issues out at the bottom of it, but at the same time carries a quantity of air with it, which is received into the pipes and forced to the innermost recesses of the coal-pits, where it issues out as if from a pair of bellows, and rarefies the body of. thick air, which would otherwise prevent the workmen from subsisting on the spot where the coals are dug."

[2] Young's ' Six Months' Tour,' vol. iii., p. 278.

marl, which was found along the sides of the canal, about a foot below the surface) that, for want of adhesion in the parts, it would not make lime. This most inventive genius happily fell upon an expedient to remedy this misfortune. He thought of tempering this earth in the nature of brick-earth, casting it in moulds like bricks, and then burning it; and the success was answerable to his wishes. In that state it burnt readily into excellent lime; and this acquisition was one of the most important that could have been made. I have heard it asserted more than once that this stroke was better than twenty thousand pounds in the Duke's pocket; but, like most common assertions of the same kind, it is probably an exaggeration. However, whether the discovery was worth five, ten, or twenty thousand, it certainly was of noble use, and forwarded all the works in an extraordinary manner." [1]

It has been stated that Brindley's nervous excitement was so great on the occasion of the letting of the water into the canal, that he took to his bed at the Wheat-sheaf, in Stretford, and lay there until all cause for apprehension was over. The tension on his brain must have been great, with so tremendous a load of work and anxiety upon him; but that he " ran away," [2] as some of

---

[1] 'Six Months' Tour,' vol. iii., p. 270-1. Mr. Hughes, C.E., says of this discovery : " The lime thus made would appear to be the first cement of which we have any knowledge in this country; since the calcareous marl here spoken of would probably produce, when burnt, a lime of strong hydraulic properties."

[2] This story was first set on foot, we believe, by the Earl of Bridge-water, in his singularly incoherent publication entitled, ' A Letter to the Parisians and the French Nation upon Inland Navigation, containing a defence of the public character of His Grace Francis Egerton, late Duke of Bridgewater. By the Hon. Francis Henry Egerton.' The first part of

this curious book (published at Paris) was dated " Hôtel Egerton, Paris, 21st Dec., 1818;" the second part was published two years later; and a third part, consisting entirely of a note about Hebrew interpretations, was published subsequently. He had in the mean time become Earl of Bridge-water, in October, 1823, having formerly been prebendary of Durham and rector of Whitchurch in Shropshire. The late Earl of Ellesmere, in his ' Essays on History, Biography,' &c., says of this nobleman that " he died at Paris in the odour of eccentricity." But this is a mild description of his lordship, who had at least a dozen distinct crazes—about canals, the Jews, punctuation, the wonderful his detractors have alleged, is at variance with the whole character and history of the man.

The Duke's canal, when finished, was for a long time regarded as the wonder of the neighbourhood. Strangers flocked from a distance to see Brindley's " castle in the air ;" and contemporary writers spoke in glowing terms of the surprise with which they saw several barges of great burthen drawn by a single mule or horse along " a river hung in the air," and over another river underneath, by the side of which some ten or twelve men might be seen slowly hauling a single barge against the stream. A lady who writes a description of the work in 1765, speaks of it as " perhaps the greatest artificial curiosity in the world ;" and she states that " crowds of people, including those of the first fashion, resort to it daily." [1] The chief value of the work, however, consisted in its uses. Manchester was now regularly and cheaply supplied with coals. The average price was at once reduced by one-half—from $7d.$ the cwt. to $3\frac{1}{2}d.$ (six score being given to the cwt.)—and the supply was regular instead of intermitting, as it had formerly been. But the full advantages of this improved supply of coals were not experienced until many years after the opening of the canal, when the invention of the steam-engine, and its extensive employment as a motive power in all manufacturing operations, rendered a cheap and abundant supply of fuel of such vital importance to the growth and prosperity of Manchester and its neighbourhood. *

---

merits of the Egertons, the proper translation of Hebrew, the ancient languages generally, but more especially about prophecy and poodle-dogs. When he drove along the Boulevards in Paris, nothing could be seen of his lordship for poodle-dogs looking out of the carriage-windows. The poodles sat at table with him at dinner, each

being waited on by a special valet. The most creditable thing the Earl did was to leave the sum of 12,000*l.* to the British Museum, and 8000*l.* to meritorious literary men for writing the well-known 'Bridgewater Treatises.' He died in February, 1829.

[1] Mr. Newbery's 'Lady's Pocket-book.'

CHAPTER IV.

THE CANAL had scarcely been opened to Manchester when we find Brindley occupied, at the instance of the Duke, in surveying the country between Stretford and the river Mersey, with the object of carrying out a canal in that direction for the accommodation of the growing trade between Liverpool and Manchester. The first boat-load of coals sailed over the Barton viaduct to Manchester on the 17th of July, 1761, and on the 7th of September following we find Brindley at Liverpool,[1] " rocconitoring ;" and, by the end of the month, he is busily engaged in levelling for a proposed canal to join the Mersey at Hempstones, about eight miles below Warrington Bridge, from whence there was a natural tideway to Liverpool, about fifteen miles distant.

The project in question was a very important one on public grounds. We have seen how the community of Manchester had been hampered by its defective road and water communications, which seriously affected its supplies of food and fuel, and, at the same time, by retarding its trade, hindered to a considerable extent the regular employment of its population. The Duke of

---

[1] It would almost seem as if the extension of the canal to the Mersey had formed part of the Duke's original plan; for Brindley was engaged in making a survey from Longford to Dunham in the autumn of the preceding year, as appears from the following account of Brindley's expenses in making the survey, preserved at the Bridgewater Canal Office at Manchester :—

" Expenses in Surveying from Longford Bridge to Dunham.

Octr 21st 1760.

Spent at Stretford	0	6
At Altringham all Night	6	0
Gave the Men to Drink that assisted	1	0
22nd		
More at Altringham	2	6
	10	0

P^d Mr. Brinley this."

Bridgewater, by constructing his canal, had opened up an abundant supply of coal, but the transport of the raw materials of manufacture was still as much impeded as before. Liverpool was the natural port of Manchester, from which it drew its supplies of cotton, wool, silk, and other produce, and to which it returned them for export when worked up into manufactured articles.

There were two existing modes by which the communication was kept up between the two places : one was by the ordinary roads, and the other by the rivers Mersey and Irwell. From a statement published in December, 1761, it appears that the quantity of goods then carried by land from Manchester to Liverpool was " upwards of forty tons per week," or about two thousand tons a year. This quantity of goods, insignificant though it appears when compared with the enormous traffic now passing between the two towns, was then thought very large, as no doubt it was when the very limited trade of the country was taken into account. But the cost of transport was the important feature ; it was not less than two pounds sterling per ton —this heavy charge being almost entirely attributable to the execrable state of the roads. It was scarcely possible to drive waggons along the ruts and through the sloughs which lay between the two places at certain seasons of the year, and even pack-horses had considerable difficulty in making the journey.

The other route between the towns was by the navigation of the rivers Mersey and Irwell. The raw materials used in manufacture were principally transported from Liverpool to Manchester by this route, at the cost of about twelve shillings a ton ; the carriage of timber and such like articles costing not less than twenty per cent. on their value at Liverpool. But the navigation was also very tedious and difficult. The boats could only pass up to the first lock at the Liverpool end with the assistance of a spring-tide ; and further up the river

there were numerous fords and shallows which the boats
could only pass in great freshes, or, in dry seasons, by
drawing extraordinary quantities of water from the
locks above. Then, in winter, the navigation was apt
to be impeded by floods, and occasionally it was stopped
altogether. In short, the growing wants of the population demanded an improved means of transit between
the two towns, which the Duke of Bridgewater now
determined to supply.

The growth of Liverpool as a seaport had been comparatively recent. At a time when Bristol and Hull
possessed thriving harbours, resorted to by foreign ships,
Liverpool was little better than a fishing-village, its
only distinction being that it was a convenient place for
setting sail to Ireland. In the war between France
and England which broke out in 1347, when Edward
the Third summoned the various ports in the kingdom
to make contributions towards the naval power according to their means, London was required to provide
25 ships and 662 men ; Bristol, 22 ships and 608 men ;
Hull, 16 ships and 466 men ; whilst Liverpool was only
asked to find 1 bark and 6 men ! In Queen Elizabeth's
time, the burgesses presented a petition to Her Majesty,
praying her to remit a subsidy which had been imposed
upon the seaport and other towns, in which they styled
their native place " Her Majesty's poor decayed town of
Liverpool." Chester was then of considerably greater
importance as a seaport. In 1634-5, when Charles I.
made his unconstitutional levy of ship-money throughout England, Liverpool was let off with a contribution
of 15*l.*, whilst Chester paid 100*l.*, and Bristol not less
than 1000*l.* The channel of the Dee, however, becoming
silted up, the trade of Chester decayed, and that of
Liverpool rose upon its ruins. In 1699 the excavation
of the old dock was commenced ; but it was used only as
a tidal harbour (being merely an enclosed space with a
small pier) until the year 1709, when an Act was

obtained enabling its conversion into a wet dock; since
which time a series of docks have been constructed,
extending for about five miles along the north shore
of the Mersey, which are among the greatest works
of modern times, and afford an unequalled amount of
shipping accommodation.

LIVERPOOL IN 1650.
[From Troughton's History of Liverpool.

From that time forward the progress of the port of
Liverpool kept steady pace with the trade and wealth
of the country behind it, and especially with the manufacturing activity and energy of the town of Manchester.
Its situation at the mouth of a deep and navigable river,
in the neighbourhood of districts abounding in coal and
iron, and inhabited by an industrious and hardy population, were unquestionably great advantages.   But these
of themselves would have been insufficient to account for
the extraordinary progress made by Liverpool within
the last century, without the opening up of the great
system of canals, which brought not only the towns of
Yorkshire, Cheshire, and Lancashire into immediate
connection with that seaport, but also the manufacturing

districts of Staffordshire, Warwickshire, and the other
central counties of England situated at the confluence of
these various navigations.[1] Liverpool thus became the
great focus of import and export for the northern and
western districts. The raw materials of commerce were
poured into it from Ireland, America, and the Indies ;
and from thence they were distributed along the canals
amongst the various seats of manufacturing industry,
returning mostly by the same route to the same port
for shipment to all parts of the world.

At the time of which we speak, however, it will be
observed that the communication between Liverpool and
Manchester was as yet very imperfect. It was not only
difficult to convey goods between the two places, but it
was also difficult to convey persons. In fine weather,
those who required to travel the thirty miles which
separated them, could ride or walk, resting at Warrington
for the night. But in winter the roads, like most of the
other country roads at the time, were simply impassable.
Although an Act had been passed as early as the year
1726 for repairing and enlarging the road from Liverpool to Prescott, coaches could not come nearer to the
town than Warrington in 1750, the road being impracticable for such vehicles even in summer.[2]

A stage-coach was not started between Liverpool and
Manchester until the year 1767, performing the journey
only three times a-week. It required six and sometimes
eight horses to draw the lumbering vehicle and its load
along the ruts and through the sloughs,—the whole day
being occupied in making the journey. The coach was
accustomed to start early in the morning from Liverpool ;

---

[1] Progress of Liverpool.

Years.	Vessels entered.	Tonnage.	Duties Paid.
1701	102	8,619	..
1760	1,245	..	£2,330
1800	4,746	450,060	23,379
1858	21,352	4,441,943	347,889

[2] Mr. Baines says : " Carriages

were then very rare, and it is mentioned as a singular fact that at the
period in question (1750) there was
but one *gentleman's* carriage in the
town of Liverpool, and that carriage
was kept by a *lady* of the name of
Clayton."—'History of Lancashire,'
vol. iv., p. 90.

it breakfasted at Prescott, dined at Warrington, and arrived at Manchester usually in time for supper. On one occasion, at Warrington, the coachman intimated his wish to proceed, when the company requested him to take another pint, as they had not finished their wine, asking him at the same time if he was in a hurry? "Oh," replied the driver, "I'm not partic'lar to an hour or so!" As late as 1775, no mail-coach ran between Liverpool and any other town, the bags being conveyed to and from it on horseback; and one letter-carrier was found sufficient for the wants of the place. A heavy stage then ran, or rather crawled, between Liverpool and London, making only four journeys a-week in the winter time. It started from the Golden Talbot, in Water-street, and was three days on the road. It went by Middlewich, where one of its proprietors kept the White Bear inn; and during the Knutsford race-week the coach was sent all the way round by that place, in order to bring customers to the Bear.

We have said that Brindley was engaged upon the preliminary survey of a canal to connect Manchester with the Mersey, immediately after the original Worsley line had been opened, and before its paying qualities could as yet be ascertained. But the Duke, having once made up his mind as to the expediency of carrying out this larger project, never halted nor looked back, but made arrangements for prosecuting a bill for the purpose of enabling the canal to be made in the very next session of Parliament. We find that Brindley's first visit to Liverpool and the intervening district on the business of the survey was made early in September, 1761. During the remainder of the month he was principally occupied in Staffordshire, looking after the working of his fire-engine at Fenton Vivian, carrying out improvements in the silk-manufactory at Congleton, and inspecting various mills at Newcastle-under-Lyme and the neighbourhood.

His only idle day during that month seems to have been the 22nd, which was a holiday, for he makes the entry in his book of " crounation of Georg and Sharlot," the new King and Queen of England.   By the 25th we find him again with the Duke at Worsley, and on the 30th he makes the entry, " set out at Dunham to Level for Liverpool."   The work then went on continuously ; the survey was completed ; and on the 19th of November he set out for London, with 7*l*. 18*s*. in his pocket.

In the course of his numerous journeys, we find Brindley carefully noting down the various items of his expenses, which were curiously small.   Although he was four or five days on the road to London, and stayed eight days in town, his total expenses, both going and returning, amounted to only 4*l*. 8*s*. ; though it is most probable that he lived at the Duke's house whilst in town.   On the 1st of December we find him, on his return journey to Worsley, resting the first night at a place called Brickhill ; the next at Coventry, where he makes the entry, " Moy mar had a bad fall in the frasst ;" the third at Sandon ; the fourth at Congleton ; and the fifth at Worsley.   He had still some inquiries to make as to the depth of water and the conditions of the tide at Hempstones ; and for three days he seems to have been occupied in traffic-taking, with a view to the evidence to be given before Parliament ; for on the 10th of December we find him at Stretford, " to count the caridgos," and on the 12th he is at Manchester for the same purpose, " counting the loded caridgos and horses."   The following bill refers to some of the work done by him at this time, and is a curious specimen of an engineer's travelling charges in those days—the engineer himself being at the same time paid at the rate of 3*s*. 6*d*. a day :—

Expenses for His Grace the Duk of Bridgwator to pay for traveling Chareges by James Brindley.

18 Novem—1761.

18 No masuring a Cros from Dunham to Warbuton Mercey and Thalwall, 3s - 11d Dunham for 2 diners 1s - 3d for the man 1s - 0d at Thalwall 1s - 2d all Night Warington	0	7	4
19 Novem Sat out from Chester for London & at Worsley Septm 5 Retorned back going to London and at London & hors back to Worsley Charged Hors & my salf	4	8	0
9 december Coming back from Ham Stone Charges at Wilderspool all Night	0	8	0
at Warington to meet Mr Ashley dining	0	4	2
10 to ataind the Turn pike Rode 2s - 6d & againe on te 12 De Rode 3s - 6d	0	6	0
21 Decm to inspect te flux and Reflux at Ham Stone 2 dayes Charges	0	6	6

26 Decr 1761. Recd the Contents of the above Bill by the Hands of John Gilbert. James Brindley £6 00 0

In the early part of the month of January, 1762, we find Brindley busy measuring soughs, gauging the tides at Hempstones, and examining and altering the Duke's paper-mills and iron slitting-mills at Worsley; and on the 7th we find this entry: " to masuor the Duks pools I and Smeaton." On the following day he makes " an ochilor survey from Saldnoor [Sale Moor] to Stockport," with a view to a branch canal being carried in that direction. On the 14th, he sets out from Congleton, by way of Ashbourne, Northampton, and Dunstable, arriving in London on the fifth day. Immediately on his arrival in town we find him proceeding to rig himself out in a new suit of clothes. His means were small, his habits thrifty, and his wardrobe scanty; but as he was about to appear in an important character, as the principal engineering witness before a Parliamentary Committee in support of the Duke's bill, he felt it necessary to incur an extra expenditure on dress for the occasion. Accordingly, on the morning of the 18th we find him expending a guinea—an entire week's pay—in the purchase of a pair of new breeches; two guineas on a coat and waistcoat of broadcloth, and six shillings for a

pair of new shoes.   The subjoined is a facsimile of the
entry in his pocketbook.

FAC-SIMILE OF BRINDLEY'S HAND-WRITING.

It will be observed that an expenditure is here entered
of nine shillings for going to "the play."   It would
appear that his friend Gilbert, who was in London with
him on the canal business, prevailed on Brindley to
go with him to the theatre to see Garrick in the play of
'Richard III.,' and he went.   He had never been to an
entertainment of the kind before; but the excitement
which it caused him was so great, and it so completely
disturbed his ideas, that he was unfitted for business for
several days after.   He then declared that no consideration
should tempt him to go a second time, and he held to

his resolution. This was his first and only visit to the
play. The following week he enters himself in his
memorandum-book as ill in bed, and the first Sunday
after his recovery we find him attending service at
" Sant Mary's Church." The service did not make him
ill, as the play had done, and on the following day he
attended the House of Commons on the subject of the
Duke's bill.

The proposed canal from Manchester to the Mersey at
Hempstones stirred up an opposition which none of the
Duke's previous bills had encountered. Its chief opponents were the proprietors of the Mersey and Irwell
navigation, who saw their monopoly assailed by the
measure ; and, unable though they had been satisfactorily
to conduct the then traffic between Liverpool and Manchester, they were unwilling to allow of any additional
water service being provided between the two towns.
Having already had sufficient evidence of the Duke's
energy and enterprise, from what he had been able to
effect in so short a time in forming the canal between
Worsley and Manchester, they were not without reason
alarmed at his present project. At first they tried to
buy him off by concessions. They offered to reduce the
rate of 3s. 4d. per ton of coals, timber, &c., conveyed
upon the Irwell between Barton and Manchester, to 6d.
if he would join their navigation at Barton and abandon
the part of his canal between that point and Manchester :
but he would not now be diverted from his plan, which
he resolved to carry into execution if possible. Again
they tried to conciliate his Grace by offering him certain
exclusive advantages in the use of their navigation. But
it was again too late; and the Duke, having a clear
idea of the importance of his project, and being assured
by his engineer of its practicability and the great commercial value of the undertaking, determined to proceed
with the measure. It offered to the public the advantages of a shorter line of navigation, not liable to be

interrupted by floods on the one hand or droughts on the other, and, at the same time, a much lower rate of freight, the maximum charge proposed in the bill being 6s. a ton against 12s., the rate charged by the Mersey and Irwell navigation between Liverpool and Manchester.

The opposition to the bill was led by Lord Strange, son of the Earl of Derby, one of the members for the county of Lancaster, who took the part of the " Old Navigators," as they were called, in resisting the bill. The question seems also to have been treated as a political one; and, the Duke and his friends being Whigs, Lord Strange mustered the Tory party strongly against him. Hence we find this entry occurring in Brindley's note-book, under date the 16th of February : " The Toores [Tories] mad had [made head] agane ye Duk." The principal objections put forward to the proposed canal were, that the landowners would suffer by it from having their lands cut through and covered with water, by which a great number of acres would be for ever lost to the public ; that there was no necessity whatever for the canal, the Mersey and Irwell navigation being sufficient to carry more goods than the then trade could supply ; that the new navigation would run almost parallel with the old one, and offered no advantage to the public which the existing river navigation did not supply ; that the canal would drain away the waters which supplied the rivers, and be very prejudicial to, if not a total obstruction of them in dry seasons ; that the proprietors of the old navigation had invested their money on the faith of Parliament, and to permit the new canal to be established would be a gross interference with their vested rights ; and so on. To these objections there were very sufficient answers. The bill provided for full compensation being made to the owners of lands through which the canal passed, and, in addition, it was provided that all sorts of manure should be carried for them ton-

nage free. It was also shown that the Duke's canal
could not abstract water from either the Mersey or the
Irwell, as the level of both rivers was considerably below
that of the intended canal, which would be supplied
almost entirely from the drainage of his own coal-mines
at Worsley; and with respect to the plea of vested
rights set up, it was shown that Parliament, in granting
certain powers to the old navigators, had regard mainly
to the convenience and advantage of the public; and
they were not precluded from empowering a new navigation to be formed if it could be proved to present a
more convenient and advantageous mode of conveyance.
And on this ground the Duke was strongly supported
by the inhabitants of the locality proposed to be served
by the intended canal. The "Junto of Old Navigators,"
as they were termed,[1] had for many years carried things
with a very high hand, extorted the highest rates, and,
in cases of loss by delay or damage to goods in transit,
refused all redress. A feeling very hostile to them and
their monopoly had accordingly grown up, which now
exhibited itself in a powerful array of petitions to Parliament in favour of the Duke's bill.

On the 17th of February, 1762, the bill came before the Committee of the House of Commons, and
Brindley gave his evidence in its support. We regret
that no copy of this evidence now exists[2] from which
we might have formed an opinion of the engineer's
abilities as a witness. Some curious anecdotes have,
however, been preserved of his appearance as a witness on canal bills before Parliament. When asked,
on one occasion, to produce a drawing of an intended

---

[1] Letter of John Hart to the Gentlemen and Tradesmen at Warrington,
Dec. 21st, 1761.
[2] Search has been made at the
Bridgewater Estate Offices at Manchester, and in the archives of the
Houses of Parliament, but no copy
can be found. It is probable that the
Parliamentary papers connected with
this application to Parliament were
destroyed by the fire which consumed
so many similar documents some
twenty years ago.

bridge, he replied that he had no plan of it on paper, but he would illustrate it by a model. He went out and bought a *large cheese*, which he brought into the room and cut into two equal parts, saying, " Here is my model." The two halves of the cheese represented the semicircular arches of his bridge ; and by laying over them some long rectangular object he could thus readily communicate to the committee the position of the river flowing underneath and the canal passing over it.[1]   On another occasion, when giving his evidence, he spoke so frequently about " puddling," describing its uses and advantages, that some of the members expressed a desire to know what this extraordinary mixture was that could be applied to so many and important purposes. Preferring a practical illustration to a verbal description, Brindley caused a mass of clay to be brought into the committee-room, and, moulding it in its raw untempered state into the form of a trough, he poured into it some water, which speedily ran through and disappeared. He then worked the clay up with water to imitate the process of puddling, and again forming it into a trough, filled it with water, which was now held in without a particle of leakage.   " Thus it is," said Brindley, " that I form a water-tight trunk to carry water over rivers and valleys, wherever they cross the path of the canal." [2] On another occasion, when Brindley was giving evidence before a committee of the House of Peers as to the lockage of his proposed canal, one of their Lordships asked him, " But what is a lock ? " on which the engineer took a piece of chalk from his pocket and proceeded to explain it by means of a diagram which he drew upon the floor, and made the matter clear at once.[3]

---

[1] Stated by Mr. Hughes, in his 'Memoir of Brindley,' as having been communicated by James Loch, Esq., M.P., the agent for the Duke's Trustees.

[2] 'Memoir of Brindley,' by S.

Hughes, C.E., in 'Weale's Papers on Civil Engineering.'

[3] As the reader may possibly desire information on the same point, we may here briefly explain the nature of a Canal Lock.   It is employed

On the day following Brindley's examination before the Committee on the Duke's bill, that is, on the 18th

as a means of carrying navigations through an uneven country, and raising the boats from one water level to another, or *vice versá*. The lock is a chamber formed of masonry, occupying the bed of the canal where the difference of level is to be overcome. It is provided with two pairs of gates, one at each end; and the chamber is so contrived that the level of the water which it contains may be made to coincide with either the higher level above, or the lower level below it. The following diagrams will explain the form and construction of the lock. A represents what is called the upper pond, B the lower, C is the left wall, and DD side culverts. When the gates at the lower end of the chamber (E) are opened, and those at the upper end (F) are closed, the water in the chamber will stand at the lower level of the canal; but when the lower gates are closed, and the upper gates are opened, the water will naturally coincide with that in the upper part of the canal. In the first case, a boat may be floated into the lock from the lower part, and then, if the lower gates be closed and

water is admitted from the upper level, the canal-boat is raised, by the depth of water thus added to the lock, to the upper level, and on the complete opening of the gates it is thus floated onward. By reversing the process, it will readily be understood how the boat may, in like manner, be lowered from the higher to the lower level. The greater the lift or the lowering, the more water is consumed in the process of exchange from one level to another; and where the traffic of the canal is great, a large supply of water is required to carry it on, which is usually provided by capacious reservoirs situated above the summit level. Various expedients are adopted for economising water : thus, when the width of the canal will admit of it, the lock is made in two compartments, communicating with each other by a valve, which can be opened and shut at pleasure; and by this means one-half of the water which it would otherwise be necessary to discharge to the lower level may be transferred to the other compartment.

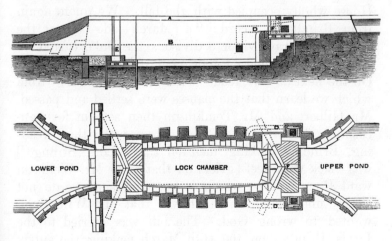

LONGITUDINAL SECTION AND PLAN OF LOCK.

of February, we find him entering in his note-book that
the Duke sent out " 200 leators " to members—friends of
the measure; doubtless containing his statement of reasons
in favour of the bill.   On the 20th Mr. Tomkinson, the
Duke's solicitor, was under examination for four hours
and a half.   Sunday intervened, on which day Brindley
records that he was " at Lord Harrington's."   On the
following day, the 22nd, the evidence for the bill was
finished, and the Duke followed this up by sending out
250 more letters to members, with an abstract of the
evidence given in favour of the measure.   On the 26th
there was a debate of eight hours on the bill, followed by
a division, in Committee of the whole House, thus
recorded by Brindley :—

> " ad a grate Division of 127 fort Duk
>   98 nos
>   ——
> for t^e Duk 29 Me Jorete "

But the bill had still other discussions and divisions
to encounter before it was safe.   The Duke and his
agents worked with great assiduity.   On the 3rd of
March he caused 250 more letters to be distributed
amongst the members ; and on the day after we find the
House wholly occupied with the bill.   We quote again
from Brindley's record : " 4 [March] ade bate at the
Hous with grate vigor 3 divisons the Duke carred by
Numbers evory time a 4 division moved but Noos yelded."
On the next day we read " wont thro the closos ;" from
which we learn that the clauses were settled and passed.
Mr. Gilbert and Mr. Tomkinson then set out for Lancashire : the bill was safe.   It passed the third reading, Brindley making mention that " Lord Strange "
was " sick with geef [grief] on that affair Mr. Wellbron
want Rong god,"—which latter expression we do not
clearly understand, unless it was that Mr. Wilbraham
wanted to wrong God.   The bill was carried to the
Lords, Brindley on the 10th March making the entry,

"Touk the Lords oath." But the bill passed the Upper
House "without opposishin," and received the Royal
Assent on the 24th of the same month.

On the day following the passage of the bill through
the House of Lords (of which Brindley makes the
triumphant entry, "Lord Strange defetted "), he set out
for Lancashire, after nine weary weeks' stay in London.
To hang about the lobbies of the House and haunt the
office of the Parliamentary agent, must have been exces-
sively irksome to a man like Brindley, accustomed to
incessant occupation and to see work growing under his
hands. During this time we find him frequently at the
office of the Duke's solicitor in "Mary Axs;" sometimes
with Mr. Tomkinson, who paid him his guinea a-week
during the latter part of his stay; and on several occasions
he is engaged with gentlemen from the country, advising
them about "saltworks at Droitwitch" and mill-arrange-
ments in Cheshire. Many things had fallen behind
during his absence and required his attention, so he at
once set out home; but the first day, on reaching Dunstable, he was alarmed to find that his mare, so long
unaccustomed to the road, had "allmost lost ye use of her
Limes" [limbs]. He therefore pushed on slowly, as
the mare was a great favourite with him—his affection
for the animal having on one occasion given rise to a
serious quarrel between him and Mr. Gilbert—and he
did not reach Congleton until the sixth day after his
setting out from London. He rested at Congleton for
two days, during which he "settled the geering of the
silk-mill," and then proceeded straight on to Worsley to
set about the working survey of the new canal.

CHAPTER V.

THE course of this important canal, which unites the trade of Manchester with the port of Liverpool, is about twenty-four miles in length.[1] From Longford Bridge, near Manchester, its course lies in a south-westerly direction for some distance, crossing the river Mersey at a point about five miles above its junction with the Irwell. At Altrincham it proceeds in a westerly direction, crossing the river Bollin about three miles further on, near Dunham. After crossing the Bollin, it describes a small semicircle, proceeding onward in the valley of the Mersey, and nearly in the direction of the river as far as the crossing of the high road from Chester to Warrington. It then bends to the south to preserve the high level, passing in a southerly direction as far as Preston, in Cheshire, from whence it again turns round to the north to join the river Mersey.

The canal lies entirely in the lower part of the new red sandstone, the principal earthworks consisting of the clays, marls, bog-earths, and occasionally the sandstones of this formation. The heaviest bog to be crossed was that of Sale Moor, on a bottom of quicksand, west of the Mersey, the construction of the canal at this part being a work of as much difficulty as the laying of the

---

[1] The following statement of the lengths of the different portions of the Duke's canal, including those originally executed, is from the map published by Brindley in 1769:—

	Miles.	furl.	chains.	
From Worsley to Longford Bridge	6	0	0	Level.
„ Longford Bridge to Manchester	4	2	0	„
„ Longford Bridge to Preston Brook	19	0	0	„
„ Preston Brook to upper part of Runcorn	4	4	0	„
„ Upper part of Runcorn to the Mersey	0	5	7	79 feet fall.

railroad upon Chat Moss some sixty years later. But Brindley, like Stephenson, looked upon a difficulty as a thing to be overcome ; and a difficulty no sooner presented itself, than he at once set his wits to work to study how it was best to be grappled with and surmounted. There were also a large number of brooks to be crossed, and two important rivers, involving a number of aqueducts, bridges, and culverts, to provide for the accommodation of the district. It will, therefore, be obvious that this undertaking was of a much more formidable character—more difficult for the engineer and much more costly to the noble proprietor—than the comparatively limited and inexpensive work between Worsley and Manchester, which we have above described.

The capital idea which Brindley early formed and determined to carry out, was to construct a level of dead water all the way from Manchester to a point as near to the junction of the canal with the Mersey as might be found practicable. Such a canal, he clearly saw, would not be so expensive to work as one furnished with locks at intermediate points. Brindley's practice of securing long levels of water in canals was in many respects similar to that of George Stephenson with reference to flat gradients upon railways ; and in all the canals that he constructed, he planned and carried them out upon this leading principle. Hence the whole of the locks on the Duke's canal were concentrated at its lower end near Runcorn, where the navigation descended by a flight of locks into the river Mersey. Lord Ellesmere has observed that this uninterrupted level of the Bridgewater Canal from Leigh and Manchester to Runcorn, and the concentration of its descent to the Mersey at the latter place, have always been considered as among the most striking evidences of the genius and skill of Brindley.

There was, as usual, considerable delay in obtaining possession of the land on which to commence the works. The tenants required a certain notice, which must neces-

sarily expire before the Duke's engineer could take possession ; and many obstacles were thrown in the way both by
tenants and landlords hostile to the undertaking. In many
cases the Duke had to pay dearly for the land purchased
under the compulsory powers of his Act. Near Lymm,
the canal passed through a little bit of garden belonging
to a poor man's cottage, the only produce growing
upon the ground being a pear-tree. For this the Duke
had to pay thirty guineas, and it was thought a very
extravagant price at that time. Since the introduction
of railways, the price would probably be considered ridi-
culously low. For the land on which the warehouses
and docks were built at Manchester, the Duke had in all
to pay the much more formidable sum of about forty
thousand pounds.

The Old Quay Navigation (by which the Mersey and
Irwell Company was called), even at this late moment,
thought to delay if not to defeat the Duke's operations,
by lowering their rates nearly one-half. Only a few days
after the Royal Assent had been given to the bill, they
published an announcement, appropriately dated the 1st
of April, setting forth the large sacrifice they were about
to make, and intimating that "from this Reduction in
the Carriage a real and permanent Advantage will arise
to the Public, and they will experience that Utility so
cried up of late, which has hitherto only existed in promises." The blow was aimed at the Duke, but he heeded
it not : he was more than ever resolved to go on with his
canal. He was even offered the Mersey navigation at
the price of thirteen thousand pounds ; but he would not
now have it at any price.

The public spirit and enterprise displayed by many
of the young noblemen of those days was truly admirable.
Brindley had for several years been in close personal
communication with Earl Gower as to the construction
of the canal intended to unite the Mersey with the Trent
and the Severn, and thus connect the ports of Liverpool,

Hull, and Bristol, by a system of inland water-communication.* With this object, as we have seen, he had often visited the Earl at his seat at Trentham, and discussed with him the plans by which this truly magnificent enterprise was to be carried out; and he had frequently visited the Earl of Stamford at his seat at Enville for the same purpose. But those schemes were too extensive and costly to be carried out by the private means of either of these noblemen, or even by both combined. They were, therefore, under the necessity of stirring up the latent enterprise of the landed proprietors in their respective districts, and waiting until they had received a sufficient amount of local support to enable them to act with vigour in carrying their great design into effect. The Duke of Bridgewater's scheme of uniting Manchester and Liverpool by an entirely new line of water-communication, cut across bogs and out of the solid earth in some places, and carried over rivers and valleys at others by bridges and embankments, was scarcely less bold or costly. Though it was spoken of as another of the Duke's "castles in the air," and his resources were by no means overflowing at the time he projected it, he nevertheless determined to enter upon the undertaking, and to go on alone with it though no one else should join him. The Duke thus proved himself a real Dux or leader of the industrial enterprise of his district; and by cutting his canal, and providing a new, short, and cheap water-way between Liverpool and Manchester, which was afterwards extended through the counties of Chester, Stafford, and Warwick, he thus unquestionably paved the way for the creation and development of the modern manufacturing system existing in the north-western counties of England. We need scarcely say how admirably he was supported throughout by the skill and indefatigable energy of his engineer. Brindley's fertility in resources was the theme of general admiration. Arthur Young, who visited the works during their progress, speaks with enthusiastic

admiration of his " bold and decisive strokes of genius,"
his " penetration which sees into futurity, and prevents
obstructions unthought of by the vulgar mind, merely
by foreseeing them : a man," says he, " with such ideas,
moves in a sphere that is to the rest of the world ima-
ginary, or at best a *terra incognita.*" *

It would be uninteresting to describe the works of the
Bridgewater Canal in detail : one part of a canal is
usually so like another, that to do so were merely to
involve a needless amount of repetition of a necessarily
dry description. We shall accordingly content ourselves
with referring to the original methods by which Brindley
contrived to overcome the more important difficulties of
the undertaking. From Longford Bridge, where the
new works commenced, the canal, which was originally
about eight yards wide and four feet deep, was carried
upon an embankment of about a mile in extent across
the valley of the Mersey. One might naturally suppose
that the conveyance of such a mass of earth must have
exclusively employed all the horses and carts in the
neighbourhood for years. But Brindley, with his usual
fertility in expedients, contrived to make the construction
of the canal itself subservient to the completion of the
remainder. He had the stuff required to make up the
embankment brought in boats partly from Worsley and
partly from other parts of the canal where the cutting
was in excess ; and the boats, filled with this stuff, were
conducted from the canal along which they had come
into caissons or cisterns placed at the point over which
the earth and clay had to be deposited.

The boats, being double, fixed within two feet of each
other, had a triangular trough supported between them
of sufficient capacity to contain about seventeen tons of
earth. The bottom of this trough consisted of a line of
trap-doors, which flew open at once on a pin being
drawn, and discharged their whole burthen into the
bed of the canal in an instant. Thus the level of

the embankment was raised to the point necessary to
enable the canal to be carried forward to the next
length. Arthur Young was of opinion that the saving
effected by constructing the Stretford embankment in
this way, instead of by carting the stuff, was equivalent
to not less than five thousand per cent.! The materials of
the caissons employed in executing this part of the work
were afterwards used in forming temporary locks across
the valley of the Bollin, whilst the embankment was
being constructed at that point by a process almost the
very reverse, but of like ingenuity.

BRINDLEY'S BALLAST BOATS.

In the same valley of the Mersey the canal had to be
carried over a large brook subject to heavy floods, by
means of a strong bridge of two arches, adjoining which
was a third, affording provision for a road. Further on,
the canal was carried over the Mersey itself upon a
bridge with one arch of seventy feet span. Westward
of this river lay a very difficult part of the work, occasioned by the carrying of the navigation over the Sale
Moor Moss. Many thought this an altogether impracticable thing; as not only had the hollow trunk of earth
in which the canal lay to be made water-tight, but to
preserve the level of the water-way it must necessarily be
raised considerably above the level of the Moor across
which it was to be laid. Brindley overcame the difficulty
in the following manner. He made a strong casing of
timber-work outside the intended line of embankment on
either side of the canal, by placing deal balks in an erect
position, backing and supporting them on the outside
with other balks laid in rows, and fast screwed together;
and on the front side of this woodwork he had his earth-

work brought forward, hard rammed, and puddled, to form the navigable canal; after which the casing was moved onward to the part of the work further in advance, and the bottom having previously been set with rubble and gravel, the embankment was thus carried forward by degrees, the canal was raised to the proper level, and the whole was substantially and satisfactorily finished.

A steam-engine of Brindley's contrivance was erected at Dunham Town Bridge to pump the water from the foundations there. The engine was called a Sawney, for what reason is not stated, and, for long after, the bridge was called Sawney's Bridge. The foundations of the under-bridge, near the same place, were popularly supposed to be set on quicksand; and old Lord Warrington, when he had occasion to pass under it, would pretend cautiously to look about him, as if to examine whether the piers were all right, and then run through as fast as he could. A tall poplar-tree stood at Dunham Banks, on which a board was nailed showing the height of the canal level, and the people long after called the place by the name of "The Duke's Folly," believing his scheme to be altogether impracticable. But the skill of the engineer baffled these and other prophets of evil; and the success of his expedients, in nearly every case of difficulty that occurred, must certainly be regarded as remarkable, considering the novel and unprecedented character of the undertaking.

Brindley invariably contrived to economise labour as much as possible, and many of his expedients with this object were very ingenious. So far as he could, he endeavoured to make use of the canal itself for the purpose of forwarding the work. He had a floating blacksmith's forge and shop, provided with all requisite appliances, fitted up in one barge; a complete carpenter's shop in another; and a mason's shop in a third; all of which were floated on as the canal advanced, and were thus always at hand to supply the requisite facilities for prosecuting the opera-

tions with economy and despatch. Where there was a
break in the line of work, occasioned, for instance, by
the erection of some bridge not yet finished, the engineer had similar barges constructed and carried by
land to other lengths of the canal which were in progress, where they were floated and advanced in like
manner for the use of the workmen. When the bridge
across the Mersey, which was pushed on with all despatch
with the object of economising labour and cost of materials, was completed, the stone, lime, and timber were
brought along the canal from the Duke's own property
at Worsley, as well as supplies of clay for the purpose
of puddling the bottom of the water-way; and thus the
work rapidly advanced at all points.

As one of the great objections made to the construction
of the canal had been the danger threatened to the surrounding districts by the bursting of the embankments,
Brindley made it his object to provide against the occurrence of such an accident by an ingenious expedient.
He had stops or floodgates contrived and laid in various
parts of the bed of the canal, across its bottom, so that,
in the event of a breach occurring in the bank and a
rush of waters taking place, the current which must
necessarily set in to that point should have the effect of
immediately raising the valvular floodgates, and so shut-
ting off the stream and preventing the escape of more
water than was contained in the division between the
two nearest gates on either side of the breach. At the
same time, these floodgates might be used for cutting
off the waters of the canal at different points, for the
purpose of making any necessary repairs in particular
lengths; the contrivance of waste tubes and plugs being
so arranged that the bed of any part of the canal, more
especially where it passed over the bridges, might be laid
bare in a few hours, and the repairs executed at once.
In devising these ingenious expedients, it ought to be
remembered that Brindley had no previous experience

to fall back upon, and possessed no knowledge of the means which foreign engineers might have adopted to meet similar emergencies. All had been the result of his own original thinking and contrivance; and, indeed, many of these devices were altogether new and original, and had never before been tried by any engineer.

It is curious to trace the progress of the works by Brindley's own memoranda, which, though brief, clearly exhibit his marvellous industry and close application to every detail of the business. He seems to have settled with the farmers for their tenant-right, sold and accounted for the wood cut down and the gravel dug out along the line of the canal, paid the workmen employed,[1] laid out the work, measured off the quantities done from time to time, planned and erected the bridges, designed the canal-boats required for conveying the earth to form the embankments, and united in himself the varied functions of land-surveyor, carpenter, mason, brickmaker, boat-builder, paymaster, and engineer. We even find him descending to count bricks and sell grass. Nothing was too small for him to attend to, or too bold for him to attempt when the necessity arose. At the same time we find him contriving a water-plane for the Duke's collieries at Worsley, and occasionally visiting Newchapel, Leek, and Congleton, in Staffordshire, for the purpose of attending to the business on which he still continued to be employed at those places.

---

[1] The following bill is preserved amongst the Bridgewater Canal papers.

Simcox was a skilled mechanic, and acted as foreman of the carpenters :—

" His Grace the Duke of Bridgewater to Sam[l] Simcox.   D[r]

		£.	s.	d.
23 Mar[h] 1760	To 12 days work at 21[d] per	1	1	0
23 Aug[t]	To 6 days more d° at d°	0	10	6
6 Sep[r]	To 8 days more d° at d°	0	14	0
		2	5	6

1 Nov[r] 1760. Rec[d] the Contents above by the Hands of John Gilbert for the Use of Sam[l] Simcox.   P[d]

JAMES BRINDLEY."

The wages of what was called a " right hand man " at that time were from 14d. to 16d. a day, and of a " left-hand man " from 1s. to 14d.

The heavy works at the crossing of the Mersey occupied him almost exclusively towards the end of the year 1763. He was there making dams and pushing on the building of the bridge. Occasionally he enters the words, "short of men at Cornbrook." Indeed, he seems then to have lived upon the works, for we find the almost daily entry of "dined at the Bull, 8*d*." On the 10th of November he makes this entry : "Aftor noon sattled about the size of the arch over the river Marsee [Mersey] to be 66 foot span and rise 16·4 feet." Next day he is "landing balk out of the ould river in to the canal." Then he goes on, "I prosceded to Worsley Mug was corking ye boats the masons woss making the senter of the waire [weir]. Whithᵉ was osing to put the lator side of the water-wheel srouds on I orderd the pit for ye spindle of ye morter-mill to be sunk level with ye canal Mr. Gilbert sade ye 20 Tun Boat should be at ye water mitang [meeting] by 7 o'clock the next morn." Next morning he is on the works at Cornhill, setting "a carpentar to make scrwos" [screws], superintending the gravelling of the towing-path, and arranging with a farmer as to Mr. Gilbert's slack. And so he goes on from day to day with the minutest details of the undertaking.

He was not without his petty "werrets" and troubles either. Brindley and Gilbert do not seem to have got on very well together. They were both men of strong tempers, and neither would tolerate the other's interference. Gilbert, being the Duke's factotum, was accustomed to call Brindley's men from their work, which the other would not brook. Hence we have this entry on one occasion,—"A meshender [messenger] from Mr G I retorned the anser No more sosiety." In fact, they seem to have quarrelled.[1]

---

[1] The Earl of Bridgewater, in his rambling 'Letter to the Parisians,' above referred to, alleges that the quarrel originated in Gilbert's horse breaking into the field where Brindley's mare was grazing—an animal of

We find the following further entries on the subject in Brindley's note-book : "Thursday 17 Novr past 7 o'clock at night   M Gilbert and sun Tom caled on mee at Gorshill and I went with them to ye Coik [sign of the Cock] tha stade all night and the had balk [blank ?] bill of parsill   18 Fryday November 7 morn I went to the Cock and Bruckfast with Gilberts he in davred to imploye ye carpinters at Cornhill in making door and window frames for a Building in Castle field and shades for the mynors in Dito and other things   I want them to Saill Moor   Hee took upon him diriction of ye back drains and likwaise such Lands as be twixt the 2 hous and ceep uper side the large farme and was displesed with such raing as I had pointed out."

Those differences between Brindley and Gilbert seem eventually to have become reconciled, most probably by the mediation of the Duke, for the services of both were alike essential to him ; and we afterwards find them working cordially together and consulting each other as

LONGFORD BRIDGE.

before on any important part of the undertaking.   At the end of the year 1763, by dint of steady work, Long-

---

which he seems to have been very fond,—and the consequence was that the engineer was for a time prevented using the animal in the pursuit of his business.   The Earl says Brindley was under the impression that Gilbert had contrived this out of spite.

ford Bridge was finished and gravelled over, and the embankment was steadily proceeding beyond the Mersey in the manner above described.

Brindley did not want for good workmen to carry out his plans. He found plenty of labourers in the neighbourhood accustomed to hard work, who speedily became expert excavators; and though there was at first a lack of skilled carpenters, blacksmiths, and bricklayers, they soon became trained into such under the vigilant eye of so expert a master as Brindley was. We find him, in his note-book, often referring to the men by their names, or rather byenames, for in Lancashire proper names seem to have been little used at that time. " Black David " was one of the foremen most employed on difficult matters, and " Bill o Toms " and " Busick Jack " seem also to have been confidential workmen in their respective departments. We are informed by a gentleman of the neighbourhood[1] that most of the labourers employed were of a superior class, and some of them were " wise " or " cunning men," blood-stoppers, herb-doctors, and planet-rulers, such as are still to be found in the neighbourhood of Manchester.[2] Their very super-

---

[1] R. Rawlinson, Esq., C.E., Engineer to the Bridgewater Canal.

[2] Whilst constructing the canal, Brindley was very intimate with one Lawrence Earnshaw, of Mottram, a kindred mechanical genius, though in a smaller way. Lawrence was a very poor man's son, and had served a seven years' apprenticeship to the trade of a tailor, after which he bound himself apprentice to a clothier for seven years; but these trades not suiting his tastes, and being of a strongly mechanical turn, he finally bound himself apprentice to a clockmaker, whom he also served for seven years. This eccentric person invented many curious and ingenious machines, which were regarded as of great merit in his time. One of these was an astronomical and geographical machine, beautifully executed, showing the earth's diurnal and annual motion, after the manner of an orrery. The whole of the calculations were made by himself, and the machine is said to have been so exactly contrived and executed that, provided the vibration of the pendulum did not vary, the machine would not alter a minute in a hundred years; but this might probably be an extravagant estimate on the part of Earnshaw's friends. He was also a musical instrument maker and music teacher, a worker in metals and in wood, a painter and glazier, an optician, a bellfounder, a chemist and metallurgist, an engraver—in short, an almost universal mechanical genius. But though he could make all these things, it is mentioned as a remarkable fact that with all his ingenuity, and after many efforts (for he made many), he never could make

stitions, says our informant, made them thinkers and calculators. The foreman bricklayer, for instance, as his son used afterwards to relate, always "ruled the planets to find out the lucky days on which to commence any important work," and he added, "none of our work ever gave way." The skilled men had their trade-secrets, in which the unskilled were duly initiated, and the following were amongst them,—simple matters in themselves, but not without use :—

A wet embankment can be prevented from slipping by dredging or dusting powdered lime in layers over the wet clay or earth.

Sand or gravel can be made water-tight by shaking it together with flat bars of iron run in some depth, say two feet, and washing down loam or soil as the bars are moved about, thus obviating the necessity for clay-puddle.

Dry-rot can be prevented in warehouses by setting the bricks opposite the ends of the main beams of the warehouse in dry sand.

As to the details of the canal works, Mr. Rawlinson observes, " All the bridges and culverts are set in the best hydraulic mortar. The plans are simple, and have special contrivances to suit peculiarities of situation, &c. As a rule, there is a vertical joint betwixt the spandrils and wing walls of bridges, to prevent injury by unequal settlements or bearings."

---

a wicker-basket! Indeed, trying to be a universal genius was his ruin. He did, or attempted to do, so much, that he never stood still and established himself in any one thing ; and, notwithstanding his great ability, he died "not worth a groat." Amongst Earnshaw's various contrivances was a piece of machinery to raise water from a coal-mine at Hague, near Mottram, and (about 1753) a machine to spin and reel cotton at one operation —in fact, a spinning-jenny—which he showed to some of his neighbours as a curiosity, but, after having convinced them of what might be done by its means, he immediately destroyed it, saying that " he would not be the means of taking bread out of the mouths of the poor." He was a total abstainer from strong drink, long before the days of Teetotal Societies. Towards the end of his life he continued on intimate terms with Brindley, holding frequent meetings with him ; and when they met they did not easily separate. Earnshaw died in 1764, at sixty years of age.

Whilst the works were in full progress, which was during several years, from three to four hundred men were regularly employed upon them, divided into gangs of about fifty, over each of which was appointed a captain and setter-out of the works. One who visited the canal while in progress thus writes to the 'St. James's Chronicle,' under date July 1st, 1765 : " I surveyed the Duke's men for two hours, and think the industry of bees or labour of ants is not to be compared to them. Each man's work seems to depend on and be connected with his neighbour's, and the whole posse appeared as I conceive did that of the Tyrians when they wanted houses to put their heads in at Carthage." At Stretford the visitor found " four hundred men at work, putting the finishing stroke to about two hundred yards of the canal, which reached nearly to the Mersey; and which, on drawing up the floodgates, was to receive a proper quantity of water and a number of loaded barges. One of these appeared like the hull of a collier, with its deck all covered, after the manner of a cabin, and having an iron chimney in the centre ; this, on inquiry, proved to be the carpentry, but was shut up, being Sabbath-day, as was another barge, which contained the smith's forge. Some vessels were loaded with soil, which was put into troughs (see Cut at p. 95) fastened together, and rested on boards that lay across two barges ; between each of these there was room enough to discharge the loading by loosening some iron pins at the bottom of the troughs. Other barges lay loaded with the foundation-stones of the canal bridge, which is to carry the navigation across the Mersey. Near two thousand oak piles are already driven to strengthen the foundations of this bridge. The carpenters on the Lancashire side were preparing the centre frame, and on the Cheshire side all hands were at work in bringing down the soil and beating the ground adjacent to the foundations of the bridge, which is designed to be covered with stone in a month, and finished in about ten days more." [1]

By these vigorous measures the works proceeded rapidly towards completion. Before, however, they had made any progress at the Liverpool end, Earl Gower, encouraged and assisted by the Duke, had applied for and obtained an Act to enable a line of navigation to be formed between the Mersey and the Trent; the Duke agreeing with the promoters of the undertaking to vary the course of his canal and meet theirs about midway between Preston-brook and Runcorn, from which point it was to be carried northward towards the Mersey, descending into that river by a flight of ten locks, the total fall being not less than 79 feet from the level of the canal to low-water of spring-tides. When this deviation was proposed, the bold imagination of Brindley projected an aqueduct across the tideway of the Mersey itself, which was there some four hundred and sixty yards wide, with the object of carrying the Duke's navigation directly onward to the port of Liverpool on the Lancashire side of the river.[2] This was an admirable idea, which, if carried out, would probably have redounded more to the fame of Brindley than any other of his works. But the cost of that portion of the canal which had already been executed, had reached so excessive an amount, that the Duke was compelled to stop short at Runcorn, at which place a dock was constructed for the accommodation of the shipping employed in the trade connected with the undertaking.

---

[1] 'A History of Inland Navigations. Particularly those of the Duke of Bridgewater in Lancashire and Cheshire.' 2nd Ed., p. 39.

[2] This bold scheme, which seems to have been earnestly proposed at the time, though never executed, was thus noticed in a Liverpool paper: "On Monday last Mr. Brindley waited upon several of the principal gentlemen of this town and others at Runcorn, in order to ascertain the expense that may attend the building of a bridge over the river Mersey at that place, which is estimated at a sum inferior to the advantages that must arise, both to the counties of Lancaster and Chester, from a communication of this sort."—Williamson's 'Liverpool Advertiser,' July 19, 1768.

From Runcorn, it was arranged that the boats should
navigate by the open tideway of the Mersey to the
harbour of Liverpool, at which place the Duke made
arrangements to provide another dock for their accommodation. Brindley made frequent visits to Liverpool
for the purpose of directing its excavation, and it still
continues devoted to the purposes of the canal navigation. It lies between the Salthouse and Albert Docks
on the north, and the Wapping and King's Docks on
the south. The Salthouse was the only public dock
near it at the time that Brindley excavated this basin.
There were only three others in Liverpool to the north,
and not one to the south; but the Duke's Dock is now
the centre of about five miles of docks, extending from
it on either side along the Lancashire shore of the
Mersey.

THE DUKE'S DOCK, LIVERPOOL.
[By E. M. Wimperis.]

CHAPTER VI.

The Duke's Difficulties—Growth of Manchester.

Long before the Runcorn locks were constructed, and before the canal from Longford Bridge to the Mersey could be made available for purposes of traffic, the Duke found himself reduced to the greatest straits for want of money. Numerous unexpected difficulties had occurred, so that the cost of the works considerably exceeded his calculations; and though the engineer carried on the whole operations with the strictest regard to economy, the expense was nevertheless almost more than any single purse could bear. The execution of the original canal from Worsley to Manchester did not cost more than about a thousand guineas a mile, to which was to be added the cost of the terminus at Manchester. There was also the outlay which had to be incurred in building the requisite boats for the canal, in opening out the underground workings of the collieries at Worsley, and in erecting various mills, workshops, and warehouses for carrying on the new business.

The Duke was enabled to do all this without severely taxing his resources, and he even entertained the hope of being able to grapple with the still greater undertaking of cutting the twenty-four miles of new canal from Longford Bridge to the Mersey. But before these works were half finished, and whilst the large amount of capital invested in them was lying entirely unproductive, he found that the difficulties of the undertaking were likely to prove almost too much for him. Indeed, it seemed an enterprise beyond the means of any private

person—more like that of a monarch with State resources
at his command, than of a young English nobleman.

But the Duke was possessed by a brave spirit. He had
put his hand to the work, and he would not look back.
He had become thoroughly inspired by his great idea,
and determined to bend his whole energies to the task of
carrying it out. He was only thirty years of age—the
owner of several fine mansions in different parts of
the country, surrounded by noble domains—he had a
fortune sufficiently ample to enable him to command the
pleasures and luxuries of life, so far as money can secure
them; yet all these he voluntarily denied himself, and
chose to devote his time to consultations with an unlet-
tered engineer, and his whole resources to the cutting of
a canal to unite Liverpool and Manchester.

Taking up his residence at the Old Hall at Worsley—
a fine specimen of the old timbered houses so common

WORSLEY OLD HALL.

[By Percival Skelton, after his original Drawing.]

in South Lancashire and the neighbouring counties,—he
cut down every unnecessary personal expense; denied
himself every superfluity, except perhaps a pipe of
tobacco; paid off his following of servants; put down
his carriages and town house; and confined himself and
his Ducal establishment to a total expenditure of 400*l.*

a-year.   A horse was, however, a necessity, for the purpose of enabling him to visit the canal works during
their progress at distant points ; and he accordingly continued to maintain one horse for himself and another for
his groom.

Notwithstanding this rigid economy, the Duke still
found his resources inadequate to the heavy cost of
vigorously carrying on the works, and on Saturday
nights he was often put to the greatest shifts to raise
the requisite money to pay his large staff of crafts-
men and labourers.   Sometimes their payment had to
be postponed for a week or more, until the cash could
be raised by sending round for contributions among the
Duke's tenantry.   Indeed, his credit fell to the lowest
ebb, and at one time he could not get a bill for 500l.
cashed in either Liverpool or Manchester.[1]

When Mr. George Rennie,* the engineer, was engaged,
in 1825, in making the revised survey of the Liverpool
and Manchester Railway, he lunched one day at Worsley
Hall with Mr. Bradshaw, manager of the Duke's property, then a very old man.   He had been a contemporary of the Duke, and knew of the monetary straits to
which his Grace had been reduced during the construction of the works.   Whilst at table, Mr. Bradshaw
pointed to a small whitewashed cottage on the Moss,
about a mile and a half distant, and said that in that
cottage, formerly a public-house, the Duke, Brindley,
and Gilbert had spent many an evening discussing the
prospects of the canal when in progress.   One of the
principal topics of conversation on those occasions was
the means of raising the necessary funds against the
next pay-night.   " One evening in particular," said Mr.

---

[1] There is now to be seen at Worsley, in the hands of a private person,
a promissory note given by the Duke,
bearing interest, for as low a sum as
five pounds.   Amongst the persons
known to be lenders of money, to
whom the Duke applied at the time,
was Mr. C. Smith, a merchant at
Rochdale ; but he would not lend a
farthing, believing the Duke to be
engaged in a perfectly ruinous undertaking.

Bradshaw, " the party was unusually dull and silent. The Duke's funds were exhausted; the canal was by no means nearly finished; his Grace's credit was at the lowest ebb; and he was at a loss what step to take next. There they sat, in the small parlour of the little public-house, smoking their pipes, with a pitcher of ale before them, melancholy and silent. At last the Duke broke the silence by asking, in a querulous tone, ' Well, Brindley, what's to be done now? How are we to get at the money for finishing this canal?' Brindley, after a few long puffs, answered through the smoke, ' Well, Duke, I can't tell; I only know that if the money can be got, I can finish the canal, and that it will pay well.' ' Ay,' rejoined the Duke, ' but *where* are we to get the money?' Brindley could only repeat what he had already said; and thus the little party remained in moody silence for some time longer, when Brindley suddenly started up and said, ' Don't mind, Duke; don't be cast down; we are sure to succeed after all!' The party shortly after separated, the Duke going over to Worsley to bed, to revolve in his mind the best mode of raising money to complete his all-absorbing project."

Still undaunted by the difficulties that beset them, the Duke and his agents exerted themselves to the utmost to find the requisite means for completing the works. Gilbert was employed to ride round among the tenantry of the neighbouring districts, and raise five pounds here and ten pounds there, until he had gathered together enough to pay the week's wages. Whilst travelling about among the farmers on one of such occasions, Gilbert was joined by a stranger horseman, who entered into conversation with him; and it very shortly turned upon the merits of their respective horses. The stranger offered to swap with Gilbert, who, thinking the other's horse better than his own, agreed to the exchange. On afterwards alighting at a lonely village inn, which he had not before frequented, Gilbert was surprised to be greeted by the landlord with mysterious marks of recognition, and still more so when he was asked if he had got a good booty.   It turned out that he had exchanged horses with a highwayman, who had adopted this expedient for securing a nag less notorious than the one which he had exchanged with the Duke's agent.[1]

At length, when the tenantry could furnish no further advances, and loans were not to be had on any terms in Manchester or Liverpool, and the works must needs come to a complete stand unless money could be raised to pay the workmen, the Duke took the road to London on horseback, attended only by his groom, to try what could be done with his London bankers.   The house of Messrs. Child and Co., Temple Bar, was then the principal banking-house in the metropolis, as it is the oldest; and most of the aristocratic families kept their accounts there.   The Duke had determined at the outset of his undertaking not to mortgage his landed property, and he had held to this resolution.   But the time arrived when he could not avoid borrowing money of his bankers on such other security as he could offer them.   He had already created a valuable and lucrative property, which was happily available for the purpose.   The canal from Worsley to Manchester had proved remunerative in an extraordinary degree, and was already producing a large annual income.   He had not the same scruples at pledging the revenues of his canal that he had to mortgage his lands; and an arrangement was concluded with the Messrs. Child under which they agreed to advance the Duke sums of money from time to time, by means of which he was eventually enabled to finish the entire canal. The books of the firm show that he obtained his first advance from them of 3800*l.* about the middle of the year 1765, at which time he was in the greatest difficulty; shortly after a further sum of 15,000*l.*; then 2000*l.*,

---

[1] The Earl of Ellesmere's ' Essays on History, Biography,' &c., p. 236.

and various other sums, making a total of 25,000*l.*; which remained owing until the year 1769, when the whole was paid off—doubtless from the profits of the canal traffic as well as the economised rental of the Duke's unburthened estates.

The entire level length of the new canal from Longford Bridge to the upper part of Runcorn, nearly twenty-eight miles in extent, was finished and opened for traffic in the year 1767, after the lapse of about five years from the passing of the Act. The formidable flight of locks,

THE LOCKS AT RUNCORN.

[By Percival Skelton, after his original Drawing.]

from the level part of the canal down to the waters of the Mersey at Runcorn, were not finished for several years later, by which time the receipts derived by the

Duke from the sale of his coals and the local traffic of
the undertaking, enabled him to complete them with comparatively little difficulty. Considerable delay was occasioned by the resistance of an obstinate landowner
near Runcorn, Sir Richard Brooke, who interposed
every obstacle which it was in his power to offer; but
his opposition too was at length overcome, and the
new and complete line of water-communication between
Manchester and Liverpool was finally opened throughout. In a letter written from Runcorn, dated the
1st January, 1773, we find it stated that "yesterday
the locks were opened, and the *Heart of Oak*, a vessel
of 50 tons burden, for Liverpool, passed through them.
This day, upwards of six hundred of his Grace's workmen were entertained upon the lock banks with an ox
roasted whole and plenty of good liquor. The Duke's
health and many other toasts were drunk with the
loudest acclamations by the multitude, who crowded from
all parts of the country to be spectators of these asto-
nishing works. The gentlemen of the country for a long
time entertained a very unfavourable opinion of this
undertaking, esteeming it too difficult to be accomplished,
and fearing their lands would be cut and defaced without producing any real benefit to themselves or the
public; but they now see with pleasure that their fears
and apprehensions were ill-grounded, and they join with
one voice in applauding the work, which cannot fail to
produce the most beneficial consequences to the landed
property, as well as to the trade and commerce of this
part of the kingdom." [1]

Whilst the canal works had been in progress, great
changes had taken place at Worsley. The Duke had
year by year been extending the workings of the coal;
and when the King of Denmark, travelling under the
title of Prince Travindahl, visited the Duke in 1768, the

---

[1] Griswell's 'Account of Runcorn and its Environs,' pp. 63-5.

tunnels had already been extended for nearly two miles
under the hill.   When the Duke began these works, he
possessed only such of the coal-mines as belonged to the
Worsley estate, but from time to time he purchased the
adjoining lands containing seams of coal which run under
the high ground between Worsley, Bolton, and Bury;
and in course of time the underground canals connecting
the different workings extended for a distance of nearly
forty miles.   Both the hereditary and the purchased mines
are worked upon two main levels, though in all there
are four different levels, the highest being a hundred and
twenty yards above the lowest.   The coals worked out
of the higher seams are shot into the boats placed below
at certain appointed places; the whole of the subter-
ranean produce being dragged to the light through the
tunnels, the entrances to which are shown in our illustration.   In opening up these underground workings
the Duke is said to have expended about 168,000*l.*; but
the immense revenue derived from the sale of the coals
by canal rendered this an exceedingly productive outlay.   Besides the extension of the canal along these
tunnels, the Duke subsequently carried a branch by the
edge of Chat-Moss to Leigh, by which means new
supplies of coal were introduced to Manchester from
that district, and the traffic was still further increased.
It was a saying of the Duke's, that " a navigation should
always have coals at the heels of it."

The total cost of completing the canal from Worsley
to Manchester, and from Longford Bridge to the Mersey
at Runcorn, amounted to 220,000*l.*   A truly magnificent undertaking, nobly planned and nobly executed.
The power conferred by wealth was probably never
more munificently exercised than in this case; for,
though the traffic proved a source of immense wealth to
the Duke, it conferred incalculable blessings upon the

population of the district. It greatly added to their
comforts, increased their employment, and facilitated
the operations of industry in all ways. The canal was
no sooner opened than its advantages were at once
felt. The charge for water-carriage between Liverpool
and Manchester was lowered one-half. All sorts of produce were brought to the latter town, at moderate rates,
from the farms and gardens adjacent to the navigation,
whilst the value of agricultural property was immediately
raised by the facilities afforded for the conveyance of lime
and manure, as well as by reason of the more ready access
to good markets which it provided for the farming
classes. The Earl of Ellesmere has not less truly than
elegantly observed, that "the history of Francis Duke
of Bridgewater is engraved in intaglio on the face of the
country he helped to civilize and enrich."

Probably the most remarkable circumstance connected
with the money history of the enterprise is this : that
although the canal yielded an income which eventually
reached about 80,000l. a year, it was planned and executed by Brindley at a rate of pay considerably less than
that of an ordinary mechanic of the present day. The
highest wage he received whilst in the employment of
the Duke was 3s. 6d. a day. For the greater part of
the time he received only half-a-crown. Brindley, no
doubt, accommodated himself to the Duke's pinched
means, and the satisfactory completion of the canal was
quite as much a matter of character with him as of pay.
Indeed, he seems to have studied economy in everything,
and strove to keep down his expenses to the very lowest
point. Whilst superintending the works at Longford
Bridge, we find him making an entry of his day's personal expenses at only 6d. for "ating and drink." On
other days his expenditure was confined to 2d. for the
turnpike. When living at "The Bull," near the works
at Throstle Nest, we find his dinner costing 8d. and his
breakfast 6d. His entire expenses were on an equally

low scale, for he studied in all ways to economize the Duke's means, that every available shilling might be expended on the prosecution of the works.

The inadequate character of his remuneration was doubtless well enough known to Brindley himself, and rendered him very independent in his bearing towards the Duke.[1] They had frequent differences as to the

---

[1] The Earl of Bridgewater, in his singular publication, the 'Letter to the Parisians,' above referred to, states that "Brindley offered to stay entirely with the Duke, and do business for no one else, if he would give him a guinea a week;" and this statement is repeated by the late Earl of Ellesmere in his 'Essays on History, Biography,' &c. But, on the face of it, the statement looks untrue; and we have since found, from Brindley's own note-book, that, on the 25th of May, 1762, he was receiving a guinea a day from the Earl of Warrington for performing services for that nobleman; nor is it at all likely that he would prefer the Duke's three-and-sixpence a day to the more adequate rate of payment which he was accustomed to charge and to receive from other employers. It is quite true, however—and the fact is confirmed by Brindley's own record—that he received no more than a guinea a week whilst in the Duke's service; which only affords an illustration of the fact that eminent constructive genius may be displayed and engineering greatness achieved in the absence of any adequate material reward. We regret to have to add that Brindley's widow (afterwards the wife of Mr. Williamson, of Longport) in vain petitioned the Duke and his representatives, as well as the above Earl of Bridgewater, for payment of a balance said to have been due to Brindley for services, at the time of the engineer's death. In her letter to Robert Bradshaw, M.P., dated the 2nd May, 1803, Mrs. Williamson says: "It will doubtless appear to you extraordinary that so very late an application should now be made . . . . but I must beg leave to state that repeated applications were made by me (after Mr. Brindley's sudden and unexpected death) to the late Mr. Thomas Gilbert and also to his brother, but without any other effect than that of constant promises to lay the matter before His Grace; and I conceive it owing to this channel of application that no settling ever took place. A letter was also written to His Grace on this subject so late as the year 1801, but no answer was received. From the year 1765 to 1772, Mr. Brindley received no money on account of his salary. At that time he was frequently in very great want, and made application to the Duke, whose answer (to use the Duke's expression) was, 'I am much more distressed for money than you; however, as soon as I can recover myself, your services shall not go unrewarded.' In consequence of this, Mr. Brindley was under the necessity of borrowing several sums to make good engagements he was then under to various canal companies. In the year 1774, two years after Mr. Brindley's death, the late Mr. John Gilbert paid my brother, Mr. Henshall, the trifling sum of 100l. on account of Mr. Brindley's time, which is all that has been received. I beg leave to suggest how small and inadequate a return this is for his services during a period of seven years. Mr. B.'s travelling expenses on His Grace's account during that time were considerable, towards which, when he had not sufficient money to carry him the whole journey, he now and then received a small sum. How far his plans and undertakings have been beneficial to His Grace's interest is well known." In

proper mode of carrying on the works; but Brindley was quite as obstinate as the Duke on such occasions, and when he felt convinced that his own plan was the right one he would not yield an inch. It is said that, after long evening discussions at the hearth of the old timbered hall at Worsley, or at the Duke's house at Liverpool, while the works there were in progress, the two would often part at night almost at daggers-drawn. But next morning, on meeting at breakfast, the Duke would very frankly say to his engineer, " Well, Brindley, I have been thinking over what we were talking about last night. I find you may be right after all; so just finish the work in your own way."

The Duke himself, to the end of his life, took the greatest personal interest in the working of his coal-mines, his canals, his mills, and his various branches of industry. These were his hobbies, and he took pleasure in nothing else. He was utterly lost to the fashionable world, and, as some thought, to a sense of its proprieties. Shortly after his canal had been opened for the convey-ance of coals, the Duke established a service of passage-boats between Manchester and Worsley, and between Manchester and a station within two miles of Warring-ton, by which passengers were conveyed at the rate of a penny a mile. The boats were fitted up like the Dutch treckschuyts, and, being found cheap as well as con-venient, were largely patronized by the public.[1] This

---

a statement of the claims of Mr. Brindley's representatives, forwarded to the Earl of Bridgewater on the 3rd of November, 1803, it is further stated that " during the period of his em-ploy under His Grace, many highly advantageous and lucrative offers were made to him, particularly one from the Prince of Hesse, in 1766, who at that time was meditating a canal through his dominions in Germany, and who offered to subscribe to any terms Mr. Brindley might stipulate. To this engagement his family strongly

urged him, but the solicitation of the Duke, in this as in every other in-stance, to remain with him, out-weighed all pecuniary considerations; relying upon such a remuneration from His Grace as the profits of his work might afterwards justify."

[1] When the Duke had put on the boats and established the service, he offered to let them for 60*l.* a year; but not being able to find any person to take them at that price, he was under the necessity of conducting the service himself, by means of an agent.

service was afterwards extended to Runcorn, and from thence to Liverpool. The Duke delighted to travel by his own boats, preferring them to any more stately and aristocratic method. He often went by them to Manchester to watch how the coal-trade was going on. When the passengers alighted at the coal-wharf, there were usually many poor people about, wheeling away their barrow-loads of coals. One of the Duke's regulations was, that whenever any deficiency in the supply was apprehended, those people who came with their wheelbarrows, baskets, and aprons for small quantities, should be served first, and waggons, carts, and horses sent away until the supply was again abundant. The numbers of small customers who thus resorted to the Duke's coal-yard rendered it a somewhat busy scene, and the Duke liked to look on and watch the proceedings. One day a customer, of the poorer sort, having got his sack filled, looked about for some one to help it on to his back. He observed a stoutish man standing near, dressed in a spencer, with dark drab smallclothes. " Heigh ! mester ! " said the man, " come, gie me a lift wi' this sack o' coal on to my shouder." Without any hesitation, the person in the spencer gave the man the required " lift," and off he trudged with the load. Some one near, who had witnessed the transaction, ran up to the man, and asked, " Dun yo know who's that yo've been speaking tull ? " " Naw ! who is he ? " " Why, it's th' Duke his-sen ! " " The Duke ! " exclaimed the man, dropping the bag of coals from his shoulder, " Hey ! what'll he do at me ? Maun a goo an ax his pardon ? " But the Duke had disappeared.

He was very fond of watching his men at work, especially when any new enterprise was on foot. When they were boring for coal at Worsley, the Duke came every morning and looked on for a long time together. The men did not like to leave off work whilst he remained there, and they became so dissatisfied at having to work so long beyond the hour at which the bell rang, that Brindley had difficulty in getting a sufficient number of hands to continue the boring. On inquiry, he found out the cause and communicated it to the Duke, who from that time made a point of immediately walking off when the bell rang, returning when the men had resumed work, and remaining with them usually until six o'clock. He observed, however, that though the men dropped work promptly as the bell rang, when he was not by, they were not nearly so punctual in resuming work, some straggling in many minutes after time. He asked to know the reason, and the men's excuse was, that though they could always hear the clock when it struck twelve, they could not so readily hear it when it struck only one. On this, the Duke had the mechanism of the clock altered so as to make it strike *thirteen* at one o'clock; which it continues to do until this day.

His time was very fully occupied with his various business concerns, to which he gave a great deal of personal attention. Habit made him a business man— punctual in his appointments, precise in his arrangements, and economical both of money and time. When it was necessary for him to see any persons about matters of business, he preferred going to them instead of letting them come to him; " for," said he, " if they come to me, they may stay as long as they please; if I go to them, I stay as long as I please." His enforced habits of economy during the construction of the canal had fully impressed upon his mind the value of money. Yet, though "near," he was not penurious, but was usually liberal, and sometimes munificent. When the Loyalty Loan was raised, he contributed to it no less a sum than

100,000*l.* in cash. He was thoroughly and strongly
national, and a generous patron of most of our public
benevolent institutions.

The employer of a vast number of workpeople, he
exercised his influence over them in such a manner as to
extort their gratitude and blessings. He did not "lord
it" over them, but practically taught them, above all
things, to help themselves. He was the pattern employer
of his neighbourhood. With a kind concern for the
welfare of his colliery workmen—then a half-savage class
—he built comfortable dwellings and established shops
and markets for them; by which he ensured that at least
a certain portion of their weekly earnings should go to
their wives and families in the shape of food and clothing,
instead of being squandered in idle dissipation and
drunkenness. In order to put a stop to idle Mondays,
he imposed a fine of half-a-crown on any workman who
did not go down the pit at the usual hour on that morning;
and hence the origin of what is called Half Crown Row
at Worsley, as thus described by one of the colliers :—
"T'ould dook fined ony mon as didn't go daown pit o'
Moonday mornin auve a craown, and abeaut thot toime
he made a new road to t'pit, so t'colliers caw'd it Auve
Craown Row." Debts contracted by the men at public-
houses were not recognised by the pay-agents. The
steadiest workmen were allowed to occupy the best and
pleasantest houses as a reward for their good conduct.
The Duke also bound the men to contribute so much of
their weekly earnings to a general sick club; and he
encouraged a religious tone of character amongst his
people by the establishment of Sunday schools, which
were directly superintended by his agents, selected from
the best available class. The consequence was, that the
Duke's colliers soon held a higher character for sobriety,
intelligence, and good conduct, than the weavers and
other workpeople of the adjacent country.

He did not often visit London, where he had long

ceased to maintain a house; but when he went there he made an arrangement with one of his friends, who undertook for a stipulated sum to provide a daily dinner for His Grace and a certain number of guests whilst he remained in town. He also made occasional visits to his fine estate of Ashridge, in Buckinghamshire, taking the opportunity of spending a few days, going or coming, with Earl Gower and his Countess, the Duke's only sister, Lady Louisa Egerton. The Countess Gower seems to have borne considerable resemblance to her brother in force of character as well as in other respects. It is related of her that when her husband was ambassador at the French Court, during the outbreak of the Revolution in 1792, the Countess was courageously kind to the then members of the French King's family who were confined in the Temple, and consequently exposed her lord to the fury of the mob. A Swiss had been the medium of these kindly acts, and the mob sought the Swiss at the Earl's hotel for the purpose of cutting off his head. The authorities offered the ambassador a guard, which he refused, on the high ground of being protected by his character; but he took care to write in large letters over his door *Hôtel de l'Ambassadeur d'Angleterre*. The Countess, describing these alarming circumstances to a friend in England at the time, concluded thus : " Now we have done all we can ; and if the mob attacks us now, it is their concern, not ours." [1] During his visits at Trentham, the Duke would get ensconced on a sofa in some distant corner of the room in the evenings, and discourse earnestly to those who would listen to him about the extraordinary advantages of canals. There was a good deal of fun made on these occasions about " the Duke's hobby." But he was always like a fish out of water until he got back to

[1] See the 'Auckland Journal and Correspondence, 1860.' The noble lady here referred to was grandmother of the late Earl of Ellesmere and the late Duke of Sutherland.

Worsley, to John Gilbert, his coal-pits, his drainage, his mills, and his canals.

No wonder he was fond of Worsley. It had been the scene of his triumphs, and the foundation of his greatness. Illustrious visitors from all parts resorted thither to witness Brindley's "castle in the air," and to explore the underground excavations beneath Worsley-hill. Frisi, the Italian, spoke of the latter with admiration when they were only a mile and a half in length; since then they have been extended to nearly forty miles. Among the numerous visitors entertained by the Duke was Fulton, the American artist, with whose speculations he was much interested.* Fulton had given a good deal of attention to the subject of canals, and was then speculating on the employment of steam power for the propulsion of canal boats. The Duke was so much impressed with Fulton's ingenuity, that he urged him to give up the profession of a painter and devote himself to that of a civil engineer. Fulton acted on his advice, and shortly after we find him residing at Birmingham—the central workshop of England—studying practical mechanics, and fitting himself for superintending the construction of canals, on which he was afterwards employed in the midland counties.[1] The Duke did not forget the idea which Fulton had communicated to him as to the employment of steam as a motive power for boats, instead of horses; and when he afterwards heard that Symington's steam-boat, *The Dundas*, had been tried successfully on the Forth and Clyde Canal, he arranged to have six canal boats constructed after Symington's model; for he was a man to shrink from no expense in carrying out an enterprise which, to use his own words, had " utility

---

[1] The treatise which Fulton afterwards published, entitled ' A Treatise on Canal Navigation, exhibiting the numerous advantages to be derived from small Canals, &c., with a description of the machinery for facili-tating conveyance by water through the most mountainous countries, independent of Locks and Aqueducts,' (London, 1796,) is well known amongst engineers.

at the heels of it." But the Duke dying shortly after, the trustees refused to proceed with the experiment, and the project consequently fell through.[1] Had the Duke lived, canal steam-tugs would doubtless have been fairly tried; and he might thus have initiated the practical introduction of steam-navigation in England, as he unquestionably laid the foundations of the canal system. He lived long enough, however, to witness the introduction of tramroads, and he saw considerable grounds for apprehension in them.* " We may do very well," he once observed to Lord Kenyon, " if we can keep clear of these ——— tram-roads."

He was an admirable judge of character, and was rarely deceived as to the men he placed confidence in. John Gilbert was throughout his confidential adviser—a practical out-doors man, full of energy and perseverance. When any proposal was made to the Duke, he would say, " Well, thou must go to Gilbert and tell him all about it ; I'll do nothing without I consult him." From living so much amongst his people, he had contracted their style of speaking, and " thee'd " and " thou'd " those whom he addressed, after the custom of the district. He was rough in his speech, and gruff and emphatic in his manner, like those amidst whom he lived ; but with the rough word he meant and did the kindly act. His early want of education debarred him in a measure from the refining influences of letters ; for he read little, except perhaps an occasional newspaper, and he avoided writing whenever he could. He also denied himself the graces of female society ; and the

---

[1] The Earl of Ellesmere, in his ' Essay on Aqueducts and Canals,' states that the Duke made actual experiment of a steam-tug, and quotes the following from the communication of one of the Duke's servants, alive in 1844 : " I well remember the steam-tug experiment on the canal. It was between 1796 and 1799. Captain Shanks, R.N., from Deptford, was at Worsley many weeks preparing it, by the Duke's own orders and under his own eye. It was set going, and tried with coal-boats ; but it went slowly, and the paddles ,made sad work with the bottom of the canal, and also threw the water on the bank. The Worsley people called it Bonaparte."—Lord Ellesmere's ' Essays,' p. 241.

seclusion which his early disappointment in love had first driven him to, at length grew into a habit. He lived wifeless and died childless. He would not even allow a woman servant to wait upon him.

In person he was large and corpulent; and the slim youth on whom the bet had been laid that he would

FRANCIS DUKE OF BRIDGEWATER (AET. 70).

[By T. D. Scott, after Picart's engraving.]

be blown off his horse when riding the race in Trentham Park so many years before, had grown into a bulky and unwieldy man. His features strikingly resembled those of George III. and other members of the royal family. He dressed carelessly, and usually wore a suit of brown

—something of the cut of Dr. Johnson's—with dark drab breeches, fastened at the knee with silver buckles. At dinner he rejected, with a kind of antipathy, all poultry, veal, and such like, calling them " white meats," and wondered that everybody, like himself, did not prefer the brown. He was a great smoker, and smoked far more than he talked. Smoking was his principal evening's occupation when Brindley and Gilbert were pondering with him over the difficulty of raising funds to complete the navigation, and the Duke continued his solitary enjoyment through life. One of the droll habits to which he was addicted was that of rushing out of the room every five minutes, with the pipe in his mouth, to look at the barometer. Out of doors he snuffed, and he would pull huge pinches out of his right waistcoat pocket and thrust the powder up his nose, accompanying the operation with sundry strong short snorts.

He would have neither conservatory, pinery, flower-garden, nor shrubbery at Worsley ; and once, on his return from London, finding some flowers which had been planted in his absence, he whipped their heads off with his cane, and ordered them to be rooted up. The only new things introduced about the place were some Turkey oaks, with which his character seemed to have more sympathy. But he took a sudden fancy for pictures, and with his almost boundless means the purchase of pictures was easy.[1] Fortunately, he was well advised as to the paintings selected by him, and the numerous fine

---

[1] Lord Ellesmere says : " An accident laid the foundation of the Bridgewater collection. Dining one day with his nephew, Lord Gower, afterwards Duke of Sutherland, the Duke saw and admired a picture which the latter had picked up a bargain, for some 10l., at a broker's in the morning. ' You must take me,' he said, ' to that d——d fellow to-morrow.' Whether this impetuosity produced any immediate result we are not informed, but plenty of such ' fellows ' were doubtless not wanting to cater for the taste thus suddenly developed ; and such advisers as Lord Farnborough and his nephew lent him the aid of their judgment. His purchases from Italy and Holland were judicious and important, and, finally, the distractions of France forcing the treasures of the Orleans Gallery into this country, he became a principal in the fortunate speculation of its purchase."—' Essays on History, Biography,' &c.

works of art which were pouring into the country at the time, occasioned by the disturbances prevailing on the Continent, enabled him to lay the foundation of the famous Bridgewater Gallery, now reckoned to be one of the finest private collections in Europe. At his death, in 1803, its value was estimated at not less than 150,000*l.*

The Duke very seldom took part in politics, but usually followed the lead of his relative Earl Gower, afterwards Marquis of Stafford, who was a Whig. In 1762 we find his name in a division on a motion to withdraw the British troops from Germany, and on the loss of the motion he joined in a protest on the subject. When the repeal of the American Stamp Act was under discussion, his Grace was found in the ranks of the opposition to the measure. He strongly supported Mr. Fox's India Bill, and generally approved the policy of that statesman.

The title of Duke of Bridgewater died with him.[1] The Earldom went to his cousin General Egerton, seventh Earl of Bridgewater, and from him to his brother the

---

[1] In 1720, when Scroop Egerton, Earl of Bridgewater, had obtained the promise of a dukedom, he acquainted his brother with the circumstance, and told him, moreover, he had so much interest he could get the dukedom settled collaterally upon him and his heirs male, in case there should happen a failure of males in his own line direct, provided his brother would pay the additional office-fees, which amounted to less than 320*l.*, for extending the patent. His brother, then Bishop of Hereford (and who, if he had lived, was to have been Archbishop of York), replied that if the Duke would consent to entail the old family estates upon the dukedom, he would consent to discharge the additional fees. To this the Duke answered that he himself had no immediate concern, and no particular interest, in the above pro- posal. He made it solely because he conceived it might be acceptable to his brother; he would bind himself, however, by no promise or condition in a matter which regarded the Bishop alone. If the Bishop thought it worth while to give about three hundred guineas for the chance, well; otherwise, the patent would stand as it was already directed to be made out. Hence the patent was not extended, and now there is a failure of males in the Duke's own line direct. The dukedom of Bridgewater is consequently become extinct, in the branch of the family of Egerton, by the death of Francis, the late Duke; and the Earldom of Bridgewater is devolved to the direct heir of the above-mentioned Henry Bishop of Hereford.—'Gentleman's Mag.' for June, 1807. Vol. 77, p. 499.

crazed Francis Henry, eighth Earl; and on his death at Paris, some thirty years since, that title too became extinct. The Duke bequeathed about 600,000*l.* in legacies to his relatives, General Egerton, the Countess of Carlisle, Lady Anne Vernon, and Lady Louisa Macdonald. He devised most of his houses, his pictures, and his canals, to his nephew George Granville (son of Earl Gower), second Marquis of Stafford and first Duke of Sutherland, with reversion to his second son, Lord Francis Egerton, first Earl of Ellesmere, who thus succeeded to the principal part of the vast property created by the Duke of Bridgewater. The Duke was buried in the family vault at Little Gaddesden, Hertfordshire, in the plainest manner, without any state, at his own express request.

The Duke was a great public benefactor. The boldness of his enterprise, and the salutary results which flowed from its execution, entitle him to be regarded as one of the most useful men of his age. A Liverpool letter of 1765 says, " The services the Duke has rendered to the town and neighbourhood of Manchester have endeared him to the country, more especially to the poor, who, with grateful benedictions, repay their noble benefactor." [1] If he became rich through his enterprise, the public grew rich with him and by him; for his undertaking was no less productive to his neighbours than it was to himself. His memory was long venerated by the people amongst whom he lived,—a self-reliant, self-asserting race, proud of their independence, full of persevering energy, and strong in their attachments. The Duke was a man very much after their own hearts, and a good deal after their own manners. In respecting him, they were perhaps but paying homage to those qualities which they most cherished in themselves. Long after the Duke had gone from amongst them, they spoke to each other

---

[1] 'History of Inland Navigation,' p. 76.

of his rough words and his kindly acts, his business zeal and his indomitable courage. He was the first great " Manchester man." His example deeply penetrated the Lancashire character, and his presence seems even yet to hover about the district. The Duke's canal still carries a large proportion of the merchandise of Manchester and the neighbouring towns; the Duke's horses [1] still draw the Duke's boats; the Duke's coals still issue from the Duke's levels; and when any question affecting the traffic of the district is under consideration, the questions are still asked of " What will the Duke say? " " What will the Duke do? "

Manchester men of this day may possibly be surprised to learn that they owe so much to a Duke, or that the old blood has helped the new so materially in the development of England's modern industry. But it is nevertheless true that the Duke of Bridgewater, more than any other single man, contributed to lay the foundations of the prosperity of Manchester, Liverpool, and the surrounding districts. The cutting of the canal from Worsley to Manchester conferred upon that town the immediate benefit of a cheap and abundant supply of coal; and when Watt's steam-engine became the great motive power in manufactures, such supply became absolutely essential to its existence as a manufacturing town. Being the first to secure this great advantage, Manchester thus got the start forward which she has never since lost.[2]

---

[1] The Duke at first employed mules in hauling the canal-boats, because of the greater endurance and freedom from disease of those animals, and also because they could eat almost any description of provender. The Duke's breed of mules was for a long time the finest that had been known in England. The popular impression in Manchester is, that the Duke's Acts of Parliament authorising the construction of his canals, forbade the use of horses, in order that men might be employed; and that the Duke consequently dodged the provisions of the Acts by employing mules. But this is not the case, there being no clause in any of them prohibiting the use of horses.

[2] The cotton trade was not of much importance at first, though it rapidly increased when the steam-engine and spinning-jenny had become generally adopted. It may be interesting to know that sixty years since it was considered satisfactory if

But, besides being a waterway for coal, the Duke's canal, when opened out to Liverpool, immediately conferred upon Manchester the immense advantage of direct connection with an excellent seaport. New canals, supported by the Duke and constructed by the Duke's engineer, grew out of the original scheme between Manchester and Runcorn, which had the further effect of placing the former town in direct water-communication with the rich districts of the north-west of England. Then the Duke's canal terminus became so important, that most of the new navigations were laid out so as to join it; those of Leigh, Bolton, Stockport, Rochdale, and the West Riding of Yorkshire, being all connected with the Duke's system, whose centre was at Manchester. And thus the whole industry of these districts was brought, as it were, to the very doors of that town.

But Liverpool was not less directly benefited by the Duke's enterprise. Before his canal was constructed, the small quantity of Manchester woollens and cottons manufactured for exportation, was carried on horses' backs to Bewdley and Bridgenorth on the Severn, from whence they were floated down that river to Bristol, then the chief seaport on the west coast. No sooner, however, was the new water-road opened out than the Bridgenorth pack-horses were taken off, and the whole export trade of the district concentrated on Liverpool. The additional accommodation required for the increased business of the port was promptly provided as occasion required. New harbours and docks were built, and before many years had passed Liverpool had shot far ahead of Bristol, and became the chief port on the west coast, if not in all England. Had Bristol been blessed with a Duke of Bridgewater, the

one cotton-flat a day reached Manchester from Liverpool. In the Duke's time the flats always "cast anchor" on their way, or at least laid up for the night, at six o'clock precisely, starting again at six o'clock on the following morning.

result might have been altogether different; and the valleys of Wilts, the coal and iron fields of Wales, and the estuary of the Severn, might have been what South Lancashire and the Mersey are now. Were statues any proof of merit, the Duke would long since have had the highest statue in Manchester as well as Liverpool erected to his memory, and that of Brindley would have been found standing by his side; for they were both heroes of industry and of peace, though even in commercial towns men of war are sometimes more honoured.

We can only briefly glance at the extraordinary growth of Manchester since the formation of the Duke's canal, as indicated by the annexed plan.

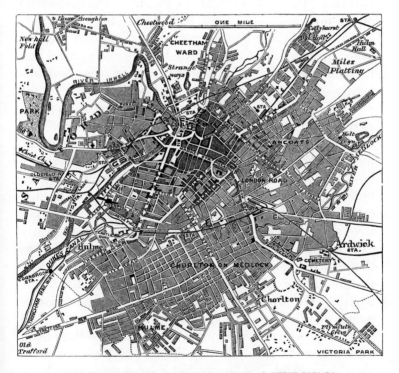

PLAN OF MANCHESTER, SHOWING ITS EXTENT AT THREE PERIODS.

(The parts printed black ⊓ represent Manchester in 1770; those dark-shaded ⌐ show its extent in 1804; and the light-shaded parts ⌐ Manchester at the present day.)

Though Manchester was a place of some importance about the middle of last century, it was altogether insignificant in extent, trade, and population compared with what it is now. It consisted of a few principal streets—narrow, dark, and tortuous—one of them, leading from the Market Place to St. Ann's Square, being very appropriately named " Dark Entry." Deansgate was the principal original street of the town, and so called because of its leading to the dean or valley along which it partly extended. From thence a few streets diverged in different directions into the open country. St. Ann's Square, the fashionable centre of modern Manchester, was in 1770 a corn-field surrounded with lofty trees, and known by the name of " Acre's Field." The cattle-fairs of the town were held there, the entrance from Deansgate being by Toll Lane, a narrow, dirty, unpaved way, so called because toll was there levied on the cattle proceeding towards the fair. The ancient seat of the Radcliffe family still stood at Pool Fold, close to the site of the modern Cross Street, and the water in the moat was used as a ducking-pond for scolds. When the pool became filled up, the ducking-pond was removed to Daub Holes, then on the outskirts of the town, where the Infirmary now stands. The site of King Street, now the very heart of Manchester, was as yet comparatively retired, a colony of rooks having established themselves in the tall trees at its upper end, from which they were only finally expelled about thirty-five years ago. Cannon Street was the principal place of business, the merchants and their families living in the comparatively humble tenements fronting the street, the equally humble warehouses in which their business was done standing in the rear. The ground on which the crowded thoroughfares of Oldham Street, London Road, Mosley Street, and their continuations, now exist, was as yet but garden or pasture-land. Salford itself was only a hamlet contained in the bend of the Irwell. It consisted of a double line of mean houses, extending from the Old Bridge (now Victoria Bridge) to about the end of Gravel Lane—a country road containing only a few detached cottages. The comparatively rural character of Manchester may be inferred from the circumstance that the Medlock and the Irk, the Tib and Shooter's Brook, were favourite fishing streams. Salmon were caught in the Medlock and at the mouth of the Irk; and the others were well stocked with trout. The Medlock and the Irk are now as black as old ink, and as thick; but the Tib and Shooter's Brook are entirely lost,—having been absorbed, like the London Fleet, in the sewage system of the town. Tib Street and Tib Lane indicate the former course of the Tib; but of Shooter's Brook not a trace is left.

The townships of Ardwick Green, Hulme, and Chorlton-upon-Medlock (formerly called Chorlton Row), were purely rural. The old rate-books of Chorlton Row exhibit some curious facts as to the transformations effected in that township. In 1720, a " lay " of 14*d.* in the pound produced a sum of 26*l.* 18*s.*, the whole disbursements for the year amounting to 28*l.* 8*s.* 5*d.* From the highway-rate laid in 1722, it appears that the contributors were only twenty persons in all, whose payments ranged from 8*d.* to 1*l.* 13*s.* 4*d.*, producing a total levy of 6*l.* 18*s.* 10*d.* for the year. From the disbursements, it appears that the regular wage paid to the workmen employed was a shilling a-day. In 1750, a lay of 3*d.* in the pound produced only 6*l.* 2*s.* 1½*d.*; so that the population and value of property in Chorlton Row had not much increased during the thirty years that had passed. In 1770, two brought in 57*l.* 8*s.* 6*d.*; and in 1794, four levies made in that year produced 208*l.* 2*s.* 4*d.*[1] Among the list of contributors in the latter year we find " Mrs. Quincey 16*s.* 6*d.*"—the mother of De Quincey,

---

[1] Recent " poor-lays " exhibit a very different result from what they did in former years. In 1860-1 the poor-rate levied in Chorlton-upon-Medlock yielded (at 2*s.* 10*d.* in the pound), 18,798*l.*; the property in the township being of the rateable value of 145,844*l.*

the English opium-eater, who was brought up in Chorlton
Row. De Quincey describes the home of his childhood
as a solitary house, " beyond which was nothing but a
cluster of cottages, composing the little hamlet of Greenhill." It was connected by a winding lane with the
Rusholme road. The house, called Greenheys—the
nucleus of an immense suburban district—built by
De Quincey's father, " was then," he says, " a clear mile
from the outskirts of Manchester," Princess Street being
the termination of the town on that side.[1] Now it is en-
veloped by buildings in all directions, and nothing of the
former rural character of the neighbourhood remains but
the names of Greenhill, Rusholme, and Greenheys.

Coming down to the second expansion of Manchester,
as exhibited on our plan, it will be observed that a considerable increase of buildings had taken place in the
interval between 1770 and 1804. The greater part of
the town was then contained in the area bounded by
Deansgate, the crooked lanes leading to Princess Street,
Bond Street, and Dand Street, to the Rochdale Canal,
and round by Ancoats Lane (now Great Ancoats Street)
and Swan Street, to Long Millgate, then a steep narrow
lane forming the great highway into North Lancashire.
Very few buildings existed outside the irregular quad-
rangle indicated by the streets we have named. The
straggling houses of Deansgate, which were principally
of timber, ended at Knott Mill. A few dye-works stood
at intervals along the Medlock, now densely occupied
with buildings for miles along both banks. Salford had
not yet extended to St. Stephen's Street in one direction,
nor above half way to Broughton Bridge in another.[2]
The comparatively limited spaces thus indicated sufficed,

---

[1] De Quincey's ' Autobiographic
Sketches,' pp. 34, 48.
[2] The corner of Irwell Street, Salford, as recently as 1828, was occupied
by an old canal " flat," tenanted by an
eccentric character, after whom it was
designated " Bannister's Ship." Opposite it was a row of cottages with
gardens in front. Oldfield and Ordsall
Lanes were country roads, and the
streets adjacent to them were not as
yet in existence.

however, for places of business and habitations for the population. Now the central districts are almost exclusively occupied for business purposes, and houses for dwellings have rapidly extended in all directions. The now populous districts of Broughton, Higher and Lower, did not exist thirty-five years ago. They contained no buildings excepting Strangeways Hall and a few cottages which lay scattered beyond the bottom of the workhouse brow.

But pastures, corn-fields, and gardens rapidly disappeared before the advance of streets and factory buildings.[1] The suburban districts of Ardwick, Hulme, and Cheetham, with their hamlets, became altogether absorbed in the great city. Stretford New Road, a broad straight street nearly a mile and a half long, forms the main highway for a district which has sprung up during the life of the present generation, and alone contains a population greater than that of many cities. Not fifty years ago, a few farm-houses and detached dwellings were all the buildings it contained, and Chester road, the principal one in the district, was a narrow winding lane, with hedges on either side. Less than thirty years since, Jackson's Lane was a mere farm-road through corn-fields. Within that time it has been converted into a spacious street, now dignified with the name of Great Jackson Street.[2] The portion of Hulme nearest Manchester was occupied by tea and ginger-beer gardens, the

---

[1] The growth of Manchester and the sister borough of Salford will be more readily appreciated, perhaps, by a glance at the population at different periods than by any other illustration :—

In 1774.	1801.	1821.	1861.
41,032	84,020	187,031	460,028

[2] A relic of rural Hulme is to be seen in a remnant of the buildings formerly known as " Jackson's Farm," which gave a name first to the " lane," and subsequently to the " street." It is a single-storey building, covered with gray flags, and standing in an oblique recess situated on the left-hand side about half-way between Chester Road and the recently formed City Road. Higher up, near the end of Upper Jackson Street, at the junction of Chapman Street with Preston Street, are still standing, but without any field near, the buildings, also covered with gray flags, which were known within a comparatively recent period as " Geary's Farm."

principal of which was at the White House. Late roy-
sterers at that place of resort, unless they could form a
party or secure the services of "the patrol," had frequently to sojourn there all night. The officers constituting the patrol [1] carried swords and horn lanterns;
and, clad as they were in heavy greatcoats with many
capes, they were by no means light of foot, or at all
formidable adversaries to the footpads who "worked"
the district.

Among the most remarkable improvements in Manchester of late years, have been the numerous spacious
thoroughfares which have been opened up in all directions. In this respect, the public spirit of Manchester
has not been surpassed by any town in the kingdom,—
the new streets being laid out on a settled plan with
a view to future extension, and executed with admirable
judgment. Narrow, dark, and crooked ways have thus
been converted into wide and straight streets, admitting
light, air, and health to the inhabitants, and affording
spacious highways for the great and growing traffic of
the town and district. The important street-improvements executed in Manchester during the last thirty
years have cost an aggregate of about 800,000*l.* The
central and oldest part of the town has thus undergone
a complete transformation. So numerous are the dark
and narrow entries that have been opened up—the obstructive buildings that have been swept away, the pro-
jecting angles that have been cut off, and the crooked ways
that have been made straight—that the denizen of a
former age would be very unlikely to recognise the Manchester of to-day, were it possible for him to revisit it.

Some of the street-improvements have their peculiar
social aspects, and call up curious reminiscences. The

---

[1] On the subject of watchmen, it
may be mentioned that the first
watchman was appointed for Chorlton-
on-Medlock in 1814. In 1832 an
Act was obtained for improving and
regulating that township, and so recently as 1833 it was first lighted
with gas. The Police Act for the
Township of Hulme was obtained in
1834.

stocks, pillory, and Old Market Cross, were removed from the Market Place in 1816. The public whipping of culprits on the pillory stage is within the recollection of the elder portion of the present inhabitants. Another " social institution," of a somewhat different character, was extinguished much more recently, by the construction of the splendid piece of terrace-road in front of the cathedral, known as the Hunt's Bank improvement. This road swept away a number of buildings, shown on the old plans of Manchester as standing on the water's edge, close to the confluence of the Irk with the Irwell. They were reached by a flight of some thirty steps, and consisted of a dye-work, employing three or four hands, two public-houses, and about a dozen cottages and other buildings. The public-houses, the ' Ring o' Bells ' and the ' Blackamoor,' particularly the former, were famous places in their day. On Mondays, wedding-parties from the country, consisting sometimes of from twenty to thirty couples, accompanied by fiddlers, visited " t' Owd Church " to get married. The ' Ring o' Bells ' was the rendezvous until the parties were duly married and ready to form and depart homewards, in a more or less orderly manner, headed by their fiddlers as they had come. The ' Ring o' Bells ' was also a favourite resort of the recruiting-serjeant, and more recruits, it is said, were enlisted there than at any other public-house in the kingdom. But these, and many other curious characteristics of old Manchester, have long since passed away; and not only the town but its population have become entirely new.

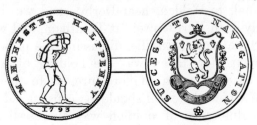

BRIDGEWATER HALFPENNY TOKEN.

CHAPTER VII.

The Grand Trunk Canal.

Long before the Duke's canal was finished, Brindley was actively employed in carrying out a still larger enterprise—no less than the canal to connect the Mersey with the Trent, and both with the Severn; thus uniting by a grand line of water-communication the ports of Liverpool, Hull, and Bristol. He had, indeed, already made a survey of such a canal, at the instance of Earl Gower, before his engagement as engineer for the Bridgewater undertaking. Thus, in the beginning of February, 1758, before the Duke's bill had been even applied for, we find him occupied for days together "a bout the novogation," and he then appears to have surveyed the country between Longbridge, in Staffordshire, and King's Mills, in Derbyshire. The enterprise, however, seems to have made little progress. It was of too formidable a character, and canals were as yet too untried in England, to be hastily entered upon. But again, in 1759, we find Brindley proceeding with his survey of the Staffordshire Canal; and in the middle of the following year he was occupied about twenty days in levelling from Harecastle, at the summit of the proposed canal, to Wilden, near Derby; and we find him frequently with Earl Gower, at Trentham, and with the Earl of Stamford, at Enville, discussing the project.

The next step taken was the holding of a public meeting at Sandon, in Staffordshire, as to the proper course in which the canal should be constructed. Considerable difference of opinion was expressed at that meeting, in consequence of which it was arranged that

Mr. Smeaton should be called upon to co-operate with Brindley in making a joint survey and a joint report. A second meeting was afterwards held at Wolseley Bridge, at which the plans of the two engineers were ordered to be engraved and circulated amongst the landowners and others interested in the project. Here the matter rested for several years more, without any action being taken. Brindley was hard at work upon the Duke's canal, and the Staffordshire projectors were disposed to wait the issue of that experiment; but no sooner had it been opened, and its extraordinary success become matter of fact, than the project of the canal through Staffordshire was again revived. The gentlemen of Cheshire and Staffordshire, especially the salt-manufacturers of the former county and the earthenware-manufacturers of the latter, now determined to enter into co-operation with the leading landowners in concerting the necessary measures with the object of opening up a line of water-communication with the Mersey and the Trent.

The earthenware manufacture, though in its infancy, had already made considerable progress, but, like every other branch of industry in England at that time, its further development was greatly hampered by the wretched state of the roads. Throughout Staffordshire they were as yet, for the most part, narrow, deep, circuitous, miry, and inconvenient; barely passable with rude waggons in summer, and almost impassable, even with pack-horses, in winter. Yet the principal materials used in the manufacture of pottery, especially of the best kinds, were necessarily brought from a great distance —flint-stones from the south-eastern ports of England, and clay from Devonshire and Cornwall. The flints were brought by sea to Hull, and the clay to Liverpool. From Hull the materials were brought up the Trent in boats to Willington; and the clay was in like manner brought from Liverpool up the Weaver to Winsford, in

Cheshire. Considerable quantities of clay were also conveyed in boats from Bristol, up the Severn, to Bridgenorth and Bewdley. From these various points the materials were conveyed by land-carriage, mostly on the backs of horses, to the towns in the Potteries, where they were worked up into earthenware and china. The manufactured articles were returned for export in the same rude way. Large crates of pot-ware were slung across horses' backs, and thus conveyed to their respective ports, not only at great risk of breakage and pilferage, but also at a heavy cost. The expense of carriage was not less than a shilling a ton per mile, and the lowest charge was eight shillings the ton for ten miles. Besides, the navigation of the rivers above mentioned was most uncertain, arising from floods in winter and droughts in summer. The effect was, to prevent the expansion of the earthenware manufacture, and very greatly to restrict the distribution of the lower-priced articles in common use.

The same difficulty and cost of transport checked the growth of nearly all other branches of industry, and made living both dear and uncomfortable. The indispensable article of salt, manufactured at the Cheshire Wiches, was in like manner carried on horses' backs all over the country, and reached almost a fabulous price by the time it was sold two or three counties off. About a hundred and fifty pack-horses, in gangs, were also occupied in going weekly from Manchester, through Stafford, to Bewdley and Bridgenorth, loaded with woollen and cotton cloth for exportation;[1] but the cost

---

[1] In a curious book published in 1766, by Richard Whitworth, of Balcham Grange, Staffordshire, afterwards Sir Richard Whitworth, member for Stafford, entitled ' The Advantages of Inland Navigation,' he points to the various kinds of traffic that might be expected to come upon the canal then proposed by him, and amongst other items he enumerates the following :—" There are three pot-waggons go from Newcastle and Burslem weekly, through Eccleshall and Newport to Bridgenorth, and carry about eight tons of pot-ware every week, at 3l. per ton. The same waggons load back with ten tons of close goods, consisting of white clay, grocery,

of the carriage by this mode so enhanced the price, that it is clear that in the case of many articles it must have acted as a prohibition, and greatly checked production and consumption. Even corn, coal, lime, and iron-stone were conveyed in the same way, and the operations of agriculture, as of manufacture, were alike injuriously impeded. There were no shops then in the Potteries, the people being supplied with wares and drapery by packmen and hucksters, or from Newcastle-under-Lyme, which was the only town in the neighbourhood worthy of the name.

The people of the district in question were quite as rough as their roads. Their manners were coarse, and their amusements brutal. Bull-baiting, cock-throwing, and goose-riding were the favourite sports. When Wesley first visited Burslem, in 1760, the potters assembled to jeer and laugh at him. They then proceeded to pelt him. " One of them," he says, " threw a clod of earth which struck me on the side of the head ; but it neither disturbed me nor the congregation." About that time the whole population of the Potteries did not amount to more than about 7000. The villages in which they lived were poor and mean, scattered up and down, and the houses were mostly covered with thatch. Hence the Rev. Mr. Middleton, incumbent of Stone—a man of great shrewdness and quaintness, distinguished for his love of harmless mirth and sarcastic humour— when enforcing the duty of humility upon his leading parishioners, took the opportunity, on one occasion,

---

and iron, at the same price, delivered on their road to Newcastle. Large quantities of pot-ware are conveyed on horses' backs from Burslem and Newcastle to Bridgenorth and Bewdley for exportation—about one hundred tons yearly, at 2l. 10s. per ton. Two broad-wheel waggons (exclusive of 150 pack-horses) go from Manchester through Stafford weekly, and may be computed to carry 312 tons of cloth and Manchester wares in the year, at 3l. 10s. per ton. The great salt-trade that is carried on at Northwich may be computed to send 600 tons yearly along this canal, together with Nantwich 400, chiefly carried now on horses' backs, at 10s. per ton on a medium."

after the period of which we speak, of reminding them
of the indigence and obscurity from which they had
risen to opulence and respectability. He said they
might be compared to so many sparrows, for that all of
them had been hatched *under the thatch*. When the
congregation of this gentleman, growing rich, bought
an organ and placed it in the church, he persisted in
calling it the hurdy-gurdy, and often took occasion to
lament the loss of his old psalm-singers.

The people further north were no better, nor were
those further south. When Wesley preached at Congleton, four years later, he said, "even the poor potters
[though they had pelted him] are a more civilized
people than the better sort (so called) at Congleton."
Arthur Young visited the neighbourhood of Newcastle-
under-Lyme in 1770, and found poor-rates high, wages
low, and employment scarce. " Idleness," said he, " is
the chief employment of the women and children. All
drink tea, and fly to the parishes for relief at the very
time that even a woman for washing is not to be had.
By many accounts I received of the poor in this neighbourhood, I apprehend the rates are burthened for the
spreading of laziness, drunkenness, tea-drinking, and
debauchery,—the general effect of them, indeed, all over
the kingdom." [1] Hutton's account of the population
inhabiting the southern portion of the same county is
even more dismal. Between Hales Owen and Stour-
bridge was a district usually called the Lie Waste, and
sometimes the Mud City. Houses stood about in every
direction, composed of clay scooped out into a tenement,
hardened by the sun, and often destroyed by the frost.
The males were half-naked, the children dirty and hung
over with rags. "One might as well look for the
moon in a coal-pit," says Hutton, " as for stays or
white linen in the City of Mud. The principal tool in

---

[1] Young's ' Six Months' Tour.'   Ed. 1770.   Vol. iii., p. 317.

business is the hammer, and the beast of burden the ass." [1]

The district, however, was not without its sprinkling of public-spirited men, who were actively engaged in devising new sources of employment for the population ; and, as one of the most effective means of accomplishing this object, opening up the communications, by road and canal, with near as well as distant parts of the country. One of the most zealous of such workers was the illustrious Josiah Wedgwood. He was one of those indefatigable workers who from time to time spring from the ranks of the common people, and by their energy, skill, and enterprise, not only practically educate the working population in habits of industry, but, by the example of diligence and perseverance which they set before them, largely influence the public activity in all directions, and contribute in a great measure to form the national character. Josiah Wedgwood was born in a very humble position in life ; and though he rose to eminence as a man of science as well as a manufacturer, he possessed no greater advantages at starting than Brindley himself did. His grandfather and granduncle were both potters, as was also his father Thomas, who died when Josiah was a mere boy, the youngest of a family of thirteen children. He began his industrial life as a thrower in a small pot-work, conducted by his elder brother ; and he might have continued working at the wheel but for an attack of virulent small-pox, which, being neglected, led to a disease in his right leg, which ended in his suffering amputation, when he became unfitted for following even that humble employment. During his illness he took to reading and thinking, and turned over in his mind the various ways of making a living by his trade, now that he could no longer work at the potter's wheel. When sufficiently recovered, he began making fancy articles out of potter's clay,

---

[1] 'History of Birmingham.' Ed. 1836, p. 24.

such as knife-hafts, boxes, and sundry curious little
articles for domestic use.   He joined in several successive
partnerships with other workmen, but made comparatively small progress until he began business on his
own account, in 1759, in a humble cottage near the
Market House in Burslem, known by the name of the
Ivy House.   He there pursued his manufacture of knife-

IVY HOUSE, BURSLEM, WEDGWOOD'S FIRST POTTERY.[1].

[From Ward's ' History of Stoke-upon-Trent.']

handles and other small wares, striving at the same time
to acquire such a knowledge of practical chemistry as
might enable him to improve the quality of his work in
respect of colour, glaze, and durability.   Success attended
Wedgwood's diligent and persistent efforts, and he proceeded from one stage of improvement to another, until
at length, after a course of about thirty years' labour, he
firmly established a new branch of industry, which not
only added greatly to the conveniences of domestic life,
but proved a source of remunerative employment to

---

[1] The Ivy House, in which Wedgwood began business on his own account, is the cottage shown on the
right-hand of the engraving.   The
other house is the old " Turk's Head."

many thousand families throughout England. His trade having begun to extend, a demand for his articles sprang up not only in London but in foreign countries. But there was this great difficulty in his way,—that the roads in his neighbourhood were so bad that he was at the same time prevented from obtaining a sufficient supply of the best kinds of clay and also from disposing of his wares in distant markets. This great evil weighed heavily upon the whole industry of the district, and Wedgwood accordingly appears to have bestirred himself at an early period in his career to improve the local communications. In conjunction with several of the leading potters he promoted an application to Parliament for powers to repair and widen the road from the Red Bull at Lawton, in Cheshire, to Cliff Bank, in Staffordshire. This line, if formed, would run right through the centre of the Potteries, open them to traffic, and fall at either end into a turnpike road. The measure was, however, violently opposed by the people of Newcastle-under-Lyme, on the ground that the proposed new road would enable waggons and packhorses to travel north and south from the Potteries without passing through their town. The Newcastle innkeepers acted as if they had a vested interest in the bad roads ; but the bill passed, and the new line was made, stopping short at Burslem. This was, no doubt, a great advantage, but it was not enough. The heavy carriage of clay, coal, and earthenware needed some more convenient means of transport than waggons and roads ; and, when the subject of water communication came to be discussed, Josiah Wedgwood at once saw that a canal was the very thing for the Potteries. Hence he immediately entered with great spirit into the movement again set on foot for the construction of Brindley's Grand Trunk Canal.

The field was not, however, so clear now as it had been before. The success of the Duke's canal led to the projection of a host of competing schemes in the county of

Cheshire, and it appeared that Brindley's Grand Trunk project would now have to run the gauntlet of a powerful local opposition. There were two other projects besides his, which formed the subject of much pamphleteering and controversy at the time, one entering the district by the river Weaver, and another by the Dee. Neither of these proposed to join the Duke of Bridgewater's canal, whereas the Grand Trunk line was laid out so as to run into his at Preston-on-the-Hill near Runcorn. As the Duke was desirous of placing his navigation —and through it Manchester, Liverpool, and the intervening districts—in connection with the Cheshire Wiches and the Staffordshire Potteries, he at once threw the whole weight of his support upon the side of Brindley's Grand Trunk. Indeed, he had himself been partly at the expense of its preliminary survey, as we find from an entry in Brindley's memorandum-book, under date the 12th of April, 1762, as follows: " Worsley—Rec^d from Mr Tho Gilbert for ye Staffordshire survey, on account, 33l. 16s. 11d." The Cheshire gentlemen protested against the Grand Trunk scheme, as calculated to place a monopoly of the Staffordshire and Cheshire traffic in the hands of the Duke; but they concealed the fact, that the adoption of their respective measures would have established a similar monopoly in the hands of the Weaver Navigation Company, whose line of navigation, so far as it went, was tedious, irregular, and expensive. Both parties mustered their forces for a Parliamentary struggle, and Brindley exerted himself at Manchester and Liverpool in obtaining support and evidence on behalf of his plan. The following letter from him to Gilbert, then at Worsley, relates to the rival schemes.

" 21 Decr. 1765

" On Tusdey Sr Georg [Warren] sent Nuton in to Manchester to make what intrest he could for Sir Georg and to gather ye old Navogtors togather to meet Sir Georg at Stoperd to make Head a ganst His Grace

" I sawe Docter Seswige who sese Hee wants to see you about pamant of His Land in Cheshire

" On Wednesday ther was not much transpired but was so dark I could carse do aneything

" On Thursdey Wadgwood of Burslam came to Dunham & sant for mee and wee dined with Lord Gree [Grey] & Sir Hare Mainwering and others Sir Hare cud not ceep His Tamer [temper] Mr. Wedgwood came to seliset Lord Gree in faver of the Staffordshire Canal & stade at Mrs Latoune all night & I whith him & on frydey sat out to wate on Mr Edgerton to seliset Him Hee sase Sparrow and others are indavering to gat ye Land owners consants from Hare Castle to Agden

" I have ordered Simcock to ye Langth falls of Sanke Navegacion.

" Ryle wants to have coals sant faster to Alteringham that Hee may have an opertunety dray of ye sale Moor Canal in a bout a weeks time.

" I in tend being back on Tusdy at fardest."

The first public movement was made by the supporters of Brindley's scheme. They held an open meeting at Wolseley Bridge, Staffordshire, on the 30th of December, 1765, at which the subject was fully discussed. Earl Gower, the lord-lieutenant of the county, occupied the chair; and Lord Grey and Mr. Bagot, members for the county,—Mr. Anson, member for Lichfield,—Mr. Thomas Gilbert, the agent for Earl Gower, then member for Newcastle-under-Lyme,—Mr. Wedgwood, and many other influential gentlemen, were present to take part in the proceedings. Mr. Brindley was called upon to explain his plans, which he did to the satisfaction of the meeting; and these having been adopted, with a few immaterial alterations, it was determined that steps should be taken to apply for a bill conferring the necessary powers in the ensuing session of Parliament. Mr. Wedgwood put his name down for a thousand pounds towards the preliminary expenses, and promised to subscribe largely for shares besides.[1] The promoters of the

---

[1] Wedgwood even entered the lists as a pamphleteer in aid of the Grand | Trunk project, and, in 1765, he and his partner, Mr. Bentley (son of Dr.

JOSIAH WEDGWOOD.

[By T. D. Scott, after Reynolds.]

measure proposed to designate the undertaking "The Canal from the Trent to the Mersey;" but Brindley, with sagacious foresight, urged that it should be called The Grand Trunk, because, in his judgment, numerous other canals would branch out from it at various points of its course, in like manner as the arteries of the human system branch out from the aorta; and we need scarcely

Bentley, archdeacon of Ely), drew up a very able statement, showing the advantages likely to be derived from the construction of the proposed canal, under the title of 'A View of the Advantages of Inland Navigation, with a plan of a Navigable Canal intended for a communication between the ports of Liverpool and Hull.' It pointed out in glowing language the advantages to be derived from opening up the internal communications of a country by means of roads, canals, &c.; and showed how the comfort and even the necessity of all classes must be so much better provided for by a reduction in the cost of carriage of useful and necessary commodities.

add that, before many years had passed, our engineer's anticipations of the progress of canal enterprise were fully justified. The Staffordshire potters were so much rejoiced at the decision of the meeting that on the following evening they assembled round a large bonfire at Burslem, and drank the healths of Lord Gower, Mr. Gilbert, and the other promoters of the scheme, with fervent demonstrations of joy.

The opponents of the measure also held meetings at which they strongly declaimed against the Duke's proposed monopoly, and set forth the superior merits of their respective schemes. One of these was a canal from the river Weaver, by Nantwich, Eccleshall, and Stafford, to the Trent at Wilden Ferry, without touching the Potteries at all. Another was for a canal from the Weaver at Northwich, passing by Macclesfield and Stockport, round to Manchester, thus completely surrounding the Duke's navigation, and preventing its extension southward into Staffordshire or any other part of the Midland districts. But there was also a strong party opposed to all canals whatever—the party of croakers, who are always found in opposition to improved communications, whether in the shape of turnpike roads, canals, or railways. These proclaimed that if the proposed canals were made, the country would be ruined, the breed of English horses would be destroyed, the innkeepers would be made bankrupts, and the packhorses and their drivers would be deprived of their subsistence. It was even said that the canals, by putting a stop to the coasting-trade, would destroy the race of seamen. It is a fortunate thing for England that it has contrived to survive these repeated prophecies of ruin. But the manner in which our countrymen contrive to grumble their way along the high road of enterprise, thriving and grumbling, is one of the peculiar features in our character which perhaps Englishmen only can understand and appreciate.

It is a curious illustration of the timidity with which
the projectors of those days entered upon canal enterprise, that one of their most able advocates, in order to
mitigate the opposition of the pack-horse and waggon
interest, proposed that " no main trunk of a canal should
be carried nearer than within four miles of any great
manufacturing and trading town; which distance from
the canal would be sufficient to maintain the same number of carriers and to employ almost the same number of
horses as before."[1]   But as none of the towns in the
Potteries were as yet large manufacturing or trading
places, this objection did not apply to them, nor prevent the canals being carried quite through the centre
of what has since become a continuous district of populous manufacturing towns and villages.   The vested
interests of some of the larger towns were, however,
for this reason preserved, greatly to their own ultimate
injury; and when the canal, to conciliate the local opposition, was so laid out as to pass them at a distance,
not many years passed before they became clamorous
for branches to join the main trunk—but not until the
mischief had been done, and a blow dealt to their own
trade, in consequence of being left so far outside the
main line of water communication, from which many of
them never after recovered.

It is not necessary to describe the Parliamentary contest upon the Grand Trunk Canal Bill.*   There was the
usual muster of hostile interests,—the river navigation
companies uniting to oppose the new and rival company; the array of witnesses on both sides,—Brindley,
Wedgwood, Gilbert, and many more, giving their evidence in support of their own scheme, and a powerful
array of the Cheshire gentry and Weaver Navigation
Trustees appearing on behalf of the others; and the
whipping-up of votes, in which the Duke of Bridgewater

---

[1] ' The Advantages of Inland Navigation,' by R. Whitworth.   1766.

and Earl Gower worked their influence with the Whig party to good purpose. Brindley's plan was, on the whole, considered the best. It was the longest and the most circuitous, but it appeared calculated to afford the largest amount of accommodation to the public. It would pass through important districts, urgently in need of an improved communication with the port of Liverpool on the one hand, and with Hull on the other. But it was not so much the connection of those ports with each other that was needed, as a more convenient means of communication between them and the Staffordshire manufacturing districts; and the Grand Trunk system—somewhat in the form of a horse-shoe, with the Potteries lying along its extreme convex part—promised effectually to answer this purpose, and to open up a ready means of access to the coast on both sides of the island.

A glance at the course of the proposed line will show its great importance. Starting from the Duke's canal at Preston-on-the-Hill, near Runcorn, it passed southwards by Northwich and Middlewich, through the great salt-manufacturing districts of Cheshire, to the summit at Harecastle. It was alleged that the difficulties presented by the long tunnel at that point were so great that it could never be the intention of the projectors of the canal to carry their "chimerical idea," as it was called, into effect. Brindley however insisted, not only that the tunnel was practicable, but that, if the necessary powers were granted, he would certainly execute it.[1]

[1] In one of the many angry pamphlets published at the time, the 'Supplement to a pamphlet entitled Seasonable Considerations on a Navigable Canal intended to be cut from the Trent to the Mersey,' &c., the following passage occurs : " When our all is at stake, these gentlemen [the promoters of the Grand Trunk Canal] must not be surprised at bold truths. We conceive more favourably of their *understanding* than of their *motive ;* we cannot suspect them of entertaining the chimerical idea of cutting through *Hare Castle !* We rather believe that they are desirous of cutting their canal at both ends, and of leaving the middle for the project of a future day. Are these projectors *jealous* of their *honour ?* Let them adopt a clause (which reason and justice strongly enforce) to restrain them from meddling with *either end* till they have finished the *great trunk.* This, and this alone, will shield them from suspicion."

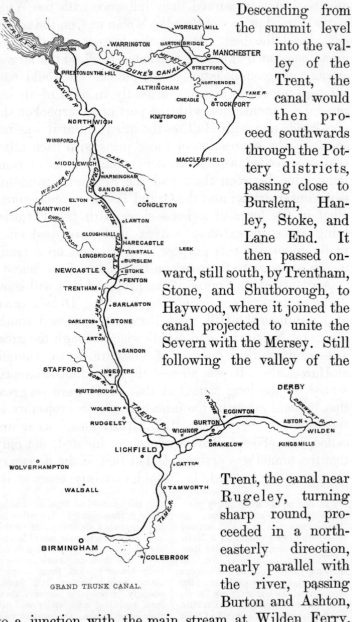

GRAND TRUNK CANAL.

Descending from the summit level into the valley of the Trent, the canal would then proceed southwards through the Pottery districts, passing close to Burslem, Hanley, Stoke, and Lane End. It then passed onward, still south, by Trentham, Stone, and Shutborough, to Haywood, where it joined the canal projected to unite the Severn with the Mersey. Still following the valley of the Trent, the canal near Rugeley, turning sharp round, proceeded in a north-easterly direction, nearly parallel with the river, passing Burton and Ashton, to a junction with the main stream at Wilden Ferry, a little above where the Derwent falls into the Trent near Derby. From thence there was a clear line of navigation, by Nottingham, Newark, and Gainsborough, to the Humber. Provided this admirable project could be carried out, it appeared likely to meet all the necessities of the case. Ample evidence was given in support of the allegations of its promoters; and the result was, that Parliament threw out the bills promoted by the Cheshire gentlemen on behalf of the old river navigation interest, and the Grand Trunk Canal Act passed into law. At the same time another important Act was passed, empowering the construction of the Wolverhampton Canal, from the river Severn, near Bewdley, to the river Trent, near Haywood Mill; thus uniting the navigation of the three rivers which had their termini at the ports of Liverpool, Hull, and Bristol, on the opposite sides of the island.

There was great rejoicing at Burslem on the news arriving at that place of the passing of the bill; and very shortly after, on the 26th of July, 1766, the first sod of the canal was cut by Josiah Wedgwood on the declivity of Bramhills, in a piece of land within a few yards of the bridge which crosses the canal at that place. Brindley was present on the occasion, and due honours were paid to him by the assembled potters. After Mr. Wedgwood had cut the first sod, many of the leading persons of the neighbourhood followed his example, putting their hand to the work by turns, and each cutting a sod or wheeling a barrow of earth in honour of the occasion. It was, indeed, a great day for the Potteries, as the event proved. In the afternoon a sheep was roasted whole in Burslem marketplace, for the good of the poorer class of potters; a *feu de joie* was performed in front of Mr. Wedgwood's house, and sundry other demonstrations of local rejoicing followed the auspicious event.

Wedgwood was of all others the most strongly impressed with the advantages of the proposed canal.

He knew and felt how much his trade had been hindered by the defective communications of the neighbourhood, and to what extent it might be increased provided a ready means of transit to Liverpool, Hull, and Bristol could be secured ; and, confident in the accuracy of his anticipations, he proceeded to make the purchase of a considerable estate in Shelton, intersected by the canal, on the banks of which he built the celebrated Etruria—the finest manufactory of the kind up to that time erected in England, alongside of which he built a mansion for himself and cottages for his workpeople. He removed his works thither from Burslem, partially in 1769, and wholly in 1771, shortly before the works of the canal had been completed.

The Grand Trunk was the most formidable undertaking of the kind that had yet been attempted in England. Its whole length, including the junctions with the Birmingham Canal and the river Severn, was $139\frac{1}{2}$ miles. In conformity with Brindley's practice, he laid out as much of the navigation as possible upon a level, concentrating the locks in this case at the summit, near Harecastle, from which point the waters fell in both directions, north and south. Brindley's liking for long flat reaches of dead water made him keep clear of rivers as much as possible. He likened water in a river flowing down a declivity, to a furious giant running along and overturning everything ; whereas (said he) " if you lay the giant flat upon his back, he loses all his force, and becomes completely passive, whatever his size may be." Hence he contrived that from Middlewich, a distance of seventeen miles, to the Duke's canal at Preston Brook, there should not be a lock ; but goods might be conveyed from the centre of Cheshire to Manchester, for a distance of about seventy miles, along the same uniform water level. He carried out the same practice, in like manner, on the Trent side of Harecastle, where he laid out the canal in as many long lengths of dead water as possible.

The whole rise of the canal from the level of the Mersey, including the Duke's locks at Runcorn, to the summit at Harecastle, is 395 feet; and the fall from thence to the Trent at Wilden is 288 feet 8 inches. The locks on the Grand Trunk proper, on the northern side of Harecastle, are thirty-five, and on the southern side forty. The dimensions of the canal, as originally constructed, were twenty-eight feet in breadth at the top, sixteen at the bottom, and four and a-half feet in depth; but from Wilden to Burton, and from Middlewich to Preston-on-the-Hill, it was thirty-one feet broad at the top, eighteen at the bottom, and five and a-half feet deep, so as to be navigable by large barges; and the locks at those parts of the canal were of correspondingly large dimensions. The width was afterwards made uniform throughout. The canal was carried over the river Dove on an aqueduct of twenty-three arches, approached by an embankment on either side—in all a mile and two furlongs in length. There were also aqueducts over the Trent, which it crosses at four different points—one of these being of six arches of twenty-one feet span each—and over the Dane and other smaller streams. The number of minor aqueducts was about 160, and of road-bridges 109.

But the most formidable works on the canal were the tunnels, of which there were five—the Harecastle, 2880 yards long; the Hermitage, 130 yards; the Barnton, 560 yards; the Saltenford, 350 yards; and the Preston-on-the-Hill, 1241 yards. The Harecastle tunnel (subsequently duplicated by Telford) was constructed only nine feet wide and twelve feet high;[1] but the others

---

[1] Brindley's tunnel had only space for a narrow canal-boat to pass through, and it was propelled by the tedious and laborious process of what is called " legging." It still continues to be worked in the same way, while horses haul the boats through the whole length of Telford's wider tunnel. The men who " leg " the boat, literally kick it along from one end to

were seventeen feet four inches high, and thirteen feet
six inches wide.* The most extensive ridge of country
to be penetrated was at Harecastle, involving by far the
most difficult work in the whole undertaking.   This
ridge is but a continuation of the high ground, forming
what is called the " back-bone of England," which extends
in a south-westerly direction from the Yorkshire moun-

NORTHERN ENTRANCE OF HARECASTLE TUNNELS.[1]

[By E. M. Wimperis, after a Sketch by the Author.]

tains to the Wrekin in Shropshire.   The flat county of
Cheshire, which looks almost as level as a bowling-green
when viewed from the high ground near New Chapel,
seems to form a deep bay in the land, its innermost point
being almost immediately under the village of Hare-

the other. They lie on their backs
on the boat-cloths, with their shoulders resting against some package,
and propel it along by means of their
feet pressing against the top or sides of
the tunnel.

[1] The smaller opening into the hill
on the right-hand of the view is
Brindley's tunnel; that on the left is
Telford's, executed some forty years
since. Harecastle church and village
occupy the ground over the tunnel
entrances.

castle; and from thence to the valley of the Trent the ridge is at the narrowest. That Brindley was correct in determining to form his tunnel at this point has since been confirmed by the survey of Telford, who there constructed his parallel tunnel for the same canal, and still more recently by the engineers of the North Staffordshire Railway, who have also formed their railway tunnel almost parallel with the line of both canals.

When Brindley proposed to cut a navigable way under this ridge, it was declared to be chimerical in the extreme. The defeated promoters of the rival projects continued to make war upon it in pamphlets, and in the exasperating language of mock sympathy proclaimed Brindley's proposed tunnel to be "a sad misfortune,"[1] inasmuch as it would utterly waste the capital raised by the subscribers, and end in the inevitable ruin of the concern. Some of the small local wits spoke of it as another of Brindley's "Air Castles;" but the allusion was not a happy one, as his first "castle in the air," despite all prophecies to the contrary, had been built, and continued to stand firm at Barton; and judging by the issue of that undertaking, it was reasonable to infer that he might equally succeed in this, difficult though it was on all hands admitted to be.

The Act was no sooner passed than Brindley set to work to execute the impossible tunnel. Shafts were sunk from the hill-top at different points down to the level of the intended canal. The stuff was drawn out of

LONGITUDINAL SECTION OF TUNNEL, SHOWING THE STRATA.

the shafts in the usual way by horse-gins; and so long as the water was met with in but small quantities, the power

---

[1] 'Seasonable Considerations,' &c.; Canal pamphlet dated 1766.

of windmills and watermills working pumps over each
shaft was sufficient to keep the excavators at work. But
as the miners descended and cut through the various strata
of the hill on their downward progress, water was met
with in vast quantities; and here Brindley's skill in
pumping-machinery proved of great value. The miners
were often drowned out, and as often set to work again
by his mechanical skill in raising water. He had a fire-
engine, or atmospheric steam-engine, of the best construction possible at that time, erected on the top of the hill, by
the action of which great volumes of water were pumped
out night and day. This abundance of water, though it
was a serious hinderance to the execution of the work,
was a circumstance on which Brindley had calculated,
and indeed depended, for the supply of water for the
summit level of his canal. When the shafts had been
sunk to the proper line of the intended waterway, the
excavation then proceeded in opposite directions, to meet
the other driftways which were in progress. The work
was also carried forward at both ends of the tunnel, and
the whole line of excavation was at length united by a
continuous driftway—it is true, after long and expensive
labour—when the water ran freely out at both ends, and
the pumping-apparatus on the hilltop was no longer
needed. At a general meeting of the Company, held
on the 1st October, 1768, after the works had been in
progress about two years, it appeared from the report of
the Committee that four hundred and nine yards of the
tunnel were cut and vaulted, besides the vast excavations
at either end for the purpose of reservoirs; and the Committee expressed their opinion that the work would be
finished without difficulty.

Active operations had also been in progress at other
parts of the canal. About six hundred men in all were
employed, and Brindley went from point to point superintending and directing their labours. A Burslem correspondent, in September, 1767, wrote to a distant friend

thus :—" Gentlemen come to view our eighth wonder of the world, the subterraneous navigation, which is cutting by the great Mr. Brindley, who handles rocks as easily as you would plum-pies, and makes the four elements subservient to his will. He is as plain a looking man as one of the boors of the Peak, or as one of his own carters; but when he speaks, all ears listen, and every mind is filled with wonder at the things he pronounces to be practicable. He has cut a mile through bogs, which he binds up, embanking them with stones which he gets out of other parts of the navigation, besides about a quarter of a mile into the hill Yelden, on the side of which he has a pump worked by water, and a stove, the fire of which sucks through a pipe the damps that would annoy the men who are cutting towards the centre of the hill. The clay he cuts out serves for bricks to arch the subterraneous part, which we heartily wish to see finished to Wilden Ferry, when we shall be able to send Coals and Pots to London, and to different parts of the globe."

In the course of the first two years' operations, twenty-two miles of the navigation had been cut and finished, and it was expected that before eighteen months more had elapsed the canal would be ready for traffic by water between the Potteries and Hull on the one hand, and Bristol on the other. It was also expected that by the same time the canal would be ready for traffic from the north end of Harecastle Tunnel to the river Mersey. The execution of the tunnel, however, proved so tedious and difficult, and the excavation and building went on so slowly, that the Committee could not promise that it would be finished in less than five years from that time. As it was, the completion of the Harecastle Tunnel occupied nine years more; and it was not finally completed until the year 1777, by which time the great engineer had finally rested from his labours.

It is scarcely necessary to describe the benefits which

the canal conferred upon the inhabitants of the districts through which it passed. As we have already seen, Staffordshire and the adjoining counties had been inaccessible during the chief part of each year. The great natural wealth which they contained was of little value, because it could with difficulty be got at; and even when reached, there was still greater difficulty in distributing it. Coal could not be worked at a profit, the price of land-carriage so much restricting its use, that it was placed altogether beyond the reach of the great body of consumers. It is difficult now to realise the condition of poor people situated in remote districts of England less than a century ago. In winter time they shivered over scanty wood-fires, for timber was almost as scarce and as dear as coal. Fuel was burnt only at cooking-times, or to cast a glow about the hearth in the winter evenings. The fireplaces were little apartments of themselves, sufficiently capacious to enable the whole family to ensconce themselves under the chimney, to listen to stories or relate to each other the events of the day. Fortunate were the villagers who lived hard by a bog or a moor, from which they could cut peat or turf at will. They ran all risks of ague and fever in summer, for the sake of the ready fuel in winter. But in places remote from bogs, and scantily timbered, existence was scarcely possible; and hence the settlement and cultivation of the country were in no slight degree retarded until comparatively recent times, when better communications were opened up.

So soon as the canals were made, and coals could be readily conveyed along them at comparatively moderate rates, the results were immediately felt in the increased comfort of the people. Employment became more abundant, and industry sprang up in their neighbourhood in all directions. The Duke's canal, as we have seen, gave the first great impetus to the industry of Manchester and that district. The Grand Trunk had precisely the same effect throughout the Pottery and other districts of Staffordshire ; and their joint action was not only to employ, but actually to civilize the people. The salt of Cheshire could now be manufactured in immense quantities, readily conveyed away, and sold at a comparatively moderate price in all the midland districts of England. The potters of Burslem and Stoke, by the same mode of conveyance, received their gypsum from Northwich, their clay and flints from the seaports now directly connected with the canal, returning their manufactures by the same route. The carriage of all articles being reduced to about one-fourth of their previous rates,[1] articles of necessity and comfort, such as had formerly been unknown except amongst the wealthier classes, came into common use amongst the people. Existence ceased to be difficult, and came to be easy. Led by the enterprise of Wedgwood and others like him, new branches of industry sprang up, and the manufacture of earthenware, instead of being insignificant and comparatively unprofitable, which it was before his time, became a staple branch of English trade. Only about ten years after the Grand Trunk Canal had been opened, Wedgwood stated in evidence before the House of Commons, that from 15,000 to 20,000

---

[1] The following comparison of the rates per ton at which goods were conveyed by land-carriage before the opening of the Grand Trunk Canal, and those at which they were subsequently carried by it, will show how great was the advantage conferred on the country by the introduction of navigable canals :—" The cost of carrying a ton of goods from Liverpool to Etruria, the centre of the Staffordshire Potteries, by land-carriage, was 50s. ; the Trent and Mersey reduced it to 13s. 4d. The land-carriage from Liverpool to Wolverhampton was 5l. a ton ; the canal reduced it to 1l. 5s. The land-carriage from Liverpool to Birmingham, and also to Stourport, was 5l. a ton ; the canal reduced both to 1l. 10s. . . . . . Thus the cost of inland transport was reduced, on the average, to about one-fourth of the rate paid previous to the introduction of canal-navigation. The advantages were enormous : wheat, for example, which formerly could not be conveyed a hundred miles, from corn-growing districts to the large towns and manufacturing districts, for less than 20s. a quarter, could be conveyed for about 5s. a quarter. These facts show how great was the service conferred on the country by Brindley and the Duke of Bridgewater."—Baines's 'History of the Commerce and Town of Liverpool.'

persons were then employed in the earthenware-manu-
facture alone, besides the large number of others employed in digging coals for their use, and the still larger
number occupied in providing materials at distant parts,
and in the carrying and distributing trade by land and
sea.   The annual import of clay and flints into Staffordshire at that time was from fifty to sixty thousand tons ;
and yet, as Wedgwood truly predicted, the trade was
but in its infancy.   The outwards and inwards tonnage
to the Potteries is now upwards of three hundred thousand
tons a-year.

The moral and social influences exercised by the canals
upon the Pottery districts were not less remarkable.*
From a half-savage, thinly-peopled district of some 7000
persons in 1760, partially employed and ill-remunerated,
we find them increased, in the course of some twenty-five
years, to about treble the population, abundantly employed, prosperous, and comfortable.[1]   Civilization is
doubtless a plant of very slow growth, and does not
necessarily accompany the rapid increase of wealth.
On the contrary, higher earnings, without improved
morale, may only lead to wild waste and gross indulgence.
But the testimony of Wesley to the improved character
of the population of the Pottery district in 1781, within
a few years after the opening of Brindley's Grand Trunk
Canal, is so remarkable, that we cannot do better than
quote it here ; and the more so, as we have already given
the account of his first visit in 1760, on the occasion of
his being pelted.   " I returned to Burslem," says Wesley ;
" how is the whole face of the country changed in about
twenty years ! Since which, inhabitants have continually
flowed in from every side.   Hence the wilderness is
literally become a fruitful field.   Houses, villages, towns,
have sprung up, and the country is not more improved
than the people."

---

[1] The population of the same district in 1861 was found to be upwards
of 120,000.

[After his work on the Manchester-Liverpool and the Grand
Trunk canals, Brindley projected and superintended numerous
other lines. One of the most important of these was the Wolverhampton Canal connecting the Trent with the Severn. Brindley
also laid out the Birmingham Canal which tied the industrial
center into the Midlands canal system. By these and other
canals, the iron manufactories gained easy access to raw materials and to market.

Smiles lamented the lack of reports and letters about Brindley's pioneer efforts in connection with his last canals. Because
Brindley could scarcely read, and wrote with difficulty, he left
limited records. Since these canals were constructed before the
significance of inland transportation was generally recognized,
few who wrote for press and publisher took note of the engineering achievements.

By the turn of the century the economic and social significance
of the canal system was more generally realized. Instead of a
decline of coastal shipping, which some predicted would result
from the spread of the canal network, the tonnage of English
ships increased threefold during the thirty years that elapsed
after the opening of the Duke's canal. This resulted in part
from the increased transit of goods from the ports to the principal inland towns. Instead of destroying the breed of horses,
the canal system greatly increased the demand for horses to
carry coal, lime, manure, and agricultural produce to and from
canal depots. By 1792 the returns upon prior canal investments
generated a speculative ferment in canal shares.

By about 1795 Brindley and the other canal builders had
opened the country by 2,600 miles of navigable canals in England, 276 miles in Ireland, and 225 miles in Scotland. Not a
place in England, south of Durham, was more than fifteen miles
from water communication.]

CHAPTER VIII.

IT will be observed that Brindley's employment as an engineer extended over a wide district. Even before his employment by the Duke of Bridgewater, he was under the necessity of travelling great distances to fit up water-mills, pumping-engines, and manufacturing machinery of various kinds, in the counties of Stafford, Cheshire, and Lancashire. But when he had been appointed to superintend the construction of the Duke's canals, his engagements necessarily became of a still more engrossing character, and he had very little leisure left to devote to the affairs of private life. He lived principally at inns, in the immediate neighbourhood of his work; and though his home was at Leek, he sometimes did not visit it for weeks together. He had very little time for friendship, and still less for courtship. Nevertheless, he did contrive to find time for marrying, though at a comparatively advanced period of his life. In laying out the Grand Trunk Canal, he was necessarily brought into close connection with Mr. John Henshall, of the Bent, near New Chapel, land-surveyor, who assisted him in making the survey. We find him visiting his house in September, 1762, and settling with him as to the preliminary operations. At these visits Brindley seems to have taken a special liking for Mr. Henshall's daughter Anne, then a girl at school, and when he went to see her father, he was accustomed to take a store of gingerbread for Anne in his pocket. She must have been a comely girl, judging by the portrait of her as a woman, which we have seen. In course of time, the liking ripened into an attachment; and shortly after the girl had left school, at the age of only nineteen, Brindley proposed to her, and was accepted. By this time he was close upon his fiftieth year, so that the union may possibly have been quite as much a matter of convenience as of love on his part. He had now left the Duke's service for the purpose of entering upon the construction of the Grand Trunk Canal, and with that object resolved to transfer his home to the immediate neighbourhood of Harecastle, as well as of his colliery at Golden Hill. Shortly after the marriage, the old mansion of Turnhurst fell vacant, and Brindley became its tenant, with his young wife. The marriage took place on the 8th December, 1765, in the parish church of Wolstanton, Brindley being described in the register as " of the parish of Leek, engineer ;" but from that time until the date of his death his home was at Turnhurst.

The house at Turnhurst was a comfortable, roomy, old-fashioned dwelling, with a garden and pleasure-ground behind, and a little lake in front. It was formerly the residence of the Bellot family, and is said to have been the last house in England in which a family fool was maintained. Sir Thomas Bellot, the last of the name, was a keen sportsman, and the panels of several of the upper rooms contain pictorial records of some of his exploits in the field. In this way Sir Thomas seems to have befooled his estate, and it shortly after became the property of the Alsager family, from whom Brindley rented it. A little summer-house, standing at the corner of the outer courtyard, is still pointed out as Brindley's office, where he sketched his plans and prepared his calculations. As for his correspondence, it was nearly all conducted, subsequent to his marriage, by his wife, who, notwithstanding her youth, proved a most clever, useful, and affectionate partner.

Turnhurst was conveniently near to the works then

BRINDLEY'S HOUSE AT TURNHURST.

in progress at Harecastle Tunnel, which was within easy walking distance, whilst the colliery at Golden Hill was only a few fields off. From the elevated ground at Golden Hill, the whole range of high ground may be seen under which the tunnel runs—the populous pottery towns of Tunstall and Burslem filling the valley of the Trent towards the south. At Golden Hill, Brindley carried out an idea which he had doubtless brought with him from Worsley. He and his partners had an underground canal made from the main line of the Harecastle Tunnel into their coal-mine, about a mile and a half in length; and by that tunnel the whole of the coal above that level was afterwards worked out, and conveyed away for sale in the Pottery and other districts, to the great profit of the owners as well as to the equally great convenience of the public.

These various avocations involved a great amount of labour as well as anxiety, and probably considerable tear and wear of the vital powers. But we doubt whether mere hard work ever killed any man, or whether Brindley's labours, extraordinary though they

were, would have shortened his life, but for the far
more trying condition of the engineer's vocation—irregular living, exposure in all weathers, long fasting, and
then, perhaps, heavy feeding when the nervous system
was exhausted, together with habitual disregard of the
ordinary conditions of physical health.   These are the
main causes of the shortness of life of most of our eminent
engineers, rather than the amount and duration of
their labours.   Thus the constitution becomes strained,
and is ever ready to break down at the weakest place.
Some violation of the natural laws more flagrant than
usual, or a sudden exposure to cold or wet, merely
presents the opportunity for an attack of disease which
the ill-used physical system is found unable to resist.
Such an accidental exposure unhappily proved fatal to
Brindley. While engaged one day in surveying a branch
canal between Leek and Froghall, he got drenched
near Ipstones, and went about for some time in his
wet clothes.   This he had often before done with im-
punity, and he might have done so again ; but, unfortunately, he was put into a damp bed in the inn at
Ipstones, and this proved too much for his constitution,
robust though he naturally was.   He became seriously
ill, and was disabled from all further work.   Diabetes
shortly developed itself, and, after an illness of some
duration, he expired at his house at Turnhurst, on the
27th of September, 1772, in the fifty-sixth year of his
age, and was interred in the burying-ground at New
Chapel, a few fields distant from his dwelling.

James Brindley was probably one of the most remarkable instances of self-taught genius to be found in
the whole range of biography.   The impulse which he
gave to social activity, and the ameliorative influence
which he exercised upon the condition of his countrymen, seem out of all proportion to the meagre intellectual culture which he had received in the course

of his laborious and active career. We must not, however, judge him merely by the literary test. It is true, he could scarcely read, and he was thus cut off, to his own great loss, from familiar intercourse with a large class of cultivated minds, living and dead; for he could not share in the conversation of educated men, nor enrich his mind by reading the stores of experience found treasured up in books. Neither could he write, except with difficulty and inaccurately, as we have shown from the extracts above quoted from his note-books still extant.

Brindley was, nevertheless, a highly-instructed man in many respects. He was full of the results of careful observation, ready at devising the best methods of overcoming material difficulties, and possessed of a powerful and correct judgment in matters of business. Where any emergency arose, his quick invention and ingenuity, cultivated by experience, enabled him almost at once unerringly to suggest the best means of providing for it. His ability in this way was so remarkable, that those about him attributed the process by which he arrived at his conclusions rather to instinct than reflection—the true instinct of genius.* " Mr. Brindley," says one of his contemporaries, " is one of those great geniuses whom Nature sometimes rears by her own force, and brings to maturity without the necessity of cultivation. His whole plan is admirable, and so well concerted that he is never at a loss ; for, if any difficulty arises, he removes it with a facility which appears so much like inspiration, that you would think Minerva was at his fingers' ends."

His mechanical genius was indeed most highly cultivated. From the time when he bound himself apprentice to the trade of a millwright—impelled to do so by the strong bias of his nature—he had been undergoing a course of daily and hourly instruction. There was nothing to distract his attention, or turn him from

pursuing his favourite study of practical mechanics.
The training of his inventive faculty and constructive
skill was, indeed, a slow but a continuous process; and
when the time and the opportunity arrived for turning
these to account—when the silk-throwing machinery of
the Congleton mill, for instance, had to be perfected and
brought to the point of effectively doing its intended
duty—Brindley was found able to take it in hand and
finish the work, when even its own designer had given
it up in despair.   But it must also be remembered that
this great facility of Brindley had been in a great measure the result of the closest observation, the most pains-
taking study of details, and the most indefatigable industry.

The same qualities were displayed in his improvements of the steam-engine, and his arrangements to
economise power in the pumping of water from drowned
mines.   It was often said of his works, as was said of
Columbus's discovery, "how easy! how simple!" but
this was after the fact.   Before he had brought his fund
of experience and clearness of vision to bear upon a
difficulty, every one was equally ready to exclaim, "how
difficult! how absolutely impracticable!"   This was the
case with his "castle in the air," the Barton Viaduct—
such a work as had never before been attempted in
England, though now any common mason would undertake it.   It was Brindley's merit always to be ready
with his simple, practical expedient; and he rarely failed
to effect his purpose, difficult although at first sight its
accomplishment might seem to be.

Like men of a similar stamp, Brindley had great confidence in himself and in his powers and resources.
Without this, it were impossible for him to have accomplished so much as he did.   It is said that the King of
France, hearing of his great natural genius, and the
works he had performed for the Duke of Bridgewater
at Worsley, expressed a desire to see him, and sent a

message inviting him to view the great canal of Languedoc. But Brindley's reply was characteristic : " I will have no journeys to foreign countries," said he, " unless to be employed in surpassing all that has been already done in them."

His observation was remarkably quick. In surveying a district, he rapidly noted the character of the country, the direction of the hills and the valleys, and, after a few journeys on horseback, he clearly settled in his mind the best line to be selected for a canal, which almost invariably proved to be the right one. In like manner he would estimate with great rapidity the fall of a brook or river while walking along the banks, and thus determined the height of his cuttings and embankments, which he afterwards settled by a more systematic survey. In these estimates he was rarely, if ever, found mistaken.

His brother-in-law, Mr. Henshall, has said of him, " when any extraordinary difficulty occurred to Mr. Brindley in the execution of his works, having little or no assistance from books or the labours of other men, his resources lay within himself.* In order, therefore, to be quiet and uninterrupted whilst he was in search of the necessary expedients, he generally retired to his bed ;¹ and he has been known to be there one, two, or three days, till he had attained the object in view. He would then get up and execute his design, without any drawing or model. Indeed, it was never his custom to make either, unless he was obliged to do it to satisfy his employers. His memory was so remarkable that he has often declared that he could remember, and execute, all the parts of the most complex machine, provided he had time, in his survey of it, to settle in his mind the

---

¹ The younger Pliny seems to have adopted almost a similar method : "Clausæ fenestræ manent. Mirè enim silentio et tenebris animus alitur. Ab iis quæ avocant abductus, et liber, et mihi relictus, non oculos animo sed animum oculis sequor, qui eadem quæ mens vident quoties non vident alia."—Epist. lib. ix., ep. 36.

several parts and their relations to each other. His method of calculating the powers of any machine invented by him was peculiar to himself. He worked the question for some time in his head, and then put down the results in figures. After this, taking it up again at that stage, he worked it further in his mind for a certain time, and set down the results as before. In the same way he still proceeded, making use of figures only at stated parts of the question. Yet the ultimate result was generally true, though the road he travelled in search of it was unknown to all but himself, and perhaps it would not have been in his power to have shown it to another." [1]

The statement about his taking to bed to study his more difficult problems is curiously confirmed by Brindley's own note-book, in which he occasionally enters the words "lay in bed," as if to mark the period, though he does not particularise the object of his thoughts on such occasions. It was a great misfortune for Brindley, as it must be to every man, to have his mental operations confined exclusively within the limits of his profession. He thought and lived mechanics, and never rose above them. He found no pleasure in anything else; amusement of every kind was distasteful to him; and his first visit to the theatre, when in London, was also his last. Shut out from the humanising influence of books, and without any taste for the politer arts, his mind went on painfully grinding in the mill of mechanics. "He never seemed in his element," said his friend Bentley, "if he was not either planning or executing some great work, or conversing with his friends upon subjects of importance." To the last he was full of projects, and full of work; and then the wheels

---

[1] 'Biographia Britannica,' 2nd Ed. Edited by Dr. Kippis. The materials of the article are acknowledged to have been obtained principally from Mr. Henshall by Messrs. Wedg-wood and Bentley, who wrote and published the memoir in testimony of their admiration and respect for their deceased friend, the engineer of the Grand Trunk Canal.

of life came to a sudden stop, when he could work no longer. It is related of him that, when dying, some eager canal undertakers insisted on having an interview with him. They had encountered a serious difficulty in the course of constructing their canal, and they *must* have the advice of Mr. Brindley on the subject. They were introduced to the apartment where he lay scarce able to gasp, yet his mind was clear. They explained their difficulty—they could not make their canal hold water. "Then puddle it," said the engineer. They explained that they had already done so. "Then puddle it again—and again." This was all he could say, and it was enough.

It remains to be added that, in his private character, Brindley commanded general respect and admiration. His integrity was inflexible ; his manner, though rough and homely, was kind ; and his conduct unimpeachable.[1] He was altogether unassuming and unostentatious, and dressed and lived with great plainness. He was the furthest possible from a narrow or jealous temper, and nothing gave him greater pleasure than to assist others with their inventions, and to train up a generation of engineers, in the persons of his pupils, able to carry out the works he had designed, when no longer able to conduct them. The principal undertakings in which he was engaged up to the time of his death were carried on by his brother-in-law, Mr. Henshall, formerly his

---

[1] It has, indeed, been stated in the crazy publication of the last Earl of Bridgewater, to which we have already alluded, that when in the service of the Duke, Brindley was "drunken." But this is completely contradicted by the testimony of Brindley's own friends; by the evidence of Brindley's note-book, from repeated entries in which it appears that his " ating and drink " at dinner cost no more than 8*d.*; by the confidence generally reposed in him, and the friendship entertained for him, by men such as Josiah Wedgwood; and by the fact of the vast amount of work that he subsequently contrived to get through. No man of " drunken " habits could possibly have done this. We should not have referred to this topic but for the circumstance that the late Mr. Baines, of Leeds, has quoted the Earl's statement, without contradiction, in his excellent ' History of Lancashire.' *

clerk of the works on the Grand Trunk Canal, and by his able pupil, Mr. Robert Whitworth, for both of whom he had a peculiar regard, and of whose integrity and abilities he had the highest opinion.

Brindley left behind him two daughters, one of whom, Susannah, married Mr. Bettington, of Bristol, merchant, afterwards the Honourable Mr. Bettington, of Brindley's Plains, Van Diemen's Land, where their descendants still live. His other daughter, Anne, died unmarried, on her passage home from Sydney, in 1838. His widow, still young, married again, and died at Longport in 1826. Brindley had the sagacity to invest a considerable portion of his savings in Grand Trunk shares, the great increase in the value of which, as well as of his colliery property at Golden Hill, enabled him to leave his family in affluent circumstances.

BRINDLEY'S BURIAL-PLACE AT NEW CHAPEL.

. . . .

JOHN RENNIE, F. R. S.

Engraved by W. Holl, after the portrait in crayons
by Archibald Skirving

LIFE OF JOHN RENNIE.

CHAPTER I.

Scotland at the Middle of last Century.

JOHN RENNIE, the architect of the three great London bridges, the engineer of the Plymouth Breakwater, of the London and East India Docks, and various other works of national importance, was born at the farm-steading of Phantassie, in East Lothian, on the 7th of June, 1761. His father was the owner of the small estate of the above

RENNIE'S NATIVE DISTRICT. [Ordnance Survey.]

name, situated about midway between Haddington and Dunbar, at the foot of the gently-sloping hills which rise from it towards the south, the village of East Linton lying close at hand on the further bank of the little river Tyne. The property had been in the family for generations, and Mr. Rennie had the reputation of being one of the best farmers in the neighbourhood. But the art of agriculture, like every thing else in Scotland, was in an incredibly backward state, compared with either England or even Ireland, at the time when our engineer was born.

The traveller through the Lothians—which now exhibit perhaps the finest agriculture in the world, where every inch of ground is turned to profitable account, and the fields are cultivated to the very hedge-roots—will scarcely believe that less than a century ago these districts were not much removed from the state in which nature had left them. In the interior there was little to be seen but bleak moors and quaking bogs. The chief part of each farm consisted of "out-field" or unenclosed land, no better than moorland, from which even the hardy black cattle could scarcely gather herbage enough to keep them from starving in winter time. The "in-field" was an enclosed patch of ill-cultivated ground, on which oats and "bear" or barley were grown; but the principal crop was weeds.

Of the small quantity of corn raised in the country nine-tenths were grown within five miles of the coast;[1] and of wheat very little was raised—not a blade north of the Lothians. When the first crop of that grain was to be seen on a field near Edinburgh, people flocked to look upon it as a wonder. Clover, turnips, and potatoes had not yet been introduced, and no cattle were fattened: it was with difficulty they could be kept alive. Mr. Rennie, the engineer's father, was one of the first to introduce turnips as a regular farmer's crop. All loads were as yet carried on horseback; but where the farm was too small, or the crofter too poor, to keep a horse, his own or his wife's back bore the load. The horse brought peats from the bog and coals from the pit, and carried the crops to market. Sacks filled with manure were also sent a-field on horseback; but the uses of manure were so little understood, that if a stream was

---

[1] Professor Forbes's 'Considerations on the Present State of Scotland,' p. 14.

near, it was thrown in and floated away, and in summer it was burnt.[1]

The towns were for the most part collections of thatched mud cottages,[2] giving scant shelter to a miserable population. The whole country was poor, desponding, gaunt, and almost haggard. The common people were badly fed and wretchedly clothed; those in the country living in despicable huts with their cattle.[3] The poor crofters were barely able to exist. Lord Kaimes says of the Scotch tenantry of the early part of last century, that they were so benumbed by oppression and poverty[4] that the most able instructors in husbandry could have made nothing of them. A writer in the Scotch 'Farmer's Magazine' sums up his account of the country at that time in these words : " Except in a few instances, it was little better than a barren waste." [5]

What will scarcely be credited, now that the industry of Scotland has become thoroughly educated by a century's discipline of work, was the inconceivable listlessness and laziness of the people at that

---

[1] 'Farmer's Magazine,' No. xxxiv., p. 200.

[2] It is stated in MacDiarmid's 'Picture of Dumfries' that at the middle of the century no lime was used in building, "except a little shell-lime, made of cockle-shells, which was burned at Colvend, and brought to Dumfries in bags." And, "in 1740, when Provost Bell (the chief magistrate or mayor of that town) built his house, the under storey was built of clay, and the upper storeys with lime brought from Whitehaven in dry-ware casks."

[3] The Rev. Dr. Playfair in 'Statistical Account of Scotland.' First edition. Vol. I., p. 513.

[4] Bad although the condition of Scotland was at the beginning of last century, there were many who believed that it would be made *worse* by the carrying of the Act of Union.

The Earl of Wigton was one of these. Possessing large estates in the county of Stirling, and desirous of taking every precaution against the impending ruin, he disposed to his tenants, on condition that they continued to pay him their then rents, low though they were, his extensive estates in the parishes of Denny, Kirkintulloch, and Cumbernauld, retaining only a few fields round the family mansion.[1] Fletcher of Saltoun equally feared the ruinous results of the Union, though he was less precipitate than the Earl of Wigton. We need scarcely say how completely all those apprehensions were falsified by the actual results.

[5] 'Farmer's Magazine,' 1803. No. xiii., p. 101.

---

[1] Farmer's Magazine, 1808, No. xxxiv., p. 193.

period.[1] They left the bog unreclaimed and the swamp undrained. They would not even be at the trouble to enclose lands easily capable of cultivation. There was no class possessed of any enterprise or wealth. A middle rank could scarcely be said to exist, or any condition between that of the starving peasantry and the impoverished proprietary, whose available means were principally expended on hard drinking:[2] Mr. Brown, an East Lothian farmer, said of the latter class, that they were still too proud, and perhaps too ignorant, to interest themselves about the amelioration of their own domains.[3] The educated class—strictly so called—was as yet extremely small, and displayed a general indifferentism on all subjects of social, political, or religious interest, which some regarded as philosophic, but which was only an exhibition in another form of the prevalent national indolence. An idea of the general poverty may be formed from the fact that about the middle of the century the whole circulating medium of the Edinburgh banks was only 200,000l., which was found amply sufficient for the requirements of trade and commerce, which had scarcely yet sprung into existence.[4] Even in East Lothian, which was probably in advance of the other Scotch counties, the ordinary wage of a day labourer was only fivepence in

---

[1] Miss Craik, in describing the difficulties which her father (William Craik, of Arbigland) had to contend against in introducing agricultural improvements in the county of Kirkcudbright, about the middle of last century, says: " For many years the indolent obstinacy of the lower class of people was almost unconquerable. Amongst other instances of their laziness, I have heard him say that, upon his first introduction of the mode of dressing the grain at night which had been thrashed during the day, all the servants in the neighbourhood refused to adopt the measure, and even threatened to destroy the houses of their employers by fire if they continued to insist upon the business. My father speedily perceived that a forcible remedy was required for the evil. He gave them their choice of removing the thrashed grain in the evening, or becoming inhabitants of Kirkcudbright jail; they preferred the former alternative, and open murmurings were no longer heard."— 'Farmer's Magazine,' No. xlvi. (June, 1811), p. 155. Art.: 'Account of William Craik, Esq., of Arbigland.'

[2] See the 'Autobiography of Dr. Alexander Carlyle,' passim.

[3] Brown on 'Rural Affairs,' Vol. I., p. 58. [4] Ibid.

winter and sixpence in summer.[1] The food of the working class was almost wholly vegetable, and even that was insufficient in quantity. The little butcher's meat consumed by the better class was salted beef and mutton, which was stored up at Ladner Time, betwixt Michaelmas and Martinmas, for the year's consumption. Mr. Buchan Hepburn says the sheriff of the county of East Lothian informed him that he remembered when not a single bullock was slaughtered in the butcher-market at Haddington for a whole year, except at the above period; and when Sir David Kinloch, of Gilmerton, sold ten wedders to an Edinburgh butcher, he stipulated for three several terms to take them away, to prevent the Edinburgh market from being overstocked with fresh butcher's meat![2]

The rest of Scotland was in no better state: in some parts it was even worse. The now rich and fertile county of Ayr, which glories in the name of "the garden of Scotland," was for the most part a wild and dreary common, with here and there a poor, bare, homely hut, where the farmer and his family were lodged.[3] There were no enclosures of land, except one or two about a gentleman's seat, and black cattle roamed at large over the face of the country.[4] More deplorable still was the

---

[1] G. Buchan Hepburn's 'General View of the Agriculture and Economy of East Lothian.' Edinburgh, 1794. P. 95.

[2] Ibid., p. 55.

[3] The Rev. Mr. Robertson, in the 'Statistical Account of Scotland.'

[4] When it was attempted, in 1723, to form enclosures in the adjoining county of Kirkcudbright, for the purpose of preventing the black cattle from straying, the poor people, who had squatted or were small tenants on the land, were turned out, and mobs assembled at different points and levelled the enclosures. "It is not pleasant," says a Kirkcudbright chronicler, " to represent the wretched

state of individuals as times then went in Scotland. The tenants in general lived very meanly, on kail, groats, milk, graddon ground in querns turned by the hand, the grain being dried in a pot, together with a crock ewe now and then about Martinmas. They were clothed very plainly, and their habitations were most uncomfortable. Their general wear was of cloth, made of waulked plaiding, black and white wool mixed, very coarse, and the cloth rarely dyed. Their hose (when they wore them) were made of white plaiding cloth, sewed together ; with single-soled shoes, and a black or blue bonnet— none having hats but the lairds, who

condition of those counties which immediately bordered the wild Highland districts, the inhabitants of which regarded the Lowlands as their lawful prey.   The only method by which security of a certain sort could be obtained for their property was by the payment of black-mail to some of the principal caterans; though this was not sufficient to protect them against the lesser marauders.   Regular contracts were drawn up between proprietors in the counties of Perth, Stirling, and Dumbarton, and the Macgregors, in which it was stipulated that if less than seven cattle were stolen—which peccadillo was styled *picking*—no redress should be required; but if the number stolen exceeded seven—such amount of theft being termed *lifting*—then the Macgregors became bound to recover.   This blackmail was regularly levied as far south as Campsie—then within six miles of Glasgow, but now almost forming part of it—down to within a few months of the outbreak of the rebellion of 1745.[1]   Under such circumstances agricultural improvement was impossible.   Another evil was, that the lawless habits of their neighbours tended to make the Lowland farmers almost as ferocious as the Highlanders themselves.   Feuds were of constant occurrence between neighbouring baronies, and even contiguous parishes; and the county fairs, which were tacitly recognised as the occasions for settling quarrels, were the scenes of as

---

thought themselves very well dressed for going to church on Sunday with a black kelt-coat of their wife's making. The distresses and poverty felt in the country continued till about the year 1735.   During these times, when potatoes were not generally raised (having been only introduced into the stewartry in 1725), there was, for the most part, a great scarcity of food, bordering on famine; for, in the whole of Kirkcudbright and Dumfries, there was not as much victual produced as was necessary for the supply of the inhabitants; and the chief part of what was required for

that purpose was brought from the Sandbeds of Esk, in tumbling cars, to Dumfries; and when the waters were high by reason of spates, and there being no bridges so that these cars could not come with the meal, I have seen the tradesmen's wives in the streets of Dumfries crying, because there was none to be got."—Letter of John Maxwell, in Appendix to MacDiarmid's 'Picture of Dumfries.'   Edinburgh, 1832.

[1] 'Farmer's Magazine:' 'Account of the Husbandry of Stirlingshire,' No. xxxiv., p. 198.

bloody fights as were ever known in Ireland, even in its worst days.

The country was as yet almost without roads, so that communication between one town and another was exceedingly difficult, especially in winter. The old track between Haddington and Edinburgh still exists as it was left when the new system of turnpike roads was introduced in Scotland. It is now used only by fox-hunters riding to cover, but it continues to bear out the description of a local writer : " Nothing," he says, " can be a greater contrast with the roads of modern times. In some places, where there was space for taking room, it was not spared. There might be seen four or five or more tracks, all collateral to one another, as each in its turn had been abandoned and another chosen, and all at last equally impassable. In wet weather they became mere sloughs, in which the carts or carriages had to slumper through in a half-swimming state, whilst in time of drought it was a continued jolting out of one hole into another." [1]

Such being the state of the highways, it will be evident that very little traffic could be conducted in vehicles of any sort. Single horse traffickers, called cadgers, plied between country towns and villages, supplying the inhabitants with salt, fish, earthenware, and articles of clothing, which they carried in sacks or creels hung across the horse's back. Even the trade between Edinburgh and Glasgow was carried on in the same primitive way. So limited was the consumption of the comparatively small population of Glasgow about the middle of last century, that most of the butter, cheese, and poultry raised within six miles of that city was carried by cadgers to Edinburgh in panniers on horse-back. On one occasion, a load of ducks, brought from Campsie to Edinburgh for sale in the Grassmarket,

---

[1] George Robertson's 'Rural Recollections,' p. 38.

finding themselves at liberty, rose upon wing and flew westward. Some of them were afterwards found at Linlithgow, and others succeeded in reaching their native "dub" at Campsie, some forty-five miles distant.[1]

It was long before travelling by coach was introduced in Scotland. When Smollett went from Glasgow to Edinburgh in 1739, on his way to London, there was neither coach, cart, nor waggon on the road. He accordingly accompanied the carriers as far as Newcastle, "sitting upon a pack-saddle between two baskets, one of which," he says, "contained my goods in a knapsack." The first vehicle which plied between the two chief cities of Scotland was not started until 1749. It was called "The Edinburgh and Glasgow Caravan," and performed the journey of forty-four miles in two days; but the packhorse continued to be the principal means of communication between the two places. Ten years later another vehicle was started, which was named "The Fly," because of its extraordinary speed, and it contrived to make the journey in rather less than a day and a half.[2] When a coach with four horses was put on between Haddington and Edinburgh, it took a full winter's day to perform the journey of sixteen miles. The effort was to reach Musselburgh in time for dinner, and go into town in the evening.[3]

In some parts of the country—as in Spain to this day —the beds of rivers served the double purpose of a river in wet, and a road in dry weather. When a common carrier began to ply between Selkirk and Edinburgh, a distance of only thirty-eight miles, he occupied a fortnight in performing the double journey. Part of the road lay along Gala Water, and in summer the carrier drove his rude cart along the bed of the stream; in winter the route was of course altogether impassable.

---

[1] 'Farmer's Magazine,' No. xxxiv., p. 200.
[2] Robertson's 'Rural Recollections.'
[3] G. Buchan Hepburn's 'Account of East Lothian.' 1794.

The townsmen of this adventurous individual, on the morning of his way-going, were accustomed to turn out and take leave of him, wishing him a safe return from his perilous journey.

The great post-road between London and Edinburgh passed close in front of the house at Phantassie in which John Rennie was born; but even that was little better than the tracks we have already described. It passed westward over Pencrake, and followed the ridge of the Garleton Hills towards Edinburgh. The old travellers had no aversion to hill tops, rather preferring them because the ground was firmer to tread on, and they could see better about them. This line of high road avoided the county town, which, lying in a hollow, was unapproachable across the low grounds in wet weather; and, of all things, swamps and quagmires were then most dreaded. A portion of this old post-road was visible until within the last few years, upon the high ground about a mile to the north of Hadding-ton. In some places it was very narrow and deep, not unlike an old broad ditch, much waterworn, and strewn with loose stones. Along this line of way Sir John Cope passed with his army, in 1745, to protect Edin-burgh against the Highland rebels; and it is related that, on marching northward to intercept them, he was com-pelled to halt for several days, waiting for a hundred horse-loads of bread required for the victualling of his army.

In 1750, a project was set on foot for improving the high road through East Lothian, and a Turnpike Act was obtained for the purpose—the first Act of the kind obtained north of the Tweed.[1] The inhabitants of the town of Haddington complained loudly of the oppres-sion practised on them, by making them pay toll for every bit of coal they burned; though before the road was made it was a good day's work for a man and

---

[1] G. Buchan Hepburn's 'Account,' p. 151.

horse to fetch a load of "divot" from Gladsmuir, or of coal from the nearest colliery, only some four miles distant. By the year 1763 this post-road must have been made practicable for wheeled vehicles; for in that year the one stage-coach, which for a time formed the sole communication of the kind between London and all Scotland, began to run; and John Rennie, when a boy, was familiar with the sight of the uncouth vehicle lumbering along the road past his door. It "set out" from Edinburgh only once a month, the journey to London occupying from twelve to eighteen days, according to the state of the roads.

Such, however, had not always been the miserable condition of Scotland. The fine old bridges which exist in different parts of the country alone serve to show that at some early period a degree of civilization and prosperity had prevailed, from which it had gradually fallen. Professor Innes has clearly pointed this out in a recent work :[1] "When we consider," he says, "the long and united efforts required in the early state of the arts for throwing a bridge over any considerable river, the early occurrence of bridges may well be admitted as one of the best tests of civilization and national prosperity." As in England itself, the original reclamation of lands, the improvement of agriculture, the making of roads, and the building of bridges throughout the Lowlands of Scotland, were for the most part due to the old churchmen; and when their ecclesiastical organization was destroyed the country again relapsed into the state from which they had raised it, and it lay in ruins almost until our own day, when it has again been rescued from barrenness, even more effectually than before, by the combined influences of education and industry.

The same "Brothers of the Bridge," who erected so many fine old bridges across the rivers of England, were

---

[1] Cosmo Innes's 'Sketches of Early Scottish History.' 1861.

equally busy beyond the Tweed, providing those essential means of intercourse for the community. Thus we find bridges early erected across most of the rapid rivers in the Lowlands, especially in those places where the ecclesiastical foundations were the richest; and to this day the magnificent old abbey or cathedral of the neighbourhood—in some corner of which the Presbyterian Church holds its worship—serves to remind one of the contemporaneous origin of both classes of structures. Thus, as early as the thirteenth century, there was a bridge over the Tay at Perth; bridges over the Esk at Brechin and Marykirk; one over the Dee at Kincardine O'Neil; one at Aberdeen; and one at the mouth of Glenmuick. The fine old bridge over the Dee, at Aberdeen, is still standing : it consists of seven arches, and, as usual, the name of a bishop—Gawin Dunbar—is connected with its erection. There is another old bridge over the Don near the same city, said to have been built by Bishop Cheyne in the time of Robert the Bruce—the famous "Brig of Balgonie," celebrated in Lord Byron's stanzas as " Balgownie Brig's black wa'." It consists of a spacious Gothic arch, resting upon the rock on either side. There was even an old bridge over the rapid Spey at Orkhill. Then at Glasgow there was a fine bridge over the Clyde, which used, in old times, to be called " the Great Bridge of Glasgow," said to have been built by Bishop Rae in 1345. Though the bridge was only twelve feet wide, it consisted of eight arches; somewhat similar to the ancient fabric which still spans the Forth under the guns of Stirling Castle. This last-mentioned bridge was, until recent times, a structure of great importance, affording almost the only access into the northern parts of Scotland for wheeled carriages.

But the art of bridge-building in Scotland, as in England, seems for a long time to have been almost entirely lost; and until Smeaton was employed to erect the bridges of Coldstream, Perth, and Banff, next to nothing was done to improve this essential part of the communications of the country. Where attempts were made by local builders to erect such structures, they very rarely stood the force of a winter's, or even a summer's, flood. "I remember," says John Maxwell, "the falling of the Bridge of Buittle, which was built by John Frew in 1722, and fell in the succeeding summer, while I was in Buittle garden seeing my father's servants gathering nettles."[1] A similar fate befell the few attempts that were made about the same time to maintain the lines of communication by replacing the old bridges where they had gone to ruin, or substituting new ones in place of fords.

The mechanical arts had indeed fallen into the very lowest state. All kinds of tools were of the most imperfect description. The implements used in agriculture were extremely rude. They were mostly made by the farmer himself, in the roughest possible style, without the assistance of any mechanic. But a plough, which was regarded as a complicated machine, was reserved for the blacksmith. It was made of young birch trees, and, if the tradesman was expert, it was completed in the course of a winter's evening.[2] This rude implement scratched, without difficulty, the surface of old crofts, but made sorry work in out-fields, where the sward was tough and stones were large and numerous. Lord Kaimes said of the harrows used in his time, that they were more fitted to raise laughter than to raise mould. Machinery of an improved kind had not yet been introduced in any department of labour. Its first application, as might be expected, was in agriculture, then the leading, and indeed almost the only, branch of industry in Scotland; and its introduction will be found to be both curious and interesting in its bearing upon the subject of our present memoir.

. . . .

_____

[1] Appendix to 'Picture of Dumfries.' By John MacDiarmid. Edinburgh, 1832.

[2] 'Farmer's Magazine,' No. xxxiv., p. 199.

CHAPTER II.

FARMER RENNIE died in the old house at Phantassie in the year 1766, leaving a family of nine children, four of whom were sons and five daughters. George, the eldest, was then seventeen years old. He was discreet, intelligent, and shrewd beyond his years, and from that

RENNIE'S BIRTHPLACE, PHANTASSIE.

[By E. M. Wimperis, after a Drawing by J. S. Smiles.]

time forward he managed the farm and acted as the head of the family. The year before his father's death he had made a tour through Berwickshire, for the purpose of observing the improved methods of farming introduced by some of the leading gentry of that county, and he returned to Phantassie full of valuable practical information. The agricultural improvements which he was shortly afterwards instrumental in introducing into East Lothian were of a highly important character ; his farm came to be regarded as a model, and his reputation as a skilled agriculturist extended far beyond the bounds of his own country, insomuch that he was resorted to for advice as to farming matters by distinguished visitors from all parts of Europe.[1]

Of the other sons, William, the second, went to sea : he was taken prisoner during the first American war, and was sent to Boston, where he died. The third, James, studied medicine at Edinburgh, and entered the army as an assistant-surgeon. The regiment to which he belonged was shortly after sent to India : he served in the celebrated campaign of General Harris against Tippoo Saib, and was killed whilst dressing the wound of his commanding-officer when under fire at the siege of Seringapatam. John, the future engineer, was the youngest son, and he was only five years old at the death of his father. He was accordingly brought up mainly under the direction of his mother, a woman possessed of many excellent practical qualities, amongst which her strong common sense was not the least valuable.

The boy early displayed his strong inclination for mechanical pursuits. When about six years old, his best loved toys were his knife, hammer, chisel, and saw, by means of which he indulged his love of construction. He preferred this kind of work to all other amusements, taking but small pleasure in the ordinary sports of boys of his own age. His greatest delight was in frequenting the smith's and carpenter's shops in the neighbouring village of Linton, watching the men use their tools, and

---

[1] Amongst Mr. Rennie's other illustrious visitors in his later years was the Grand Duke Nicholas (afterwards Emperor) of Russia, who stayed several nights at Phantassie, and during the time was present at the celebration of a " hind's wedding."

trying his own hand when they would let him. But his favourite resort was Andrew Meikle's millwright's shop, down by the river Tyne, only a few fields off. When he began to go to the parish school, then at Prestonkirk, he had to pass Meikle's shop daily, going and coming ; and he either crossed the river by the planks fixed a little below the mill, or by the miller's boat when the waters were high. But the temptations of the millwright's workshop while passing to school in the mornings not unfrequently proved too great for him to resist, and he played truant ; the delinquency being only discovered by the state of his fingers and clothes on his return home, when an interdict was laid against his " idling " away his time at Andrew Meikle's shop.*

The millwright, on his part, had taken a strong liking for the boy, whose tastes were so congenial to his own. Besides, he was somewhat proud of his landlady's son frequenting his house, and was not disposed to discourage his visits. On the contrary, he let him have the run of his workshop, and allowed him to make his miniature water-mills and windmills with tools of his own. The river which flowed in front of Houston Mill was often swollen by spates or floods, which descended from the Lammermoors with great force ; and on such occasions young Rennie took pleasure in watching the flow of the waters, and following the floating stacks, field-gates, and other farm wreck along the stream, down to where the Tyne joined the sea at Tyningham, about four miles below. Amongst his earliest pieces of workmanship was a fleet of miniature ships. But not finding tools to suit his purposes, he contrived, by working at the forge, to make them for himself ; then he constructed his fleet, and launched his ships, to the admiration and astonishment of his playfellows. This was when he was about ten years old. Shortly after, by the advice and assistance of his friend Meikle, who took as much pride in his performances as if they had been his own, Rennie made a model of a windmill, another of a fire-engine (or steam-engine), and a third of Vellore's pile-engine, displaying upon them a considerable amount of manual dexterity; some of these early efforts of the boy's genius being still preserved.

Though young Rennie thus employed so much of his time on this amateur work in the millwright's shop, he was not permitted to neglect his ordinary education at the parish school. That of Prestonkirk was kept by a Mr. Richardson, who seems to have well taught his pupil in the ordinary branches of education; but by the time he had reached twelve years of age he seems almost to have exhausted his master's store of knowledge, and his mother then thought the time had arrived to remove him to a seminary of a higher order.[1]   He was accordingly

---

[1] Though a poor country, as we have seen, Scotland was already rich in parish and burgh schools; the steady action of which upon the rising generation was probably the chief cause of that extraordinary improvement in all its branches of industry to which we have above alluded. John Knox—himself a native of East Lothian—explicitly set forth in his first 'Book of Discipline'—"That every several kirke have ane schoolmaister appointed," "able to teach grammar and the Latin tongue, if the town be of any reputation;" and if an upland town, then a reader was to be appointed, or the minister himself must attend to the instruction of the children and youth of the parish. It was also enjoined that "provision be made for the attendance of those that be poore, and not able by themselves nor their friends to be sustained at letters;" "for this," it was added, "must be carefully provided, that no father, of what estate or condition that ever he may be, use his children at his own fantasie, especially in their youthhead; but all must be compelled to bring up their children in learning and virtue." During the troubles in which Scotland was involved, almost down to the Revolution of 1688, although attempts were made to establish a school in every parish, they seem to have been attended with comparatively small results; but at length, in 1696, the Scottish Parliament was enabled, with the concurrence of William of Orange, to put in force the Act of that year, which is regarded as the charter of the parish-school system in Scotland. It is there ordained "that there be a school settled and established, and a schoolmaster appointed in every parish not already provided, by advice of the heritors and minister of the parish." The consequence was, that the parish schools of Scotland, working steadily upon the rising generation, all of whom passed under the hands of the parish teachers during the preceding half-century, had been training a population whose intelligence was greatly in advance of their material condition as a people; and it is in this circumstance, we apprehend, that the true explanation is to be found of the rapid start forward which the whole country now took, dating more particularly from the year 1745. Agriculture was naturally the first branch of industry to exhibit signs of decided improvement; to be speedily followed by like advances in manufac-

taken from the parish school at that age, but his friends had not made up their minds as to the steps they were to adopt with reference to his further education. The boy, however, found abundant employment for himself with his tools, and went on model-making; but feeling that he was only playing at work, he became restless and impatient, and entreated his mother that he might be allowed to go to Andrew Meikle's to learn to be a millwright. This was agreed to, and he was sent to Meikle's accordingly, where he worked for two years, during which period he learnt one of the most valuable parts of education—the use of his hands. He seemed to overflow with energy, and was ready to work at anything — at smith's work, carpenter's work, or millwork; taking most pleasure in the latter, in which he shortly acquired considerable expertness. Having the advantage of books—limited though the literature of mechanism was in those days—he studied the theory as well as the practice of mechanics, and the powers of his mind became steadily strengthened and developed with application and self-culture.

At the end of two years his friends determined to send him to the burgh school of Dunbar, one of that valuable class of seminaries directed and maintained by the magistracy, which have been established for the last hundred years and more in nearly every town of any importance in Scotland.[1] Dunbar High School was

---

tures, commerce, and shipping. Indeed, from that time, the country never looked back, but her progress went on at a constantly accelerating rate, issuing in results as marvellous as they have probably been unequalled.

[1] The origin of what are technically termed "Grammar Schools" in Scotland, is involved in considerable obscurity. They are, for the most part, of ancient foundation, and are supposed to have been endowed by generous individuals, who vested in some public body, usually the borough corporation, sums of money for the purpose of educating the youth of the towns in which they are established. The money or property so devised was legally termed a "mortification." Many of such bequests were made in the remote times when Scotland was a Catholic nation. John Knox himself was educated at the Grammar School of Haddington, near to which town he was born and brought up, and there he says he learnt the elements of the Latin language.

then a seminary of considerable celebrity. Mr. Gibson, the mathematical master, was an excellent teacher, full of love and enthusiasm for his profession ; and it was principally for the benefit of his discipline and instruction that young Rennie was placed under his charge. The youth, on entering this school, possessed the advantage of being fully impressed with a sense of the practical value of intellectual culture. His two years' service in Meikle's workshop, while it trained his physical powers had also sharpened his appetite for knowledge, and he entered upon his second course of instruction at Dunbar with the disciplined powers almost of a grown man. He had also this advantage, that he prosecuted his studies there with a definite aim and purpose, and with a determinate desire to master certain special branches of education required for the successful pursuit of his intended business. Accordingly, we are not surprised to find that in the course of a few months he outstripped all his schoolfellows and took the first place in the school. A curious record of his proficiency as a scholar is to be found in a work by one Mr. David Loch, Inspector-General of Fisheries, published in 1779. It was his duty to hold a court of the herring skippers of Dunbar, then the principal fishing-station on the east coast ; and it appears that at one of his visits to the town he attended an examination of the burgh schools, and was so much pleased with the proficiency of the pupils that he makes special mention of it in his book.[1]

---

[1] After speaking of the teachers of Latin, English, and arithmetic, he goes on to say: " But Mr. Gibson, teacher of mathematics, afforded a more conspicuous proof of his abilities, by the precision and clearness of his manner in stating the questions which he put to the scholars; and their correct and spirited answers to his propositions, and their clear demonstrations of his problems, afforded the highest satisfaction to a numerous audience. And here I must notice in a particular manner the singular proficiency of a young man of the name of Rennie : he was intended for a millwright, and was breeding to that business under the famous Mr. Meikle, at Linton, East Lothian ; he had not attended Mr. Gibson for the mathematics, &c., much more than six months, but on his examination he discovered such amazing power of genius, that one would have imagined him a second Newton. No problem was too hard for him to de-

Rennie remained with Mr. Gibson for about two years. During that period he went as far in mathematics and natural philosophy as his teacher could carry him, after which he again proposed to return to Meikle's workshop. But at this time the mathematical master was promoted to a higher charge—the rectorship of the High School of Perth—and a question arose as to the appointment of his successor. The loss to the town was felt to be great, and Mr. Gibson was pressed by the magistrates to point out some person whom he thought suitable for the office. The only one he could think of was his favourite pupil; and though not yet quite seventeen years old, he strongly recommended John Rennie to accept the appointment. The young man, how-ever, already beginning to be conscious of his powers, had formed more extensive views of life, and could not entertain the idea of settling down as the " dominie " of a burgh school, respectable and responsible though that office must be held to be. He accordingly declined the honour which the magistrates proposed to confer upon him, but agreed to take charge of the mathematical classes until Mr. Gibson's successor could be appointed. He continued to carry on the classes for about six weeks, and conducted them so satisfactorily that it was matter of much regret when he left the school and returned to his family at Phantassie for the purpose of prosecuting his intended profession.

At home he pursued the study of his favourite branches of instruction, more particularly mathematics, mechanics, and natural philosophy, frequenting the work-shop of his friend Meikle, assisting him with his plans,

---

monstrate. With a clear head, a de-cent address, and a distinct delivery, his master could not propose a ques-tion, either in natural or experimental philosophy, to which he gave not a clear and ready solution, and also the reasons of the connection between causes and effects, the power of gravi- tation, &c., in a masterly and con-vincing manner, so that every person present admired such an uncommon stock of knowledge amassed at his time of life. If this young man is spared, and continues to prosecute his studies, he will do great honour to his country."

and taking an especial interest in the invention of the thrashing-machine, which Meikle was at that time employed in bringing to completion. He was also entrusted to superintend the repairs of corn-mills in cases where Meikle could not attend to them himself; and he was sent, on several occasions, to erect machinery at a considerable distance from Prestonkirk. Rennie thus gained much valuable experience, at the same time that he acquired confidence in his own powers; and before the end of a year he began to undertake millwork on his own account. His brother George was already well known as a clever farmer, and this connection helped the young millwright to as much employment in his own neighbourhood as he desired. Meikle was also ready to recommend him in cases where he could not accept the engagements offered in distant counties; and hence, as early as 1780, when Rennie was only nineteen years of age, we find him employed in fitting up the new mills at Invergowrie, near Dundee. He designed the machinery as well as the buildings for its reception, and superintended them to their completion. His next work was to prepare an estimate and design for the repairs of Mr. Aitcheson's flour-mills at Bonnington, near Edinburgh. Here he employed cast iron pinions, instead of the wooden trundles formerly used: one of the first attempts made to introduce iron into this portion of the machinery of mills.

These, his first essays in design, were considered very successful, and they brought him both fame and emolument. Business flowed in upon him, and before the end of his nineteenth year he had as much employment as he could comfortably get through. But he had no intention of confining himself to the business of a country millwright, however extensive, aiming at a higher professional position and a still wider field of work. Desirous, therefore, of advancing himself in scientific culture and prosecuting the studies in mechanical philosophy which he had begun at Phantassie and pursued in the burgh school at Dunbar, he determined to place himself at the University of Edinburgh, then a seminary of rising celebrity. In taking this step he formed the resolution—by no means unusual amongst young men of his country inspired by a laudable desire for self-improvement—of supporting himself at college entirely by his own labour. He was persuaded that by diligence and assiduity he would be enabled to earn enough during the summer months to pay for his winter's instruction and maintenance; and his habits being frugal, and his style of living very plain, he was enabled to prosecute his design without difficulty.

He accordingly matriculated at Edinburgh in November, 1780, and entered the classes of Dr. Robison, Professor of Natural Philosophy, and of Dr. Black, Professor of Chemistry; both men of the highest distinction in their respective walks. Robison was an eminently prepossessing person, frank and lively in manner, full of fancy and humour, and, though versatile in talent, a profound and vigorous thinker. His varied experience of life, and the thorough knowledge which he had acquired of the principles as well as the practice of the mechanical arts, proved of great use to him as an instructor of youth. The state of physical science was then at a very low ebb in this country, and the labours of Continental philosophers were but little known even to those who occupied the chairs in our universities; the results of their elaborate researches lying concealed in foreign languages, or being known, at most, to a few inquirers more active and ardent than their fellows; whilst the general student, mechanic, and artisan were left to draw their principal information from the ancient but ordinary springs of observation and daily experience.*

Under Dr. Robison the study of natural philosophy became invested with unusual significance and importance.

The range of his knowledge was most extensive : he was familiar with the whole circle of the accurate sciences, and in imparting information his understanding seemed to work with extraordinary energy and rapidity.   The labours of others rose in value under his hands, and new views and ingenious suggestions never failed to enliven his prelections on mechanics, hydrodynamics, astronomy, optics, electricity, and magnetism, the principles of which he unfolded to his pupils in language at once fluent, elegant, and precise.   Lord Cockburn remembers him as somewhat remarkable for the humour in which he indulged in the article of dress.   " A pigtail so long and thin that it curled far down his back, and a pair of huge blue worsted hose, without soles, and covering the limbs from the heel to the top of the thigh, in which he both walked and lectured, seemed rather to improve his wise elephantine head and majestic person." He delighted in holding familiar intercourse with his pupils, whom he charmed and elevated by his brilliant conversation and his large and lofty views of life and philosophy. Rennie was admitted freely to his delightful social influence, and to the close of his career he was accustomed to look back upon the period which he spent at Edinburgh as amongst the most profitable and instructive in his life.

During his college career Rennie carefully read the works of Emerson, Switzer, Maclaurin,* Belidor, and Gravesande, allowing neither pleasure nor society to divert him from his line of study.   As a relief from graver topics, he set himself to learn the French and German languages, and was shortly enabled to read both with ease.   His recreation was mostly of a solitary character, and, having a little taste for music, he employed some of his leisure time in learning to play upon various instruments.   He acquired considerable proficiency on the flute and the violin, and he even went so far as to buy a pair of bagpipes and learn to play upon them, though the selection of such an instrument probably does not say much for his musical taste. When he had left Edinburgh, however, and entered seriously upon the business of life, the extensive nature of his engagements so completely occupied his time that in a few years, flute, fiddle, and bagpipes were laid aside altogether.

During the three years that he attended college, our student was busily occupied in the summer vacation—extending from the beginning of May to the end of October in each year—in executing millwork in various parts of the country. Amongst the undertakings on which he was thus employed, may be mentioned the repair or construction of the Kirkaldy and Bonnington Flour Mills, Proctor's Mill at Glammis, and the Carron Foundry Mills. When not engaged on distant works, his brother George's house at Phantassie was his headquarters, where he prepared his designs and specifications. He had the use of the workshop at Houston Mill for making such machinery as was intended for erection in the neighbourhood; but when he was employed at some distant point, the work was executed in the most convenient places he could find for the purpose. There were as yet no large manufactories in Scotland where machinery of an important character could be turned out as a whole; the millwright being under the necessity of sending one portion to the blacksmith, another to the founder, another to the brass-smith, and another to the carpenter,—a state of things involving a great deal of trouble and often risk of failure, but which was eminently calculated to familiarize our young engineer with the details of every description of work required in the practice of his profession.

His college training having ended in 1783, and being desirous of acquiring some knowledge of English engineering practice, Rennie set out upon a tour in the manufacturing districts. Brindley's reputation attracted him first towards Lancashire for the purpose of inspecting the works of the Bridgewater Canal. There being no stage coaches convenient for his purpose, he travelled on horseback, and in this way was enabled readily to diverge from his route for the purpose of visiting any structure more interesting than ordinary. At Lancaster he inspected the handsome bridge across the Lune, then in course of construction by Mr. Harrison, afterwards more celebrated for his fine work of Chester Gaol. At Manchester he examined the works of the Bridgewater Canal; and at Liverpool he visited the docks there in progress.

Proceeding by easy stages to Birmingham, then the centre of the mechanical industry of England, and distinguished for the ingenuity of its workmen and the importance of its manufactures in metal, he took the opportunity of visiting the illustrious Boulton and Watt at Soho. His friend, Dr. Robison, had furnished him with a letter of introduction to James Watt, who received the young engineer kindly and showed him every attention; and a friendship then began which lasted until the close of Watt's life.

The condensing engine had by this time been brought into an efficient working state, and was found capable not only of pumping water—almost the only purpose to which it had formerly been applied—but of driving machinery, though whether with advantageous results was still a matter of doubt. Thus, in November, 1782, Watt wrote to his partner Boulton, "There is now no doubt but that fire-engines will drive mills, but I entertain some doubts whether anything is to be got by them." [1] About the beginning of March, 1783, however, a company was formed in London for the purpose of erecting a large corn-mill, to be driven by one of Boulton and Watt's steam-engines, and the work was in progress at the time Rennie visited Soho. Watt had much conversation with his visitor on the subject of corn-mill machinery, and was gratified to learn the extent and

---

[1] Muirhead's 'Origin and Progress of the Mechanical Inventions of James Watt,' vol. ii., p. 165.

accuracy of his information. He seems to have been provoked beyond measure by the incompetency of his own workmen. "Our millwrights," he wrote to his partner, "have kept working, working, at the corn-mill ever since you went away, and it is not yet finished; but my patience being exhausted, I have told them that it must be at an end to-morrow, done or undone. There is no end of millwrights once you give them leave to set about what they call machinery; here they have multiplied wheels upon wheels until it has now almost as many as an orrery." [1]

Watt himself had but little knowledge of millwork, and stood greatly in need of some able and intelligent millwright to take charge of the fitting up of the Albion Mills. Young Rennie seemed to him at the time to be a very likely person; but, with characteristic caution, he said nothing of his intentions, but determined to write privately to his friend Robison upon the subject, requesting particularly to know his opinion as to the young man's qualifications for taking the superintendence of such important works. Dr. Robison's answer was most decided; his opinion of Rennie's character and ability was so favourable, and expressed in so confident a tone, that Watt no longer hesitated; and he shortly after wrote to the young engineer, when he had returned home, inviting him to undertake the supervision of the proposed mills, so far as concerned the planning and erection of the requisite machinery.

Watt's invitation found Rennie in full employment again. He was engaged in designing and erecting mills and machinery of various kinds. Amongst his earlier works, we also find him, in 1784, when only in his twenty-third year, occupied in superintending the building of his first bridge—the humble forerunner of a series of structures which have not been surpassed

---

[1] Muirhead, vol. ii., p. 177.

in any age or country.   His earliest building of this
kind was erected for the trustees of the county of
Mid-Lothian, across the Water of Leith, near Steven-
house Mill, about two miles west of Edinburgh.   It is
the first bridge on the Edinburgh and Glasgow turnpike
road.

RENNIE'S FIRST BRIDGE.

[By R. P. Leitch, after a Sketch by J. S. Smiles.]

Notwithstanding the extent of his engagements and
the prospect of remunerative employment which was
opening up before him, Rennie regarded the invitation
of Watt as too favourable an opportunity for enlarging
his experience to be neglected, and, after due deliberation, he wrote back accepting the appointment.   He
proceeded, however, to finish the works he had in hand ;
after which, taking leave of his friends and home at
Phantassie, he set out for Birmingham on the 19th
of September, 1784.   He remained there two months,
during which he enjoyed the closest personal intercourse with Watt and Boulton, and was freely admitted
to their works at Soho, which had already become the
most important of their kind in the kingdom.   Birmingham was then the centre of the mechanical industry of
England.   For many centuries, working in metals had
been the staple trade of the place.   Swords were made

there in the time of the ancient Britons. In the reign of Henry VIII., Leland found "many smythes in the town that use to make knives and all manner of cutting tools, and many loriners that make bittes, and a great many nailers; so that a great part of the town is maintained by smythes who have their iron and sea-coal out of Staffordshire."

The artisans of the place thus had the advantage of the training of many generations; aptitude for handicraft, like every other characteristic of a people, descending from father to son like an inheritance. There was then no town in England where mechanics were to be found so capable of satisfactorily executing original and unaccustomed work, nor has the skill yet departed from them. Though there are now many districts in which far more machinery is manufactured than in Birmingham, the workmen of that place are still superior to most others in executing machinery requiring manipulative skill and dexterity out of the common track, and especially in carrying out new designs. The occupation of the people gave them an air of quickness and intelligence which was quite new to strangers accustomed to the quieter aspects of rural life. When Hutton entered Birmingham, he was especially struck by the vivacity of the persons he met in the streets. "I had," he says, "been among dreamers, but now I was among men awake. Their very step showed alacrity. Every man seemed to know and prosecute his own affairs." He also adds, that men whose former disposition was idleness no sooner breathed the air of Birmingham than diligence became their characteristic.

Rennie did not stand in need of this infection being communicated to him, yet he was all the better for his contact with the population of the town. He made himself familiar with their processes of handicraft, and, being able to work at the anvil himself, he could fully appreciate the skill of the Birmingham artisans. The manufacture of steam-engines at Soho chiefly attracted his notice and his study. He had already made himself acquainted with the principles as well as the mechanical details of the steam-engine, and was ready to suggest improvements, in a very modest way, even to Watt himself, who was still engaged in perfecting his wonderful machine. The partners thought that they saw in him a possible future competitor in their trade; and in the agreement which they entered into with him as to the erection of the Albion Mills, they sought to bind him, in express terms, not only to abstain from interfering in any way with the construction and working of the steam-engines required for the mills, but to prohibit him from executing such work upon his own account at any future period. Though ready to give his word of honour that he would not in any way interfere with Watt's patents, he firmly refused to bind himself to such conditions; being resolved in his own mind not to be debarred from making such improvements in the steam-engine as experience might prove to be desirable. And on this honourable understanding the agreement was concluded; nor did Rennie ever in any way violate it, but retained to the last the friendship and esteem of both Watt and Boulton.

On the 24th of November following, after making himself fully acquainted with the arrangements of the engines by means of which his machinery was to be driven, our engineer set out for London to proceed with the designing of the millwork. It was also necessary that the plans of the building—which had been prepared by Mr. Samuel Wyatt, an architect of reputation in his day—should undergo revision; and, after careful consideration, Rennie made an elaborate report on the subject, recommending various alterations, which were approved by Boulton and Watt, and forthwith ordered to be carried into effect.

CHAPTER III.

WHEN Rennie arrived in London in 1785, the country was in a state of serious depression in consequence of the unsuccessful termination of the American War. Parliament was engaged in defraying the heavy cost of the recent struggle with the revolted colonies. The people were ill at ease, and grumbled at the increase of the debt and taxes. The unruly population of the capital could with difficulty be kept in order. The police and local government were most inefficient. Only a few years before, London had, during the Gordon riots, been for several days in the hands of the mob, and blackened ruins in different parts of the city still marked the track of the rioters. Though the largest city in Europe, the population was scarcely more than a third of what it is now ; yet it was thought that it had become so vast as to be unmanageable. Its northern threshold was at Hicks's Hall, in Clerkenwell. Somers Town, Camden Town, and Tyburnia were as yet green fields ; and Kensington, Chelsea, Marylebone, and Bermondsey were outlying villages. Fields and hedgerows led to the hills of Highgate and Hampstead. The West End of London was a thinly-inhabited suburb, Fitzroy Square having only been commenced in 1793. The westernmost building in Westminster was Millbank, a wide tract of marshy ground extending opposite Lambeth. Executions were conducted in Tyburn fields, long since covered with handsome buildings, down to 1783. Oxford Street, from Princes Street eastward as far as High Street, St. Giles's, had only a few houses on the north side. "I remember it," says Pennant, "a deep hollow road and full of sloughs, with here and there a ragged house, the lurking-place of cut-throats; insomuch that I was never taken that way by night, in my hackney-coach, to a worthy uncle's who gave me lodgings at his house in George Street, but I went in dread the whole way." Paddington was "in the country," and the communication with it was kept up by means of a daily stage—a lumbering vehicle, driven by its proprietor—which was heavily dragged into the city in the morning, down Gray's Inn Lane, with a rest at the Blue Posts, Holborn Bars, to give passengers an opportunity of doing their shopping. The morning journey was performed in two hours and a half, "quick time," and the return journey in the evening in about three hours.

Heavy coaches still lumbered along the country roads at little more than four miles an hour. A new state of things had, however, been recently inaugurated by the starting of the first mail-coach on Palmer's plan, which began running between London and Bristol on the 24th of August, 1784, and the system was shortly extended to other places. Numerous Acts were passed by Parliament authorising the formation of turnpike-roads and the erection of bridges.[1] The general commerce of the country was also making progress. The application of recent inventions in manufacturing industry gave a stimulus to the general improvement, and this was further helped by a succession of favourable harvests. The India Bill had just been renewed by Pitt, and trade with India was brisk. Besides, a commercial treaty with France was on foot, from which great things were ex-

---

[1] In the interval between 1784 and 1792, not fewer than 302 Acts were passed authorising the construction of new roads and bridges, 64 authorising the formation of canals and harbours, and still more numerous Acts for carrying out measures of drainage, enclosure, paving, and other local improvements—a sufficient indication of the industrial activity of the nation at the time.

pected; although the outbreak of the Revolution, which shortly after took place, put an end for a time to those hopes of fraternity and peaceful trade in which it had originated. The Government boldly interposed to check smuggling, and Pitt sent a regiment of soldiers to burn the smugglers' boats laid up on Deal beach by the severity of the winter, so that the honest traders might have the full benefit of the treaty with France which Pitt had secured. Increased trade flowed into the Thames, and ministers and monarch indulged in drawing glowing pictures of prosperity. When Pennant visited London in 1790, he found the river covered with shipping, presenting a double forest of masts, with a narrow avenue in mid-channel. The smaller vessels discharged directly at the warehouses along the banks of the river, whilst the India ships of large burden mostly lay down the river as far as Blackwall, and discharged into lighters, which floated up their cargoes to the city wharves. London as yet possessed no public docks,—only a few private ones of very limited extent,—although Pennant speaks of Mr. Perry's dock and ship-yard at Blackwall, on the eastern side of the Isle of Dogs, as "the greatest private dock in all Europe!" Another was St. Saviour's, denominated by Pennant "the port of Southwark," though it was only thirty feet wide, and used for discharging barges of coal, corn, and other commodities. There was also Execution Dock at Wapping, which witnessed the occasional despatch of seagoing criminals, who were hanged on a gallows at low-water mark, and left there until the tide flowed over their dead bodies.

Among the commercial enterprises to which the increasing speculation of the times gave birth, was the erection of the Albion Mills. For the more convenient transit of corn and flour, as well as to secure a plentiful supply of water for engine purposes, it was determined to erect the new mills on the banks of the Thames, near the south-east end of Blackfriars Bridge. Hand-mills, which had in the first place been used for pounding wheat into flour, had long since been displaced by water-mills and windmills; and now a new agency was about to be employed, of greater power than either—the agency of steam. Fire-engines had heretofore been employed almost exclusively in pumping water out of mines; but the possibility of adapting them to the driving of machinery having been suggested to the inventive mind of James Watt, he set himself at once to the solution of the problem, and the result was the engines for the Albion Mills—the most complete and powerful which he had until then turned out of the Soho manufactory. They consisted of two double-acting engines, of the power of 50 horses each, with a pressure of steam of five pounds to the superficial inch—the two engines, when acting together, working with the power of 150 horses. They drove twenty pairs of millstones, each four feet six inches in diameter, twelve of which were usually worked together, each pair grinding ten bushels of wheat per hour, by day and night if necessary. The two engines working together were capable of grinding, dressing, &c., complete, 150 bushels an hour—by far the greatest performance achieved by any mill at that time, and probably not since surpassed, if equalled. But the engine power was also applied to a diversity of other purposes, then altogether novel—such as hoisting and lowering the corn and flour, loading and unloading the barges, and in the processes of fanning, sifting, and dressing—so that the Albion Mills came to be regarded as among the greatest mechanical wonders of the day. The details of these various ingenious arrangements were entirely worked out by Mr. Rennie himself, and they occupied him nearly four years in all, having been commenced in 1784, and set to work in 1788. Mr. Watt was so much satisfied with the result of his employment of Rennie, that he wrote to Dr. Robison, thanking him for his recommendation of his young friend, and speaking in the highest terms of the ability with which he had designed and executed the millwork and set the whole in operation.

THE ALBION MILLS.

Amongst those who visited the new mills and carefully inspected them was Mr. Smeaton, the engineer, who pronounced them to be the most complete, in their arrangement and execution, which had yet been erected in any country; and though naturally an undemonstrative person, he cordially congratulated Mr. Rennie on his success. The completion of the Albion Mills, indeed, marked an important stage in the history of mechanical improvements; and they may be said to have effected an entire revolution in millwork generally. Until then, machinery had been constructed almost entirely of wood, and it was consequently exceedingly clumsy, involving great friction and much waste of power. Mr. Smeaton had introduced an iron wheel at Carron in 1754, and afterwards in a mill at Belper, in Derbyshire —mere rough castings, imperfectly executed, and neither chipped nor filed to any particular form; and Mr. Murdock (James Watt's ingenious assistant) had also employed cast iron work to a limited extent in a mill erected by him in Ayrshire; but these were very inferior speci-

mens of iron work, and exercised no general influence
on mechanical improvement.   Mr. Rennie's adoption of
wrought and cast iron wheels, after a system, was of
much greater importance, and was soon adopted generally
in all large machinery.   The whole of the wheels and
shafts of the Albion Mills were of these materials, with the
exception of the cogs in some cases, which were of hard
wood, working into others of cast iron ; and where the
pinions were very small, they were of wrought iron.   The
teeth, both wooden and iron, were accurately formed by
chipping and filing to the form of epicycloids.   The
shafts and axles were of iron and the bearings of brass,
all accurately fitted and adjusted, so that the power employed worked to the greatest advantage and at the least
possible loss by friction.   The machinery of the Albion
Mills, as a whole, was regarded as the finest that had
been executed to that date, forming a model for future
engineers to work by ; and although Mr. Rennie executed many splendid specimens in his after career,[1] he

---

[1] Shortly after the completion of
these mills, Mr. Rennie was largely
consulted on the subject of machinery
of all kinds.   The Corporations of
London, Edinburgh, Glasgow, Perth,
and other places, took his advice as to
flour-mills.   Agriculturists consulted
him about thrashing-mills, millers
about grinding-mills, and manufacturers and distillers respecting the
better arrangement of their works.   He
supplied plans for a steel lead-rolling-
mill for Messrs. Locke and Co., at
Newcastle-on-Tyne ; he was called in
to remedy the defective boiler-arrange-
ments at Meux· and Co.'s brewery ;
he advised the Government as to the
power for working their small-arms
manufactory at Enfield, and the Navy
Board respecting the apparatus for
blowing the forge at Portsmouth.   In
1792 he invented the depressing
sluice for water-mills, which a Government engineer, a Mr. Lloyd, afterwards brought out (in 1807) as his
own invention.   The Don Navigation

Company's mills at Doncaster were
entirely rebuilt after his designs ; he
sent plans of large flour-mills to one
Don Diego at Lisbon, and of the extensive saw-mills erected at Arch-
angel in Russia.   In July, 1798, he
was called upon to examine the machinery and arrangements at the
Royal Mint on Tower Hill.   The result was, the construction of an entire
new mint, worked by steam-power,
with improved rolling, cutting-out,
and stamping machinery, after Mr.
Rennie's designs.   The new machinery
was introduced between the years
1806 and 1810.   Although it has
now been in · use for half a century,
it continues in as efficient a working state as in the year it was
erected.   It is still capable of turning out from the metal, in each
day of twelve hours, two and a half
tons of copper, and a ton each of
gold and silver coin.   The whole
process, as carried out by this apparatus, is extremely beautiful and effi-

himself was accustomed to say that the Albion Mill machinery was the father of them all.

As a commercial enterprise, the mills promised to be perfectly successful : they were kept constantly employed, and were realising a handsome profit to their proprietors, when unhappily they were destroyed by fire on the 3rd of March, 1791, only three years after their completion.* Their erection had been viewed with great hostility by the trade, and the projectors were grossly calumniated on the ground that they were establishing a monopoly injurious to the public, which was sufficiently disproved by the fact that the mills were the means of considerably reducing the price of flour while they continued in operation. The circumstances connected with the origin of the fire were never cleared up, and it was generally believed at the time that it was the work of an incendiary. During the night in which the buildings were destroyed, Mr. Rennie, who lived near at hand, felt unaccountably anxious. A presentiment as of some great calamity hung over him, which he could not explain to himself or to others. He went to bed at an early hour, but could not sleep. Several times he went off in a doze, and suddenly woke up, having dreamt that the mills were on fire! He rose, looked out, and all was quiet. He went to bed again, and at last fell into a profound sleep, from which he was roused by the cry of " Fire !" under his windows, and the rumble of the fire-engines on their way to the mills! He dressed hastily, rushed out, and to his dismay found his chef-d'œuvre wrapt in flames, which brightened the midnight sky. The engi-

---

cient. The cutting-out and stamping-machines were the invention of the late Matthew Boulton, of Soho, but the machinery was by Mr. Rennie. On one occasion, in 1819, a million of sovereigns were turned out in eight days! During the great silver coinage in 1826, the eight presses turned out, for nine months, not less than 247,696 pieces per day, the rolling going on day and night, and the stamping for fifteen hours out of every twenty-four. Mr. Rennie also supplied the machinery for the mints at Calcutta and Bombay ; that erected at the former place being capable of turning out 200,000 pieces of silver in every eight hours.

neer was amongst the foremost in his efforts to extin-
guish the conflagration; but in vain. The fire had made
too great progress, and the Albion Mills, Rennie's pride,
were burnt completely to the ground, and never rebuilt.

The Albion Mills, however, established Mr. Rennie's
reputation as a mechanical engineer, and introduced him
to extensive employment. His practical knowledge of
masonry and carpentry also served to point him out
as a capable man in works of civil engineering, which
were in those days usually entrusted to men bred to
practical mechanics. There was not as yet any special
class trained to this latter profession, the number of
persons who followed it being very small; and these
were usually determined to it by the strong instinct of
constructive genius. Hence the early engineers were
mainly self-educated—Smeaton, like Watt, being originally a mathematical instrument maker, Telford a stonemason, and Brindley and Rennie millwrights; force of
character and bent of genius enabling each to carve out
his career in his own way. The profession of engineering being still in its infancy in England, there was very
little previous practice to serve for their guide, and they
were called upon in many cases to undertake works of
an entirely new character, in which, if they could not find
a road, they had to make one. This threw them upon
their own resources and compelled them to be inventive:
it practised their powers and disciplined their skill, and
in course of time the habitual encounter with difficulties
brought fully out their character as men as well as their
genius as engineers.

When the ruins of the Albion Mills had been cleared
away, Mr. Rennie obtained leave from the owners to
erect a workshop upon a part of the ground, wherein he
continued for the rest of his life to carry on the business
of a mechanical engineer. But from an early period the
civil branch of the profession occupied a considerable
share of his attention, and eventually it became his

principal pursuit; though down to the year 1788 he was
chiefly occupied in designing and constructing machinery
for dye-works, water-works (at London Bridge amongst
others), flour-mills, and rolling-mills, in all of which
Boulton and Watt's engine was the motive power employed.

[Working in London, Rennie attracted the attention of the
eccentric Earl Stanhope.* The Earl enjoyed the society of
clever mechanics and was a first-rate workman skilled as carpenter, blacksmith, and millwright. He communicated to Rennie
his ingenious scheme for the application of the steam engine to
navigation as early as 1790, but no significant practical achievements resulted from their projects and experiments.

After 1790 Rennie was frequently consulted about canal
undertakings. He worked on the survey, designs for the viaducts
and bridges, as well as on the execution of the Kennet and
Avon Canal. He was also engaged on the Rochdale Canal,
constructed to provide direct transportation between the manufacturing districts of West Yorkshire and South Lancashire.
Rennie, as company engineer, worked on the Lancaster Canal,
projected in part to provide navigable communication between
the coal fields near Wigan and the area's lime districts. Though
he had a vested interest in canals, he appreciated the potential
of the railway as at least an adjunct to the canal system.]

CHAPTER IV.

MR. RENNIE'S DRAINAGE OF THE LINCOLN AND CAMBRIDGE FENS.

NOTWITHSTANDING all that had been done for the drainage of the Fens, as described in the early part of this work, large districts of reclaimable lands in Lincoln still lay waste and unprofitable. As early as 1789 Mr. Rennie's attention was drawn to the drowned state of the rich low-lying lands to the south of Ely; and having become impressed with a conviction of the extensive uses to which his friend Watt's steam-engine might be applied, he recommended it for pumping the water from the Botteshaw and Soham Fens, which contained about five thousand acres of what was commonly called "rotten land," because of the rot which infected the sheep depastured upon it. But he found the prejudices in favour of drainage by the old method of windmills, imported from Holland, too strong to be uprooted; and it was not until many years after, that his recommendation was adopted and the steam-engine was applied to pump the water from low-lying swamps which could not otherwise be cleared. The results were so successful that the same agency became generally employed for the purpose, not only in England but in Holland itself, where the forty-five thousand acres of Haarlemer Meer have since been effectually drained by the application of the steam-engine.

One of the most important works of thorough drainage carried out by Mr. Rennie was in that extensive district of South Lincolnshire which extends along the south verge of the Wolds, from near the city of Lincoln eastward to the sea. It included Wildmore Fen, West Fen, and East Fen, and comprised about seventy-five thousand acres of land which lay under water for the greater part of every year, and was thus comparatively useless either for grazing or tillage. The only crop grown there was tall reeds, which were used as thatch for houses and barns, and even for churches. The river Witham, which flows by Lincoln, had been grievously neglected and allowed to become silted up, its bottom being in many places considerably above the level of the land on either side. Hence, bursting of the banks frequently occurred during floods, causing extensive inundation of the lower levels, only a small proportion of the flood-waters being able to force their way to the sea. The wretched state of these lands may be inferred from the fact that, about seventy years since, a thousand acres in Blankney Fen, constituting part of " the Dales "—now one of the most fertile parts of the district between Lincoln and Tattershall—were let annually by public auction at Harecastle, and the reserved bid was only 10*l.* for the entire area! [1] It is stated that, about the middle of last century, there were not two houses in the whole parish of Dogdyke communicable with each other for whole winters round except by boat; this being also the only means by which the Fen-slodgers could get to church. Hall, the Fen Poet, speaks of South Kyme, where he was born, as a district in which, during the winter season, nothing was to be seen—

> " But naked flood for miles and miles."

The entire breadth of Lincolnshire north of Boston often lay under water for months together :—

> " 'Twixt Frith bank and the wold side bound,
> I question one dry inch of ground.
> From Lincoln all the way to Bourne,
> Had all the tops of banks been one,
> I really think they all would not
> Have made a twenty acre spot."

---

[1] 'Journal of Royal Agricultural Society,' 1847, vol. viii., p. 124.

Until as recent a date as forty years back, the rich and fertile district of Waldersea, about eight thousand acres in extent, was, as its name imports, a sea in winter. Well might Roger Wildrake describe his paternal estate of " Squattlesea Mere " as being in the " moist county of Lincoln ! "

Arthur Young visited this district in 1793, and found the freeholders of the high lands adjoining Wildmore and West Fens depasturing their sheep on the drier parts during the summer months ; but large numbers of them were dying of the rot. " Nor is this," he adds, " the only evil, for the number stolen is incredible. They are taken off by whole flocks, as so wild a country (whole acres being covered with thistles and nettles four feet high and more) nurses up a race of people as wild as the fen." The few wretched inhabitants who con- trived to live in the neighbourhood for the most part sheltered themselves in huts of rushes or lived in boats. They were constantly liable to be driven out of their cabins by the waters in winter, if they contrived to survive the attacks of the ague to which they were perennially subject.

The East Fen was the worst of all. It was formerly a most desolate region, though it now presents probably the richest grazing land in the kingdom. Being on a lower level than the West and Wildmore Fens, and the natural course of the waters to the sea being through it to Wainfleet Haven, it was in a much more drowned state than those to the westward. About two thousand acres were constantly under water, summer and winter. One portion of it was called Mossberry or Cranberry Fen, from the immense quantities of cranberries upon it. A great part of the remainder of the East Fen consisted of shaking bog, so treacherous and so deep in many places that only a desperate huntsman would ven- ture to follow the fox when he took to it, and then he must needs be well acquainted with the ground.

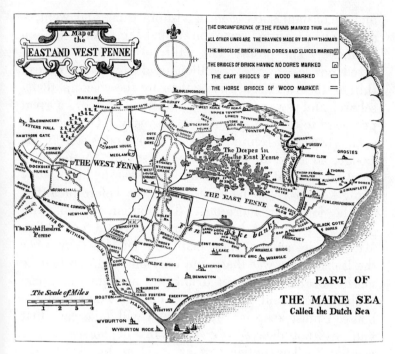

THE LINCOLNSHIRE FENS. [Before their Drainage by Mr. Rennie.]

Matters were in this state when Sir Joseph Banks, then President of the Royal Society, endeavoured to stir up the landowners to undertake the drainage of the district. He was the proprietor of a good estate at Revesby, near Tattershall; and his mansion of Abbot's Lodge, standing on an elevated spot, overlooked the waste of the East and West Fens, of which it commanded an extensive view. Sir Joseph spent a portion of every year at Revesby, as he did at his other mansions, leaving each at special times appointed beforehand, almost with the regularity of clockwork. He was a popular and well-known man, jolly and good-humoured, full of public spirit, and, though a philosopher, not above taking part in the sports and festivities of the neighbourhood in which he resided. While Sir Joseph lived at Revesby he used to keep almost open house,

and a constant succession of visitors came and went,—
some on pleasure, some on friendship, and some on
business. The profuse hospitality of the place was enjoyed not less by the postillions and grooms who drove
thither the baronet's guests than by the visitors themselves ; and it was esteemed by the hotel postboys a great
privilege to drive a customer to Revesby. On one occasion, when Mr. Rennie went to dine and sleep at the
Lodge, he took an opportunity of saying to the principal butler that he hoped he would see to his postboy
being kept *sober*, as he wished to leave before breakfast
on the following morning. The butler replied, with
great gravity, that he was sorry he could not oblige
Mr. Rennie, as the same man had left Revesby sober
the last time he was there, but only on condition that he
might be allowed to get drunk the next time he came.
" Therefore," said the butler, " for the honour of the
house, I must keep my word ; but I will take care that
you are not delayed for the want of horses and a
postboy." The butler was as good as his word : the
man got drunk, the honour of Revesby was saved, and
Mr. Rennie was enabled to set off in due time next
morning.

From an early period Sir Joseph Banks entertained
the design of carrying out the drainage of the extensive
fen lands lying spread out beneath his hall window,
and making them, if possible, a source of profit to the
owners, as well as of greater comfort and better subsistence for the population. Indeed, the reclamation of
these unhealthy wastes became quite a hobby with him ;
and when he could lay hold of any agricultural im-
prover, he would not let him go until he had dragged him
through the Fens, exhibited what they were, and demonstrated what fertile lands they might be made. When
Arthur Young visited Revesby about 1799, Sir Joseph
immediately started his favourite topic. " He had the
goodness," says Young, in his Report on Lincolnshire,

"to order a boat, and accompanied me into the heart
of East Fen, which had the appearance of a chain of
lakes, bordered by great crops of reed." Sir Joseph
was a man of great public spirit and determination : he
did not allow the matter to sleep, but proceeded to
organize the ways and means of carrying his design
into effect. His county neighbours were very slow to
act, but they gradually became infected by his example,
and his irresistible energy carried them along with him.
The first step taken was to call meetings of the proprietors in the several districts adjoining the drowned
and "rotten lands." Those of Wildmore Fen met at
Horncastle on the 27th of August, 1799, and resolu-
tions were adopted authorizing the employment of Mr.
Rennie to investigate the subject and report to a future
meeting.*

One reason, amongst others, which weighed with Sir
Joseph Banks in pressing on the measure was the
scarcity of corn, which about that time had risen almost
to a famine price. There was also great difficulty in
obtaining supplies from abroad, in consequence of the
war which was then raging. Sir Joseph entertained
the patriotic opinion that the best way of providing for
the exigency was to extend the area of our English food-
ground by the reclamation of the waste lands ; and hence
his determination to place under tillage, if possible, the
thousands of acres of rich soil, equal to the area of some
English counties, lying under water almost at his own
door. A few years' zealous efforts, aided by the skill
of his engineer, produced such results as amply to
justify his anticipations, and proved his patriotism to
be as wise as his public spirit was beneficent.

The manner in which Mr. Rennie proceeded to work
out the problem presented to him was thoroughly characteristic of the man. Most of the drainage attempted
before his time was of a very partial and inefficient
character. It was enough if the drainers got rid of the

surplus water anyhow, either by turning it into the
nearest river, or sending it in upon a neighbour. What
was done in one season was very often undone, or
undid itself, in the next. The ordinary drainer did not
care to look beyond the land immediately under his own
eyes. Mr. Rennie's practice, on the contrary, was founded
on a large and comprehensive view of the whole subject. He was not bounded by the range of his physical vision, but took into account the whole contour of
the country; the rainfall of the districts through which
his drains were to run, as well as of the central
counties of England, whose waters flowed down upon
the Fens; the requirements of the lands themselves as
regarded irrigation and navigation; and the most effectual method not only of removing the waters from
particular parts, but of providing for their effectual
discharge by proper outfalls into the sea.

What was the problem now to be solved by our
engineer? It was how best to carry out to sea the
surplus waters of a district extending from the eastern
coast to almost the centre of England. Various streams
descending from the Lincolnshire wolds flowed through
the level, whilst the Witham brought down the rainfall
not only of the districts to the north and east of Lincoln, but of a large part of the central counties of
Rutland and Leicester. It was therefore necessary to
provide for the clear passage of these waters, and also to
get rid of the drainage of the Fens themselves, a considerable extent of which lay beneath the level of the
sea at high water. It early occurred to Mr. Rennie
that, as the waters of the interior for the most part came
from a higher level, their discharge might be provided
for by means of distinct drains, and prevented from at
all mingling with those of the lower lying lands. But
would it be possible to "catch" these high land waters
before their descent upon the Fens, and then to carry
them out to sea by means of independent channels?

He thought it would; and with this leading idea in his mind he proceeded to design his plan of a great "catch-water drain," extending along the southern edge of the Lincolnshire wolds.

But there were also the waters of the Fens themselves to be got rid of, and how was this to be accomplished? To ascertain the actual levels of the drowned land, and the depth to which it would be necessary to carry the outfall of his drains into the sea, he made two surveys of the district, the first in October, 1799, and the second in March, 1800,—thus observing the actual condition of the lands both before and after the winter's rains. At the same time he took levels down to the sea outfalls of the existing drains and rivers. He observed that the Wash, into which the Fen waters ran, was shallow and full of shifting sands and silt. He saw that during winter the rivers were loaded with alluvial matter held in suspension, and that at a certain distance from their mouths the force of the inland fresh and the tidal sea waters neutralized each other, and there a sort of stagnant point was formed, at which the alluvium was no longer held in suspension by the force of the current. Hence it became precipitated in the channels of the rivers, and formed banks or bars in the Wash outside their mouths, which proved alike obstructive to drainage and navigation.

It required but little examination to detect the utter inadequacy of the existing outfalls to admit of the discharge of the surplus waters of so extensive a district. The few sluices which had been provided had been badly designed and imperfectly constructed. The levels of the outfalls were too high, and the gowts and sluices too narrow, to accommodate the drainage in flood-times. These outfalls were also liable, in dry summers, to become choked up by the silt settling in the Washes; and when a heavy rain fell, down came the waters from the high lands of the interior, and, unable to find an outlet, they burst the defensive banks of the rivers, and an amount of mischief was thus done which the drainage of all the succeeding summer failed to repair. Accordingly, the next essential part of Mr. Rennie's scheme was the provision of more effectual outfalls ; with which object he designed that they should be cut down to the lowest possible level of low water, whilst he arranged that at the points of outlet they should be mounted with strong sluices, opening outwards ; so that, whilst the fresh waters should be allowed freely to escape, the sea should be valved back and prevented flowing in upon the land. The third and last point was to provide for the drainage of the Fen districts themselves by means of proper cuts and conduits for the voidance of the Fen waters.

Such were the general conclusions formed by Mr. Rennie after a careful consideration of the circumstances of the case, which he embodied in his report to the Wildmore Fen proprietors[1] as the result of his investigations. The two great features of his plan, it will be observed, were (1) his intercepting or catchwater drains, and (2) his cutting down the outfalls to lower levels than had ever before been proposed. Simple though his system appears, now that its efficacy has been so amply proved by experience, it was regarded at the time as a valuable discovery in the practice of fen-draining, and indeed it was nothing less. There were, however, plenty of detractors, who alleged that it was nothing of the kind. Any boy, they said, who has played at dirt pies in a gutter, knows that if you make an opening sufficiently low to let the whole contained water escape, it will flow away. Very true ; yet the thing had never been done until Mr. Rennie proposed it, and, simple though the method was, it cost him many years of arguing, illustration, and enforcement, before he could induce intelligent men in other districts to adopt the

---

[1] Report, dated the 7th of April, 1800.

simple but thoroughly scientific method which he thus invented for the effectual discharge of the drainage of the Fens. And even to this day there are whole districts in which the stubborn obstinacy of ignorant obstructives still continues to stand in the way of its introduction. The Wildmore Fen proprietors, however, had the advantage of being led by a sagacious, clear-seeing man in Sir Joseph Banks, who cordially supported the adoption of the proposed plan with all the weight of his influence, and Mr. Rennie was eventually empowered to carry it into execution.

In laying out the works, he divided them according to their levels, placing Wildmore and West Fen in one plan, and East Fen in another. In the drainage of the former, the outlet was made by Anton's Gowt, about two miles and a half above Boston, and by Maud Foster, a little below that town. But both of these, being found too narrow and shallow, were considerably enlarged and deepened, and provided with double sluices and lifting gates : one set pointing towards the Witham in order to keep out the tides and river-floods; the other to the land, in order to prevent the water in summer from draining too low, and thereby hindering navigation as well as the due irrigation of the lands. An extensive main drain was also cut through the Wildmore and West Fens to the river Witham, about twenty-one miles long and from eighteen to thirty feet wide, the bottom being an inclined plane falling six inches in the mile.

The level of the East Fen being considerably lower than that of the Fens to the westward, it was necessary to provide for its separate drainage, but on precisely the same principles. From the levels which were taken, it appeared that the bottom of " the Deeps," which formed part of the East Fen, was only two feet six inches above the cill of Maud Foster Sluice, thirteen miles distant; whereas its highest parts were but eight feet above the same point, giving a fall of only an inch and eight tenths per mile at low water of neap tides. From some of the more distant parts of the same Fen, sixteen miles from the outfall, there would only have been a fall of five tenths of an inch per mile at low water. It was clear, therefore, that even the higher levels of the East Fen could not be effectually drained by the outfall at Anton's Gowt or Maud Foster; and hence arose the necessity for cutting an entirely separate main drain, with an outfall at a point in the Wash outside the mouth of the river Witham.[1] This east main cut, called the Hobhole Drain, is about eighteen miles long and forty feet wide, diminishing in breadth according to its distance from the outfall; the bottom being an inclined plane falling four inches in the mile towards the sluice at Hobhole in the Wash. This drain is an immense work, defended by broad and lofty embankments extending inland from its mouth, to prevent the contained waters flooding the surrounding lands. It is protected at its sea outlet by a strong sluice, consisting of three openings of fifteen feet each. When the tide rises, the gates, acting like a valve, are forced back and hermetically closed; and when it falls, the drainage waters, which have in the mean time accumulated, force open the gates again, and the waters flow away down to the level of low water. A connection was also formed between the main drains emptying themselves at Maud Foster (three miles higher up the Witham) and the Hobhole Drain, the flow being regulated by a gauge; so that, during heavy floods, not only the low land waters of the East Fen districts were effectually discharged at Hobhole, but also a considerable portion of the drainage of the West and Wildmore Fens.

An essential part of the scheme was the cutting of the catchwater drains, which were carried quite round the

---

[1] See the map on p. 226.

base of the high lands skirting the Fens; beginning with a six feet bottom, and widening out towards their embouchures to sixteen feet. The principal work of this kind commenced near Stickney, and was carried eastward towards Wainfleet, to near the Steepings river. It was connected at Cowbridge with the main Hobhole Drain, into which the high land waters brought down by the catchwater drain were thus carried, without having been allowed, until reaching that point, to mix with the Fen drainage at all. It would be tedious to describe the works more in detail; and perhaps the outline we have given, aided by the maps of the district, will enable the reader to understand the leading features of Mr. Rennie's comprehensive design. The works were necessarily of a very formidable character, the extent of the main and arterial drains cut during the seven or eight years they were under execution being upwards of a hundred miles. They often dragged for want of funds, and encountered considerable opposition in their progress; though the wisdom of the project was in all respects amply justified by the result.[1]

The drainage of Wildmore and West Fens was first finished, when forty thousand acres of valuable land were completely reclaimed, and in a few years yielded

---

[1] The following letter, written by a Lincolnshire gentleman, in January, 1807, appears in the 'Farmer's Magazine' of February in that year:— "Our fine drainage works begin now to show themselves, and in the end will do great credit to Mr. Rennie, the engineer, as being the most complete drainage that ever was made in Lincolnshire, and perhaps in England. I have been a commissioner in many drainages, but the proprietors never would suffer us to raise money sufficient to dig deep enough through the old enclosures into the sea before; and, notwithstanding the excellency of Mr. Rennie's plan, we have a party of uninformed people, headed by a little parson and magistrate, who keep publishing letters in the newspapers to stop the work, and have actually petitioned Sir Joseph Banks, the lord of the manor, against it; but he answered them with a refusal, in a most excellent way. . . . I think Mr. Rennie's great work will promote another general improvement here, which is, to deepen and enlarge the river Witham from the sea, through Boston and Lincoln, to the Trent, so as to admit of a communication for large vessels, as well as laying the water so much below the surface of the land as to do away with the engines. We have got an estimate, and find the cost may be about 100,000*l*."

heavy crops of grain. East Fen was attacked the last,
the difficulties presented by its formidable chain of lakes
being much the greatest; but the prize also was by far
the richest. When the East Fen waters were drained
off, the loose black mud settled down into fertile soil.
Boats, fish, and wildfowl disappeared, and the plough
took their place. After being pared and burnt, the
land in the East Fen yielded two and even three crops
of oats in succession, of not less than ten quarters to
the acre. The cost of executing the drainage had no
doubt been very great, amounting to about 580,000*l.*
in all, inclusive of expenditure on roads, &c.; against
which had to be set the value of the lands reclaimed.
In 1814 Mr. Anthony Brown, surveyor and valuer, estimated their improved rental at 110,561*l.*; and allowing
five per cent. on the capital expended on the works, we
thus find the increased net value of the drained lands to
be not less than 81,000*l.* per annum, which, at thirty
years' purchase, gives a total increased value of nearly
two millions and a half sterling![1]

It was a matter of great regret to Mr. Rennie that his
design was not carried out as respected the improved
outfall of the Witham. It was an important part of his
original plan that a new and direct channel should be
cut for this river from Boston down to deep water at
Clayhole, where the tide ebbed out to the main sea
level, and there was little probability of the depth being

---

[1] Mr. Brown's estimate was as follows :—

Land reclaimed in the East Fen	..	12,664 acres worth 40s. per ann.	£25,328
,, ,, West Fen	..	17,044 ,, 50s. ,,	42,610
,, ,, Wildmore	..	10,773 ,, 42s. ,,	22,623
Adjoining low lands improved	..	20,000 ,, 20s. ,,	20,000

Total acreage of improved and drained lands	60,481	Annual value	£110,561

Less capital expended on drainage ...	£433,905	
,, ,, roads, &c...	146,800	

                £580,705 at 5 per cent.   29,035

                                    £81,526

Which at 30 years' purchase gives £2,445,780.

materially interfered with by silting for many years to come. This new channel would have enabled all the waters—low land as well as high land—to be discharged into the sea with the greatest ease and certainty. It would also have completely restored the navigation of the river, which had become almost entirely lost through the silting up of its old winding channel. But the Witham was under the jurisdiction of the corporation of Boston, who were staggered by the estimated cost of executing the proposed works, though it amounted to only 50,000*l.* Accordingly, nothing was done to carry out this part of the design, and the channel continued to get gradually worse, until at length it was scarcely possible even for small coasters to reach Boston Quay. As late as the year 1826 the water was so low that little boys were accustomed to amuse themselves by wading across the river below the town even at high water of neap tides. The corporation were at last compelled to bestir themselves to remedy this deplorable state of affairs, and they called in Sir John Rennie to advise them in their emergency. The result was, that as much of the original plan of 1800 [1] was carried out as the state of their funds would permit: the lower part of the channel was straightened, and the result was precisely

---

[1] In his admirable Report, dated the 6th October, 1800, Mr. Rennie pointed out that the lines of direction in which the rivers Welland and Witham entered the Wash tended to the silting-up of the channels of both, and he suggested that the two river outlets should be united in one, and diverted into the centre of the Wash, at Clayhole, which would at the same time greatly increase the depth, and enable a large area of valuable land to be reclaimed for agricultural purposes. This suggestion has since been elaborated by Sir John Rennie, whose plan of 1837, when fully carried out, will have the effect of greatly improving the outfalls of all the rivers entering the Great Wash—the Ouse, the Nene, the Welland, and the Witham—and the drainage of the low level lands depending upon them, comprising above a million of acres, and ultimately gaining from the Wash between 150,000 and 200,000 acres of rich new land, or equal to the area of a good-sized county. In the Wash of the Nene, called Sutton Wash, 4000 acres have already been reclaimed after this plan—the land, formerly washed by the sea at every tide, being now covered with rich cornfields and comfortable farmsteads. It was at this point that King John's army was nearly destroyed when crossing the sands before the advancing tide.

that which the engineer had more than thirty years before anticipated. The tide returned to the town, the shoals were removed, and vessels drawing from twelve to fourteen feet water could again come up to Boston Quays at spring tides.

Mr. Rennie was equally successful in carrying out drainage works in other parts of the Fens, and on the same simple but comprehensive principles.[1] He thus drained the low lands of Great Steeping, Thorpe, Wainfleet, All Saints, Forsby, and the districts thereabout, converting the Steepings river into a catchwater drain, and effectually reclaiming a large acreage of highly valuable land. He was also consulted as to the better drainage of the North Level, the Middle Level, South Holland, and the Great Bedford Level ; and his valuable reports on these subjects, though not carried out at the time, for want of the requisite means, or of public spirit on the part of the landowners, laid the foundations of a course of improvement which has gone on until the present day. It is much to be regretted that his grand plan of 1810 for the drainage of the Great Level, by means of more effectual outfalls and a system of intercepting catchwater drains, was not carried out; for there is every reason to believe that it would have proved as completely successful as his drainage of the Fens of Lincolnshire. But the only part of this scheme that was executed in his time was the Eau Brink Cut, for the purpose of securing a more effectual outfall of the river into the Wash near King's Lynn.

The necessity for this work will be more clearly under-

---

[1] Among other important works of the same kind executed by Mr. Rennie, but which it would be tedious to describe in detail, was the reclamation (in 1807) of 23,000 acres of fertile land in the district of Holderness, near Hull. He was extensively employed to embank lands exposed to the sea, and succeeded (in 1812) in effectually protecting the thirty miles of coast extending from Wainfleet to Boston, and thence to the mouths of the rivers Welland and Glen. Two years later (in 1814) he, in like manner, furnished a plan, which was carried out, for protecting the Earl of Lonsdale's valuable marsh land on the south shore of the Solway Frith.

stood when we explain the circumstances under which its construction was recommended. It will be observed from the Ordnance Survey map covering the Fen district, that the river Ouse flows into the shallows of the Wash near the town of King's Lynn, charged with the waters of the Great Bedford Level as well as those of Huntingdon, Bedford, and Cambridge, and of the high lands of the western parts of Norfolk and Suffolk. Immediately above Lynn the old river made an extensive bend of about five miles in extent, to a point called German's Bridge. This channel was of very irregular breadth and full of great sand beds which were constantly shifting. In some places it was as much as a mile in width, and divided into small streams which varied according as the tidal or the fresh waters were for the time being most powerful. During floods, the flow of the river was so much obstructed that the waters could not possibly get away out to sea during the ebb, so that at the next rise of the tide they were forced back into the interior, and thus caused serious inundations in the surrounding country.[1] The fresh waters were in this way penned up within the land to the extent of about seven feet; and over an extensive plain, such as the Bedford Level, where a few inches of fall makes all the difference between land drained and land drowned, it is clear how seriously this obstruction of the Ouse outfall must have perilled the agricultural operations of the district. Until,

---

[1] When Mr. Rennie was first consulted respecting the drainage of the Great Level, he found that much good land which had been formerly productive had become greatly deteriorated, or altogether lost for purposes of agriculture. Some districts were constantly flooded, and others were so wet that they were rapidly returning to their original state of reeds and sedge. In the neighbourhood of Downham Eau, the harvestmen were, in certain seasons, obliged to stand upon a platform to reap their corn, which was carried to and from the drier parts in boats; and some of the farmers, in like manner, rowed through their orchards in order to gather the fruit from the trees. A large portion of Littleport Fen, in the South Level, was let at 1s. an acre, and, in the summer-time, stock were turned in amongst the reed and "turf-bass," and not seen for days together. In Marshland Fen, the soil was so soft that wooden shoes, or flat boards, were nailed on the horses' feet over the iron ones, to prevent them from sinking into the soil.

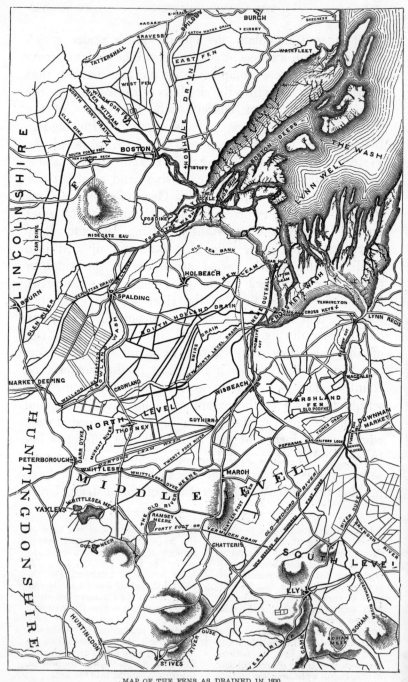

MAP OF THE FENS AS DRAINED IN 1830.

[After Telford's Plan and the Ordnance Survey.]

therefore, this great impediment to the drainage of the
Level could be removed, it was clear to Mr. Rennie's
mind that no inland works could be of any permanent
advantage.   The remedy which he proposed was, to cut
off the great bend in the Ouse by making a direct new
channel from Eau Brink, near the mouth of Marshland
Drain, to a point in the river a little above the town of
Lynn, as shown in the following plan :—

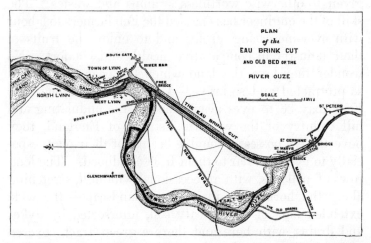

The cut was to be about three miles in length, and of
sufficiently capacious dimensions to contain the whole
body of the river.   By thus shortening the line of the
stream, Mr. Rennie calculated that the channel would be
kept clear of silt by the greater velocity of the current,
and that the fresh waters would at the same time be able
to force their way out to sea without difficulty.   An Act
was accordingly obtained enabling the Eau Brink to
be cut ; but some years passed before any steps were
taken to carry out the works, which were not actually
begun until the year 1817, when Mr. Rennie was formally
appointed the chief engineer.   After about four years'
labour the cut was finished and opened, and its immediate effect was to give great relief to the whole of the
district watered by the Ouse.   An extra fall of not less
than five feet and a half was obtained at St. German's,

by which the surface of the waters throughout the whole
of the Middle and South Levels was reduced in proportion. Thus the pressure on all the banks along the
rivers of the Level was greatly relieved, whilst inundations were prevented, and the sluices provided for the
evacuation of the inland waters were enabled effectually
to discharge themselves.

By labours such as these an immense value has been
given to otherwise worthless swamps and wastes.* The
skill of the engineer has enabled the Fen farmers to labour
with ever-increasing profit, and to enjoy the fruits of
their industry in comparative health and comfort. No
wonder they love the land which has been won by toil
so protracted and so brave. Unpicturesque though the
Fens may be to eyes accustomed to the undulating and
hill country of the western districts of England, they
nevertheless possess a humble beauty of their own, especially to eyes familiar to them from childhood. The long
rows of pollards, with an occasional windmill, stretching
along the horizon as in a Dutch landscape—the wide
extended flats of dark peaty soil, intersected by dykes
and drains, with here and there a green tract covered
with sleek cattle—have an air of vastness, and even
grandeur, which is sometimes very striking. To this we
may add, that the churches of the district, built on sites
which were formerly so many oases in the watery de-
sert, loom up in the distance like landmarks, and are
often of remarkable beauty of outline.

It has been said of Mr. Rennie that he was the greatest
"slayer of dragons" that ever lived,—this title being
given in the Fens to persons who, by skill and industry,
have perfected works of drainage, and thereby removed
the causes of sickness and disease, typified in ancient
times as dragons or destroyers.[1] In this sense, certainly,
Mr. Rennie is entitled, perhaps more than any other
man, to this remarkable appellation.

---

[1] Thompson's 'History of Boston,' 1856, p. 639.

CHAPTER V.

Mr. Rennie's Bridges.

THE bridges erected by our engineer are amongst the finest of his works, and sufficient of themselves to stamp him as one of the greatest masters of his profession. We have already given a representation of his first bridge, erected over the Water of Leith, near Edinburgh, the forerunner of a series of similar structures unrivalled for solidity and strength, contrived with an elegance sometimes ornate, but for the most part of severe and massive simplicity.

Unlike some of his contemporaries, Mr. Rennie did not profess a disregard for theory; for he held that true practice could only be based on true theory. Taken in the sense of mere speculative guessing, however ingenious, he would have nothing to do with it; but as matter of inference and demonstration from fixed principles, he held by theory as his sheet-anchor. His teacher, Professor Robison, had not failed to impress upon him its true uses in the pursuit of science and art; and he never found reason to regret the fidelity with which he carried out his instructions in practice. In 1793 he had the advantage of much close personal intercourse with his old friend the Professor, who paid him a visit at his house in London for the express purpose of conferring with him upon mechanical subjects. In the letter announcing the object of his visit, Dr. Robison candidly avowed that it was in order " that he might extract as much information from him as possible." The Doctor had undertaken to prepare the articles on Mechanics for the third edition of the ' Encyclopedia

Britannica,' and he believed he should be enabled to impart an additional value to his writings by throwing upon
them the light of Rennie's strong practical judgment.
He proposed to take a lodging in the immediate neighbourhood of Rennie's house, then in the Great Surrey
Road, and to board with him during the day ; but Rennie
would not listen to this proposal, and insisted on being
the Professor's entertainer during the period of his visit.

One of the points which he particularly desired to
discuss with Mr. Rennie was the theory of the equilibrium of arches—a subject at that time very imperfectly understood, but which the young engineer had
studied with his usual energy and success. He had
clearly proved that the proper proportion and depth
of the key-stone to that of the extrados (or exterior
curve) should be in proportion to the size and form
of the arch and the materials of which it was composed ; and he had also established the ratio in which the
arch-stones should increase from the key-stone to the
piers or abutments. Up to this time there had been no
rules laid down for the guidance of the engineer or architect, who worked very much in the dark as to principles ;
and it was often a matter entirely of chance whether a
bridge stood or fell when the centres were removed.
According to the views of Hutton and Attwood, the
weight upon the haunches and abutments, to put the
arch in a state of equilibrium so that it should stand, was
unlimited ; whereas Mr. Rennie established the limit to
which the countervailing force or weight on the extrados
should be confined. Hence he adopted the practice of
introducing a flat inverted arch between the extrados of
each two adjoining arches, (at the same time increasing
the width of the abutment,)—the radii of the voussoirs or arch-stones being continued completely through
them. And in order to diminish the masonry, the
lower or foundation course was inclined also,—thus
combining the work more completely together, and

enabling it better to resist the lateral thrust. Dr. Robison had much discussion with Mr. Rennie on these and many other points, and the information he obtained was shortly after worked up into numerous original contributions of great value ; amongst which may be mentioned his articles in the ' Encyclopedia ' on the Arch, Carpentry, Roof, Waterworks, Resistance of Fluids, and Running of Rivers [1]—on all of which subjects Mr. Rennie had much original information to impart. It may readily be imagined that the evenings devoted by Dr. Robison to conversation and discussion on such topics at Rennie's house were of interest and advantage to both ; and when the Doctor returned to his Edinburgh labours, he carried with him the cordial affection and respect of the engineer, who continued to keep up a correspondence with him until the close of his life.

In the early part of his career Mr. Rennie was called upon to furnish designs of many bridges, principally in Scotland, which, however, were not carried out, in most cases because the requisite funds could not be raised to build them. Thus, in 1798, he designed one of eight cast iron arches to span the river Don at Aberdeen. Four years later he was called upon to furnish further designs, when he supplied three several plans, two of granite bridges ; but the structures were of too costly a character for the people of Aberdeen then to carry out. The first important bridge which Mr. Rennie was authorised to execute was that across the Tweed at Kelso, and it afforded a very favourable specimen of his skill as an architect. It was designed in 1799 and opened in 1803. It consists of five semi-elliptical arches of 72 feet span, each rising 28 feet, and four piers each 12 feet thick,

---

[1] Dr. Robison was the first contributor to the ' Encyclopedia ' who was really a man of science, and whose articles were above the rank of mere compilations. He sought information from all quarters—searched the works of foreign writers, and consulted men of practical eminence, such as Rennie, to whom he could obtain access,—and extraordinary value was thus imparted to his articles.

with a level roadway 23 feet 6 inches wide between the
parapets, and 29 feet above the ordinary surface of the
river.   The foundations were securely laid upon the solid
rock in the bed of the Tweed, by means of coffer-dams,
and below the deepest part of the river.   The piers and
abutments were ornamented with three-quarter columnar
pilasters of the Roman Doric order, surmounted by a plain
block cornice and balustrade of the same character.   The
whole of the masonry was plain rustic coursed work, and
in style and execution it was long regarded as one of the
most handsome and effective structures of its kind.   It
may almost be said to have formed the commencement
of a new era of bridge-building in this country.   The
semi-elliptical arches, the columnar pilasters on the piers,
the balustrade, and the level roadway, are the same
as in Waterloo Bridge, except as regards size and character ; so that Kelso Bridge may be regarded as the
model of the greater work.   We believe it was one of
the first bridges in this country constructed with a *level*

KELSO BRIDGE   [By Percival Skelton.]

MUSSELBURGH BRIDGE.

[By E. M. Wimperis, after a Drawing by J. S. Smiles.]

roadway.   Some of the old-fashioned bridges were ex-
cessively steep, and to get over them was like climbing
the roof of a house.   There was a heavy pull on one
side and a corresponding descent on the other.   The old
bridge across the Esk at Musselburgh, forming part of
the high road between Edinburgh and London, was
of this precipitous character.   It was superseded by a
handsome and substantial bridge, with an almost level
roadway, after a design by Rennie.   When the engineer
was taking the work off the hands of the contractor, one
of the magistrates of the town, who was present, asked
a countryman who was passing at the time with his cart
how he liked the new brig?   "Brig!" said the man,
"it's nae brig ava! ye neither ken whan ye're on't, nor
whan ye're aff't!"

Mr. Rennie's boldness in design grew with experience,
and when consulted as to a bridge near Paxton, over
the Whitadder (a rapid stream in Berwickshire), he

proposed, in lieu of the old structure, which had been
carried away by a flood, a new one of a single arch, of
150 feet span; but unhappily the road trustees could
not find the requisite means for carrying it into effect.
Another abortive but grand design was proposed by
him in 1801. He had been requested by the Secretary
of State for Ireland to examine the road through North
Wales to Holyhead, with the object of improving the
communication with Ireland, which was then in a
wretched state. The connection of the opposite shores
of the Menai Strait by means of a bridge was considered an indispensable part of any improvement of
that route; and Mr. Rennie proposed to accomplish this
object by a single great arch of cast-iron 450 feet in
span,—the height of its soffit or crown to be 150 feet
above high water at spring tides.[1] A similar bridge,
of 350 feet span, having its crown 100 feet above the
same level, was also proposed by him for the crossing
of Conway Ferry. These bridges were to be manufactured after a plan invented by Mr. Rennie in 1791,
and communicated by him to Dr. Hutton in 1794;
and he was strongly satisfied of its superiority to all
others that had been proposed. The designs were alike

---

[1] The great arch of 450 feet was to
be supported on two stone piers, each
75 feet thick, the springing to be 100
feet above high water. There were to
be arches of stone on the Caernarvon
side to the distance of about 156
yards, and on the Anglesea side to
the distance of about 284 yards;
making the total length of the bridge,
exclusive of the wing walls, about
640 yards. The estimated cost of
the whole work and approaches was
268,500l. The point at which the
bridge was recommended to be thrown
across was, either opposite Inys-y-
Moch island, on which one of the
main piers would rest, or at the
Swilly rocks, about 800 yards to the
eastward; but, on the whole, he preferred the latter site. He also sent
in a subsequent design, showing an
iron arch on each side of the main
one of 350 feet span, in lieu of masonry, with other modifications, by
which the dimensions of the main
piers were reduced, and the estimate
somewhat lessened. Other plans were
prepared and submitted, embodying
somewhat similar views, the prominent idea in all of them being the
spanning of the strait by a great cast-
iron arch, the crown of which was to
be 150 feet above the sea at high-
water. The plans and evidence on
the subject are to be found set forth
in the 'Reports from Committees of
the House of Commons on Holyhead
Roads' (1810-22), ordered to be
printed 25th July, 1822.

bold and skilful, and it is to be regretted that they
were not carried out; for their solidity would not only
have proved sufficient for the purposes of a roadway,
but probably also of a locomotive railway. In that
case, however, we should have been deprived of the
after-display of much engineering ability in bridging
the straits at Menai and the ferry at Conway. But the
plans were thought far too daring for the time, and the
expense too great. The whole subject was therefore
allowed to sleep for many years, until eventually Telford
spanned both these straits with suspension road bridges,
and Robert Stephenson afterwards with tubular railway
bridges, at a total cost of about a million sterling.

BOSTON BRIDGE. [By Percival Skelton.]

The first bridge constructed by Mr. Rennie in England, and the earliest of his cast iron bridges, was that
erected by him over the Witham, in the town of Boston,
Lincolnshire, in 1803. It consists of a single arch of
iron ribs, forming the segment of a circle, the chord of

which is 80 feet. It is simple yet elegant in design;
its flatness and width contributing to render it most
convenient for the purpose for which it was intended
—that of accommodating the street-traffic of one of the
most prosperous and busy towns in the Fens.

Mr. Rennie's reputation as an engineer becoming well
established by these and other works, he was, during the
remainder of his professional career, extensively consulted on this branch of construction;[1] and many solid
memorials of his skill in bridge-work are to be found in
different parts of the kingdom. But the finest of the
buildings of this character which were erected by him are
unquestionably those which grace the metropolis itself.

---

[1] Among his minor works may be mentioned the bridge over the stream which issues out of Virginia Water and crosses the Great Western Road (erected in 1805); Darlaston Bridge across the Trent, in Staffordshire (1805); the timber and iron bridge over the estuary of the Welland at Fossdyke Wash, about nine miles below Spalding (1810); the granite bridge of three arches at New Galloway, on the line of the Dumfries and Portpatrick Road (1811); a bridge of five arches across the Cree at Newton Stewart (1812); the cast iron bridge over the Goomtee at Lucknow, erected after his designs in 1814, and frequently referred to in the military operations for the relief of that city a few years ago; Wellington Bridge, over the Aire, at Leeds (1817); Isleworth Bridge (1819); a bridge of three elliptical arches of 75 feet span each, at Bridge of Earn, Perthshire (1819); Cramond Bridge, of eight semi-circular arches of 50 feet span, with the roadway 42 feet above the river (1819); and Ken Bridge, New Galloway, of five stone arches, the centre 90 feet span (1820). An adventure of some peril attended Mr. Rennie's erection of the bridge at Newton Stewart. He happened to visit the works on one occasion during a heavy flood, which swept down the valley with great fury; and the passage of the ferry was thus completely interrupted. Mr. Rennie and his son (the present Sir John) were consequently unable to cross over to Newton Stewart, on the further side of the river, and they were under the necessity of spending the night in a miserable public-house on the eastern bank. About 11 P.M. the violence of the storm had somewhat abated, and the moon came out, though obscured by the clouds which drifted across her face. Mr. Rennie went out at that late hour to look at the bridge works, and even to try whether he might not reach the other side by crossing the timber platform by means of which the works were being carried on. There was a gangway of only two planks from pier to pier on the eastern side, and this he safely crossed. The torrent was still raging furiously beneath, shaking the frail timbers of the scaffolding. As Mr. Rennie was about to place his foot on the plank which led to the third pier, his son observed the framework tremble, and pulled his father back, just in time to see the whole swept into the stream with a tremendous crash. Fortunately the planking still stood across which they had passed, and they succeeded in retracing their steps in safety. The bridge was finished and opened during the summer of 1814.

The project of erecting a new bridge to connect the Strand, near Somerset House, with the Surrey side of the Thames at Lambeth, was started by a Bridge Company in 1809—a year distinguished for the prevalence of one of those joint-stock fevers which periodically seize the moneyed classes of this country. The first plan considered was the production of Mr. George Dodds, a well-known engineer of the time. The managing committee were not satisfied with the design, and referred it to Mr. Rennie and Mr. Jessop for their opinion. It was found to be for the most part a copy of M. Peyronnet's celebrated bridge of Neuilly, with modifications rendered necessary by the difference of situation and the greater width of the river to be spanned. It showed a bridge of nine arches of 130 feet span; each being a compound curve, the interior an ellipsis, and the face or exterior a segment of a circle, as in the bridge at Neuilly.[1] The reporting engineers pointed out various

---

[1] In their report on this design, Mr. Rennie and his colleague observed: " We should not have thought it necessary to quote the production of a foreign country for the sake of showing the practicability of constructing arches of 130 feet span, had we not been led to it by the exact similarity of the designs, and by the principle which is therein adopted of the compound curve; because our own country affords examples of greater boldness in the construction of arches than that of Neuilly. There is a bridge over the river Taff, in the county of Glamorgan, of upwards of 135 feet span, with a rise not exceeding 32 feet, and what is more remarkable is, that the depth of the arch-stones is only 30 inches; so that in fact that bridge far exceeds in boldness of design that of Neuilly." [See our Memoir of William Edwards in Vol. I. of this work.] After some observations as to the importance and necessity of making a bridge in such a situation, at the bend of the river, with as large arches as possible, to accommodate the navigation and present as little obstruction as possible to the rise and fall of the water, they proceed: " We confess we do not wholly approve of M. Peyronnet's construction as adapted for the intended situation. It is complicated in its form, and, we think, wanting in effect. The equilibrium of the arches has not been sufficiently attended to; for when the centres of the bridge at Neuilly were struck, the top of the arches sank to a degree far beyond anything that has come to our knowledge, whilst the haunches retired or rose up, so that the bridge as it now stands is very different in form from what it was originally designed. No such change of shape took place in the bridge over the Taff (Pont-y-Prydd); the sinking after the centres were struck did not amount to one-half of that at Neuilly, although the one was designed and built under the direction of the first engineer of France, without regard to expense, whilst the other was designed and built by a country mason with par-

objections to Mr. Dodds's design, as well as to the plan
proposed by him for founding the piers; and they
showed that his estimate of cost was altogether insufficient.   The result was, that no further steps were taken
with Mr. Dodds's plan; but when the Act authorising the
construction of the bridge had been obtained, the committee again applied to Mr. Rennie; and on this occasion
they requested him to furnish them with the design of a
suitable structure.[1]   The first step which he took was
to prepare an entirely fresh chart of the river and the
adjacent shores, after a careful and accurate survey
made by Mr. Francis Giles.   In preparing his plan, he
kept in view the architectural elegance of the structure
as well as its utility; and while he designed it so as to
enhance the beauty of the fine river front of Somerset
House, by contriving that the face of the northern
abutment should be on a line with its noble terrace, he
laid out the roadway so that it should be as nearly upon
a level with the great thoroughfare of the Strand as
possible,—the rise from that street to the summit on
the bridge being only 1 in 250, or about two feet in all.
Two designs were prepared—one of seven equal arches,
the other of nine; and the latter being finally approved
by the committee as the less costly, it was ordered to be
carried into effect.

The structure as executed is an elegant and substantial
bridge of nine arches of 120 feet span, with piers 20 feet
thick; the arches being plain semi-ellipses, with their
soffits or crowns 30 feet above high-water of ordinary
spring tides.   Over the points of each pier are placed

---

simonious economy.   Our opinion
therefore is, that the arches of the
bridge over the Thames should either
be plain ellipses, without the slanting
off in the haunches so as to deceive
the eye by an apparent flatness which
does not in reality exist, or they should
be of a flat segment of a circle formed
in such a manner as to give the re-

quisite room for the passage of the
current and barges under it."

[1] In June, 1810, we find him accepting the direction of the new
bridge at 1000*l.* a year for himself
and assistants, or 7*l.* 7*s.* a day and
expenses; but on no account were any
of his people to have to do with the
payment or receipt of moneys.

two three-quarter Doric column pilasters, after the design
of the temple of Segesta in Sicily. These pilasters are
5 feet 8½ inches diameter at the base, and 4 feet 4 inches
at the under side of the capital, forming recesses in the
roadway 17 feet wide and 5 feet deep. The depth of
the arch-stones at the crown is 4 feet 6 inches, and they
increase regularly to 10 feet at the haunches. Between
each pair of arches, at the level of 19 feet above the
springing, there is an inverted arch, the stones of which

SECTION OF WATERLOO BRIDGE.

are 4 feet 6 inches deep at the crown, and decrease
regularly on each side as they unite and abut against the
extrados or backs of the voussoirs of the main haunches.
The abutments are 40 feet in thickness at the base, and
decrease to 30 feet at the springing. The cope of the
arches and piers is surmounted by a Grecian Doric
block-cornice and entablature, upon which is placed a
balustrade parapet 5 feet high. The total width of the
bridge from outside to outside of the parapets is 45 feet.
The footpaths on each side are 7 feet wide, and the
roadway for carriages 28 feet. There are four sets of
landing-stairs—two to each abutment; and the arrangement of this part of the work has been much admired,
on account of its convenience for public uses as well as
its architectural elegance.

In the construction of this bridge, there are four features of distinctive importance to be noted :—1st. The
employment of coffer-dams in founding piers in a great
tidal river—an altogether new use of that engineering
expedient, though now become customary. 2nd. The
ingenious method employed for constructing, floating,
and fixing the centres; since followed by other engineers
in works of like magnitude. 3rd. The introduction and
working of granite stone to an extent before unknown,

and in much larger and more substantial pieces of masonry than had previously been practised. 4th. The adoption of elliptical stone arches of an unusual width, though afterwards greatly surpassed by the same engineer in his New London Bridge.

Mr. Rennie invariably took the greatest pains in securing the most solid foundations possible for all his structures, and especially of his river works, laying them far below the scour of the river, at a depth beyond all probable reach of injury from that cause. The practice adopted in founding the piers of the early bridges across the Thames, was to dredge the bottom to a level surface, and build the foundations on the bed of the river, protecting them outside by rubble, by starlings, or by sheet-piling. Mr. Dodds had proposed to follow the method employed by Labelye at Westminster Bridge, of founding the piers by means of caissons; but Mr. Rennie insisted on the total insufficiency of this plan, and that the most effectual method was by means of coffer-dams. This would no doubt be more costly in the first instance, but vastly more secure; and he foresaw that the inevitable removal of the piers of Old London Bridge, by increasing the current of the river, would severely test the foundations of all the bridges higher up the stream—which proved to be the case. Having already extensively employed coffer-dams in getting in the foundations of the London and East India Dock walls, he had no doubt as to their success in this case; and they were adopted accordingly.[1]

---

[1] The coffer-dams in which the foundations of the abutments were built, were formed by driving two rows of piles 13 by 6½ inches each, with a counter or abutting pile at every 12 feet, 12 by 12, driven in the form of an ellipsis, and strongly cemented together, at low-water and high-water levels, by double horizontal walings or bracks, having a space of about 8 inches clear between them for the intermediate or half piles. The whole were driven close together from 15 to 20 feet deep into the ground, well caulked, so as to be water-tight, and all connected firmly together by strong wrought iron bars and bolts, besides shores and intermediate braces. The spaces between the two rows of piles were then rammed close with well-tempered clay, so that they formed, as it were,

Mr. Rennie also introduced a practice of some novelty and importance in the centering upon which the arches of the bridge were built.   He adopted the braced prin-

CENTERING OF ARCH, WATERLOO BRIDGE. [After E. Blore.]

ciple.   The centres spanning the whole width of the arch were composed of eight ribs each, formed in one piece, resting upon the same number of solid wedges,

a solid vat or tub impermeable to water ; and within these, when pumped clear of water, the excavation was made to the proper depth, and in the space so dug out the building operations proceeded. The coffer-dams for the piers were formed in a similar manner, with modifications according to circumstances. By this means the bed of the river, where the piers were to be erected, was exposed and dug out to the proper depth, and the foundations were commenced from a level nine feet at least below low water mark. The foundations there rested upon timber piles from 20 to 22 feet long, driven into the solid bed of the river. Upon the heads of these piles half-timber planking was spiked, and on this the solid masonry was built—every stone being fitted, mortared, and laid with studious accuracy and precision. The whole work was done with such solidity that, after the lapse of fifty years, the foundations have not yielded by a straw's breadth at any point.

supported by inclined tressels placed upon longitudinal
bearers, firmly fixed to the offsets of the piers and abutments. At the intersecting point of the bearers or
braces in each rib, there was a cast iron box with two
holes or openings in it, so that the butt-ends rested firmly
against the metal; and to prevent them from acting like
so many wedges to tear the rib to pieces when the vertical weight of the arch began to act upon them, pieces
of hard wood were driven firmly into the holes above
described, to check the effect of the bearers or strutts of
the ribs; and this arrangement proved completely successful. The eight ribs were firmly connected together by
braces and ties, so as to form one compact frame, and the
curve or form of the arch was accurately adjusted by
means of transverse timbers, 12 inches wide and 6 inches
thick, laid across the whole of the ribs, set out to the
exact form of the curve by ordinates from the main or
longitudinal axis of the ellipsis; and in proportion as
the voussoirs or arch-stones were carried up from the
adjoining piers, the weight which had been laid upon
the top of the centre to keep it in equilibrium according
to the form of the arch during construction, was gradually removed as it advanced towards completion.
When the arch was about two-thirds completed, a small
portion of it was closed with the centre, and the remaining part of each side was brought forward regularly
by offsets to the crown until the whole was finished.
Each key-stone was accurately fitted to its respective
place, and the last portion of each, for the space of about
eighteen inches, was driven home by a heavy wooden
ram or pile-engine, so as almost to raise the crown of
the arch from the centre.

About ten days after the main arches had been
completed, and the inverts and spandrel walls between them carried up to the proper height, the arches
were gently slackened, to the extent of about two
inches, so as to bring each to its bearing to a cer-

tain extent. This was effected by driving back the wedges upon which the ribs of the centres rested, by means of heavy wooden rams attached to them, so that they could swing backwards and forwards with great facility when any external force was applied to them; and this was done by ropes worked by hand-labour. After the first striking or slackening, the arches were allowed to stand for ten days, when the wedges were driven back six inches further. After ten days more the wedges were driven back sufficiently to render the arch altogether clear of the centering. By this means the mortar was firmly imbedded into all the joints, and the arch came gradually to its ultimate bearing without any undue crushing. In order to ascertain whether any change of form took place, three straight lines were drawn in black chalk on the extreme face of the arch previous to commencing the operation of striking the centre,—one horizontally in the centre of the voussoirs forming the crown, and two from the haunches of the arch, each intersecting the first line at about 25 feet on each side of the keystone; so that if there had been any derangement of the curve or irregular sinking, it would at once have been clearly apparent. After the centres had been removed, it was found that the sinking of the arches varied from $2\frac{1}{2}$ to $3\frac{1}{4}$ inches, which was as nearly as possible the allowance made by the engineer in designing the work; the whole plan being worked out with admirable precision and accuracy.

The method of fixing and removing the centres was entirely new; being precisely the same as was afterwards followed by Mr. Robert Stephenson in fixing the wrought iron ribs of the Conway and Britannia bridges,—that is, by constructing them complete on a platform adjacent to the river, and floating them between the piers on barges expressly contrived for the purpose.* They were then raised into their proper places by four strong screws, 8 inches in diameter and 4 feet long,

fixed in a strong cast iron box firmly bedded in the solid
floor of the barge. The apparatus worked so well and
smoothly, that the whole centre, consisting of eight ribs,
each weighing about fifty tons, was usually placed within
the week.

The means employed by Mr. Rennie for forming his
road upon the bridge were identical with those adopted
by Mr. Macadam at Bristol some six years later. But the
arrangement constituted so small a part of our engineer's
contrivances, that, as in many other cases, he made no
merit of it. When the clay puddle placed along the
intended roadway was sufficiently hard, he spread a
stratum of fine screened gravel or hoggins, which was
carefully levelled and pressed down upon the clay. This
was then covered over with a layer of equally broken
flints, about the size of an egg ; after which the whole
was rolled close together, and in a short time formed
an admirable "macadamized" road. Mr. Rennie had
practised the same method of making roads over his
bridges long before 1809; and he continued to adopt
it in all his subsequent structures.

The whole of the stone required for the bridge
(excepting the balustrades, which were brought ready
worked from Aberdeen) was hewn in some fields adjacent to the erection on the Surrey side. It was
transported on to the work upon trucks drawn along railways, in the first instance over temporary bridges of
wood; and it is a remarkable circumstance that nearly
the whole of the material was drawn by one horse,
called "Old Jack"—a most sensible animal, and a great
favourite. His driver was, generally speaking, a steady
and trustworthy man, though rather too fond of his dram
before breakfast. As the railway along which the stone
was drawn passed in front of the public-house door, the
horse and truck were usually pulled up while Tom
entered for his "morning." On one occasion the driver
stayed so long that "Old Jack," becoming impatient,

poked his head into the open door, and taking his master's coat-collar between his teeth, though in a gentle sort of manner, pulled him out from the midst of his companions, and thus forced him to resume the day's work.

WATERLOO BRIDGE.

[By Percival Skelton, after his original Drawing.]

The bridge was opened with great ceremony by His Royal Highness the Prince Regent, attended by the Duke of Wellington and many other distinguished personages, on the 18th of June, 1817. It was originally named the Strand Bridge; but after that date the name was altered to that of "Waterloo," in honour of the

Duke. At the opening, the Prince Regent offered to confer the honour of knighthood on the engineer, who respectfully declined it. Writing to his friend Whidbey, he said, "I had a hard business to escape knighthood at the opening." He was contented with the simple, unadorned name of John Rennie, engineer and architect of the magnificent structure which he had thus so successfully brought to completion. Waterloo Bridge is indeed a noble work, and probably has not its equal for magnitude, beauty, and solidity. Dupin characterised it as a colossal monument worthy of Sesostris or the Cæsars; and what most struck Canova during his visit to England was, that the trumpery Chinese bridge, then in St. James's Park, should be the production of the Government, whilst Waterloo Bridge was the enterprise of a private company. Like all Rennie's works, it was built for posterity. That it should not have settled more than a few inches—not five in any part—after the centres were struck, is an illustration of solidity and strength probably without a parallel. We believe that to this day not a crack is visible in the whole work.

The necessity for further bridges across the Thames increased with the growth of population on both sides of the river; and in the year 1813 a Company was formed to provide one at some point intermediate between Blackfriars and London Bridge, of which Mr. Rennie was appointed the engineer. The scheme was at first strongly opposed by the Corporation on the ground of the narrowness of the river at the point at which it was proposed to erect the new structure; but the public demands being urgent, they at length gave way and allowed the necessary Act to pass, insisting, however, on the provision of a very large waterway, so that the least obstruction should be offered to the navigation. Mr. Rennie prepared a design to meet the necessities of the case, and in order to secure the largest possible waterway, he projected his well-known Southwark Bridge, extending from Queen Street, Cannon Street, to Bridge Street, Southwark. It consists of three cast iron arches, with two stone piers and abutments. The arches are flat segments of circles, the centre one being not less than 240 feet span (or 4 feet larger than Sunderland Bridge, the largest cast iron arch that had until then been erected), rising 24 feet, and springing 6 feet above high water of spring tides. The two side arches are of 210 feet span, each rising 18 feet 10 inches, and springing from the same level. The two piers were 24 feet wide each at the springing, and 30 feet at the base.

The works commenced with the coffer-dam of the south pier on the Southwark side, and the first stone was laid by Admiral Lord Keith about the beginning of 1815. All the centering for the three arches was fixed by the autumn of 1817, and the main ribs were set by the end of April, 1818. The centres were struck by the end of the month of June following, and completely removed by the middle of October; and the bridge was opened for traffic in March, 1819.

In the course of this work great precautions were used in securing the foundations of the piers. The river was here at its narrowest and deepest point, the bed being 14 feet below low water of ordinary spring tides. The coffer-dams were therefore necessarily of great depth and strength to resist the pressure of the body of water, as well as the concussion of the barges passing up and down the river, which frequently drove against them. Hence the dams were constructed in the form most capable of resisting external pressure, and yet suitable to the dimensions of the foundations.[1] The masonry

---

[1] They were made elliptical, and consisted of two main rows of piles 14 inches square, placed 6 feet apart. These were cemented together on each side at the level of low water at half-tide, and 2 feet above high water of spring tides, with horizontal walings or braces, also 14 inches square, firmly secured to every tenth pile by wrought iron screw bolts $2\frac{1}{2}$ inches diameter passing through each row of piles, and fixed at each end by

and iron work of the bridge were erected with great care and completeness. The blocks of stone in the piers were accurately fitted to their places by moulds, and driven down by a heavy wooden ram. The least possible quantity of finely tempered mortar was used, so that

---

large bracket-pieces of timber and wrought iron plates and nuts, so that the whole could be firmly screwed up and braced together. The piles were well headed and shod with wrought iron, and driven from 15 to 20 feet into the solid bottom of the river. On the outside of the two main rows of piles there was another placed six feet from the inner, and driven to the same depth, but their heads only extended up to the level of half-tide. They were tied together with horizontal braces and wrought iron screw bolts, in the same manner as the two main rows. The joints of all were well caulked, and the spaces between the three rows were filled with well-puddled clay, so as to be completely impervious to water. On each dam there was a trunk three feet square, fitted with a valve, which could be lowered or raised at pleasure, so that the water could be let off to the level of low tide, or filled at any time, in the event of any accident occurring in the building of the piers. The abutment coffer-dams were constructed in a semi-circular form, and consisted of two rows of piles of like dimensions, similarly rammed between with clay to keep out the water. The enclosed spaces were pumped out by means of steam-engines of 20-horse power. The engines worked double pumps of 15 inches diameter by an arrangement of slide-rods. When necessary, as at the City abutment, where the soil was more porous, an additional engine with pumps was employed. By these complete methods the water was kept under until the foundations were got in at the great depth we have above stated. The piers and abutments rest on solid platforms of piles, cills, and planking about 2 feet 6 inches thick. The piles, of fir, elm, and beech,

20 feet long and 12 inches in diameter, shod with wrought iron, were driven in regular rows three feet apart, until a ram of 15 cwt. falling from a height of 28 feet moved them downward only half an inch at a blow. Their heads were then cut off level; the earth and clay removed from about them to the depth of a foot, and the space filled in with Kentish rag-stone, well rammed and grouted together with lime and gravel. Cills of fir, beech, and elm, 12 inches thick and 14 inches wide, were then accurately fitted and spiked to each pile-head, in the transverse direction of the piers and abutments, with wrought iron jagged spikes 18 inches long and six-eighths of an inch square. The spaces between the cills were solidly filled with brickwork, and another row of cills of the same dimensions, laid at right angles to those below, was fitted and spiked to them over each row of piles, in the same manner as above explained. The spaces were then filled with brick-work; the whole surface was covered with a solid flooring of elm-plank 6 inches thick, well bedded in mortar, and spiked down to each cill with wrought iron jagged spikes 10 inches long and five-eighths of an inch square. It may here be observed, as to the use of timber at this depth, that when it is exposed to an equable degree of moisture it is found almost imperishable—timber having been taken up, as fresh as when laid down, from the foundations of structures laid in water for more than a thousand years. Besides, timber is found, better than any other material, capable of distributing the pressure over the whole surface, as well as binding all the parts of the foundations together.

every part should have a perfectly true permanent bearing. Great care was also taken in the selection of the blocks. The exterior of the piers was constructed of hard silicious stone brought from Craigleith Quarry near Edinburgh, and Dundee; the interior, from the bottom

SOUTHWARK BRIDGE.

[By Percival Skelton, after his original Drawing.]

of the foundations to the springing of the arches, of hard Yorkshire grit; while that part of the piers and abutments from which the arches spring consisted of the hardest and closest blocks of Cornish and Aberdeen granite: in fine, it may be affirmed that a more

solid piece of masonry does not exist than Southwark
Bridge.

The iron work consists of eight arched ribs, the main
strength of the arches being embodied in their lower
parts, which are solid.   The lower or main arch is
divided into thirteen pieces, with a rib 5½ inches thick
at the top and bottom, and 2½ inches in the centre.
The joints radiate outwards from the lower edge, and
form so many cast iron instead of stone voussoirs, from
6 to 8 feet deep and 13 feet long.   At the junction of
each of these main rib pieces there are transverse
plates of the same depth, having flanges cast upon
them on both sides in a wedge form, so that the ends
of the main rib piers fit into them on one side, whilst
on the other there is a cast iron wedge, driven in
between the rib and flange piece, and enabling the whole
to be accurately adjusted and connected together.   In
addition to this, each rib piece had a flange, cast at each
end with a certain number of holes three quarters of
an inch in diameter, into which wrought iron screw
bolts were introduced to connect the whole firmly together in the direction of the arch.   These rib pieces
were also of great importance during construction, the
chief dependence being placed upon their lateral thrust
in holding the arches together.   At each pier and
abutment there was a similar cast iron bed or abutting
plate, let accurately 1½ inch into the stone ; but between
the end of each main rib which sprang from this plate
there was a groove cut out of the solid stone behind
the springing plates and main iron ribs of the arches,
18 inches wide, 3 inches thick at the top, and 2 inches
at bottom.   This groove was accurately dressed and
polished.   Three cast iron wedges, 9 feet long, 6 inches
wide, and 3¼ inches thick at top and 2 inches at bottom,
were then made and most accurately chipped and filed,
so as to fit exactly the groove above mentioned to within
12 inches of its bottom.   When the whole of these wedges

at both ends of the arch had been put into their places, they were carefully driven home to the bottom of the grooves at the same time by heavy wooden rams, by which means the ribs of the arches were relieved from the centres and took their own bearing. In other words, the arches were keyed from the abutments only, instead of from the centre, as is usual in bridges of stone. This was an extremely delicate and nice process, as it required that the variations of the thermometer should be carefully observed, in order that each operation should be carried on at as nearly as possible the same degree of temperature, otherwise the form of the arch would have been distorted, the vertical and lateral pressure of the different parts would have been affected, and an undue strain thrown upon the abutments as well as the different parts of the arch. But so nicely was the whole operation arranged and adjusted, that nothing of the kind occurred : the parts remained in perfect equilibrium ; not a bolt was broken, and not the smallest derangement was found in the structure after the process had been completed.

The spandrel pieces attached to the top of the main ribs were cast in the form of open diamonds or lozenges, connected together in the transverse direction by two tiers of solid crosses laid nearly horizontally—all closely wedged and firmly bolted together. In addition to the transverse connecting plates cast in open squares, there were also diagonal braces of cast iron, commencing at the extremity of the outer rib of each arch and inter-secting each other so as to form a diamond-shaped space in the centre. These were also secured at their ends by wedges and bolts, like the main rib pieces. After the main ribs of the arches were relieved from the centres, and had taken their bearing, before the centres were removed from beneath them, experiments were made how far they might be affected by expansion and con-traction, in proportion to the different degrees of tempera-ture to which the bridge might be exposed ; and for this

purpose different gauges were made, of brass, iron, and
wood. These gauges were firmly attached to the middle
or crown of the wooden centres, and divided into
sixteenths of an inch, and at each a Fahrenheit thermometer was placed; so that, the ends of the arch being
fixed, the variation in the temperature would be indicated
by the rise and fall in the centre. The observations
were made daily—in the morning, at midday, and at
sunset—for several months during summer and winter,
when it was ascertained that the arches rose and fell
about one-tenth of an inch for every 10 degrees of temperature, more or less.

The whole iron work is covered with solid plates,
having flanges cast on their upper side. These plates
are laid in the transverse direction and on the top of the
spandrel walls, so that they form a solid and compact
cast iron floor to support the roadway. The cornice,
which is cast hollow, is of the plain Roman-Doric order,
and is secured to the roadway-plates by strong stays and
bolts at proper intervals. The parapet consists of a
plinth, also cast hollow, with a groove at the top to
receive the railing, which is cast in the form of open
diamonds corresponding with the spandrels. The roadway is 42 feet wide from outside to outside, and formed
in the same manner as that over Waterloo Bridge,
already described. The total quantity of cast iron in
the bridge is 3620 tons, and of wrought iron 112 tons.
It has been said that an unnecessarily large quantity
of material has been employed; and no doubt a lighter
structure would have stood. But looking at the imper-
fections of workmanship and possible flaws in the
castings, Mr. Rennie was probably justified in making
the strengths such as he did, in order to ensure the
greatest possible solidity and durability—qualities which
eminently characterize his works, and perhaps most of
all, his majestic metropolitan bridges. Although the
Southwark Bridge was built before the Railway era,

which has given so great an impetus to the construction
of iron bridges, it still stands pre-eminent in its class,
and is a model of what a bridge should be.   Its design
was as bold as its execution was masterly.   Mr. Robert
Stephenson has well said of it that, "as an example of
arch-construction, it stands confessedly unrivalled as
regards its colossal proportions, its architectural effect,
and the general simplicity and massive character of its
details." [1]

[1] Article on Iron Bridges in 'Encyclopedia Britannica.'

WATERLOO BRIDGE.   [By R. P. Leitch.]

CHAPTER VI.

Mr. Rennie's Docks and Harbours.

THE growth of the shipping business, and the increase in our home and foreign commerce, led to numerous extensive improvements in the harbours of Britain about the beginning of the present century. The natural facilities of even the most favourably situated ports, though to some extent improved by art, no longer sufficed for the accommodation of their trade. Comparatively little had as yet been done to improve the port of London itself, the great focus of the maritime and commercial industry of Britain.[1] It is true its noble river the Thames provided a great amount of shipping room and a vast extent of shore convenience between Millwall and London Bridge; but the rise and fall of the tide twice in every day, and the great exposure of the vessels lying in the river to risks of collisions, and other drawbacks, were felt to be evils which the shipping interest found it necessary to remedy. Besides the crowding of the river by ships and lighters—the larger vessels having to anchor in the middle of the stream as low as Blackwall, from which their cargoes were lightered to the warehouses higher up the Thames—the warehouse

---

[1] The increase in the trade of London is exhibited by the following abstract of vessels entered at the port at different periods since the beginning of last century :—

YEARS.	BRITISH.		FOREIGN.		COLLIERS AND COASTERS.	
	Vessels.	Tonnage.	Vessels.	Tonnage.	Vessels.	Tonnage.
1702	839	80,040	496	76,995	No return.	No return.
1751	1498	198,023	184	36,346	Do.	Do.
1798	1649	397,096	1771	229,991	10,133	1,250,449
1860	6320	1,828,911	4857	1,152,499	18,346	3,152,853

accommodation was found very inadequate in extent as well as difficult of access. There· was also a regular system of plunder carried on in the conveyance of the merchandise from the ship's side to the warehouses, the account of which, given in Mr. Colquhoun's work on the 'Commerce and Police of the Thames,' affords a curious contrast to the security and regularity with which the shipping operations of the port are carried on at the present day. Lightermen, watermen, labourers, sailors, mates and captains occasionally, and even the officers of the revenue, were leagued together in a system of pilfering valuables from the open barges. The lightermen claimed as their right the perquisites of "wastage" and "leakage," and they took care that these two items should include as much as possible. There were regular establishments on shore for receiving and disposing of the stolen merchandise.[1] The Thames Police was established, in 1798, for the purpose of checking this system of wholesale depredation ; but, so long as the goods were conveyed from the ship's side in open lighters, and the open quays formed the principal shore accommodation—sugar hogsheads, barrels, tubs, baskets, boxes, bales, and other packages, being piled up in confusion on every available foot of space—it was clear that mere police regulations would be unequal to meet the difficulty. It was also found that the confused manner in which the imports were brought ashore led

---

[1] Mr. Colquhoun, the excellent Police Magistrate, estimated that, in 1798, the depredations on the foreign and coasting trade amounted to the almost incredible sum of 506,000l., and on the West India trade to 232,000l.—together 738,000l.! He stated the number of depredators—including mates, inferior officers, crews, revenue-officers, watermen, lightermen, watchmen, &c.—to be 10,850 ; and the number of opulent and inferior receivers, dealers in old iron, small chandlers, publicans, &c., interested in the plunder, to be 550 ! Colquhoun's book, and its descriptions of the lumpers, scuffle-hunters, long-apron men, bumboat men and women, river-pirates, light-horsemen, and other characters who worked at the water-side, with their skilful appendages of jiggers, bladders with nozzles, pouches, bags, sacks, pockets, &c., form a picture of life on the Thames sixty years since worthy to rank with Mayhew's 'London Labour and London Poor' of the present day.

to a vast amount of smuggling, by which the honest merchant was placed at a disadvantage at the same time that the revenue was cheated. The Government, therefore, for the sake of its income, and the traders for the security of their merchandise, alike desired to provide an effectual remedy for these evils.

Mr. Rennie was consulted on the subject in 1798, and requested to devise a plan. Before that time various methods had been suggested, such as quays and warehouses, with jetties, along the river on both sides; but all these eventually gave place to that of floating docks or basins communicating with the river, surrounded with quays and warehouses, shut in by a lofty enclosure-wall, so that the whole of the contained vessels and their merchandise should be placed, as it were, under lock and key. By such a method it was believed the goods could be loaded and unloaded with the greatest economy and despatch, whilst the Customs duties would be levied with facility, at the same time that the property of the merchants was effectually protected against depredation. At the beginning of the century a small dock had existed on the Thames, called the Greenland Dock; but it was of very limited capacity, and only used by whaling vessels. Docks had existed at Liverpool for a considerable period, which had been greatly extended of recent years; so that there was no novelty in the idea of providing accommodation of a similar kind on the Thames, though it is certainly remarkable that, with the extraordinary trade of the metropolis, the expedient should not have been adopted at a much earlier period.

The first of the modern floating docks actually constructed on the Thames was the West India, occupying the isthmus that formerly connected the Isle of Dogs with Poplar, and of which Mr. Jessop [1] was the

---

[1] Mr. William Jessop, C.E., was among the most eminent engineers of his day. His father was engaged under Smeaton in the building of the Eddystone Lighthouse; and, dying in 1761, he left the guardianship of his

engineer. At the same period, in 1800, a company was formed by the London merchants for the purpose of constructing docks at a point as near the Exchange as might be practicable, for the accommodation of general merchandise, and of this scheme Mr. Rennie was appointed the engineer. He proposed several designs for consideration on a scale more or less extensive, adopting his usual course of submitting alternative plans, from which practical men might make a selection of the one most suitable for the purposes of their business; at the same time inviting suggestions, which he afterwards worked up into his more complete designs. As the future trade of London was an unknown quantity, he wisely provided for the extension of the docks as circumstances might afterwards require.

In carrying out the London Docks it was deemed advisable, in the first instance, to limit the access to the present Middle River Entrance at Bell Dock, 150 feet

---

family to that engineer, who adopted William as his pupil, and carefully brought him up to the same profes-

WILLIAM JESSOP, C.E.

sion. With Smeaton Jessop continued for ten years; and, after leaving him, he was engaged successively on the Aire and Calder, the Calder and Hebble, and the Trent Navigations.

He also executed the Cromford and the Nottingham Canals; the Loughborough and Leicester, and the Horncastle Navigations; but the most extensive and important of his works of this kind was the Grand Junction Canal, by which the whole of the north-western inland navigation of the kingdom was brought into direct connection with the metropolis. He was also employed as engineer for the Caledonian Canal, in which he was succeeded by Telford, who carried out the work. He was the engineer of the West India Docks (1800-2) and of the Bristol Docks (1803-8), both works of great importance. He was the first engineer who was employed to lay out and construct railroads as a branch of his profession; the Croydon and Merstham Railroad, worked by donkeys and mules, having been constructed by him as early as 1803. He also laid down short railways in connection with his canals in Derbyshire, Yorkshire, and Nottinghamshire. During the later years of his life he was much afflicted by paralysis, and died in 1814.

long and 40 feet wide, with the cill laid five feet below
low water of spring tides. The entrance lock communicated with a capacious entrance basin, called the
Wapping Basin, covering a space of three acres, and this
again with the great basin called the Western Dock,
1260 feet long and 960 feet wide, covering a surface of
20 acres. The bottom of the dock was laid 20 feet
below the level of high water of an 18 feet tide. The
quays next to the river were five feet above high water,
increasing to nine feet at the Great Dock. From the
east side of the latter, it was ultimately proposed to make
two or more docks, communicating with each other
and with a larger and deeper entrance lower down the
river at Shadwell; all of which works have since been
carried out.

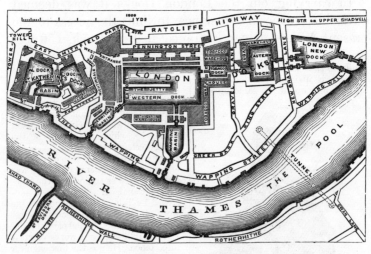

PLAN OF LONDON DOCKS.

As the site of the Docks was previously in a great
measure occupied by houses, considerable time necessarily elapsed before these could be purchased and
cleared away; so that the works were not commenced
until the spring of 1801, when two steam-engines were
erected, of 50 horse power each, for pumping the water,

and three minor engines for other purposes, such as grinding mortar, working the pile-engine, and landing materials from the jetty—an application of steam power as an economist of labour which Mr. Rennie was among the first to introduce in the execution of such works. The coffer-dam for the main entrance, and the excavation of the Docks, were begun in the spring of 1802;[1] after which time the works were carried forward with great vigour until their completion on the 30th of January, 1805, when they were opened with considerable ceremony.

At a subsequent period Mr. Rennie designed the present westernmost or Hermitage entrance lock and basin, the former of which is 150 feet long and 38 feet wide, with the cill laid two feet below low water of spring tides; the basin and main dock covering a surface of

---

[1] The locks were founded upon piles driven firmly into the soil, with rows of grooved and long-end sheeting piles in front of and behind the gates, in order to prevent the water from getting under them. The chamber between the lock-gates was formed by an inverted arch of masonry 2 feet 6 inches thick, strongly embedded in brickwork. The side-walls of the lock recesses and chamber were 7 feet thick, with strong counterfoots behind at the proper intervals. The whole of the locks and chambers were built of fine masonry, composed principally of hard blue sandstone from Dundee. All the retaining walls of the basins and docks were made curvilinear in the face, drawn from a radius of 80 feet, the centre being level with the top of the wall, and the bottom being inclined at the same angle as the radius. The wall was of a parallel thickness of 6 feet, except three or four footings at the back, where there were also counterfoots 3 feet 4 inches square, 15 feet asunder. The dock walls were founded generally upon a strong bed of gravel, which rendered piling unnecessary, and were built upon a flooring of beech and elm plank 6 inches thick. Under the front and back of this flooring ran a strong cill 12 inches square, to which the planks were firmly spiked. The walls and counterfoots were built of brickwork, the front, for 14 inches inwards, being formed of vitrified pavier bricks, and the remainder of good hard burned stock, the joints being a quarter of an inch thick in front and three-eighths thick at the back; the whole well bedded in excellent mortar made by a mill. There were two through or binding courses of stone, 14 inches thick in front, increasing in thickness backwards according to the radius of the front of the wall. The whole of the locks were furnished with cast iron turning or swivel-bridges erected across them. The works were generally done by contract, but the locks, which required greater care, accuracy, and completeness, were executed by daywork, under the engineer's immediate direction. [For further particulars as to these docks see Sir John Rennie's able work on 'British and Foreign Harbours;' Art. London Docks.]

one acre and a quarter. Another small dock of one acre
was afterwards added on the north-east side of the
Great Basin, exclusively devoted to the tobacco trade;
and it was ultimately extended to the Thames at Shadwell, as contemplated in the original design.

After the Docks had been opened for trade, Mr.
Rennie gave his careful attention to the working details,
and he was accustomed from time to time to make suggestions with a view to increased despatch and economy
in the conduct of the business. Thus, in 1808, he recommended that the whole of the lifting cranes in the
Docks should be worked by the power of a steam-engine
instead of by human or horse labour. He estimated
that the saving thus effected, in the case of only twenty-
six cranes, would amount to at least 1500*l.* a year,
besides ensuring greater regularity and despatch of
work; and, if applied to the whole of the cranes along
the Docks and in the warehouses, a much greater annual
saving might be anticipated. It was, however, regarded
as too bold an innovation for the time; and we believe
the suggestion has not been carried out to this day,
the cranes in the London Docks being still worked
by hand labour, at a great waste of time and money,
as well as loss of business. Another of Mr. Rennie's
valuable suggestions, with a view to greater economy,
was the adoption of tramways all round the quays, provided with trucks, by means of which the transfer of
goods from one part of the Dock to another might be
effected with the greatest ease and in the least possible
time. But this too was disregarded. Labour-saving
processes were not then valued as they now are. The
application and uses of machinery were as yet imperfectly understood, and there were in most quarters
powerful prejudices to be overcome before it could be
introduced. To this day the goods in the London Docks
are hauled in trollies, waggons, or hand-barrows from
ship to ship, or from the vessels to the respective

bonded warehouses; and it still remains matter of surprise that a system so clumsy, so wasteful of time, so obstructive to rapid loading and unloading in dock, should be permitted to continue.

Shortly after these works were set on foot, and when the great importance and economy of floating docks began to be recognised by commercial men, another project of a similar character was started, to provide accommodation exclusively for vessels of the East India Company, of from 1000 to 1800 tons burden. A company was formed for the purpose, and an Act was obtained in 1803, the site selected being immediately to the west of the river Lea, at the point at which it enters the Thames, and where at that time there were two small floating basins or docks, provided with wooden locks, and surrounded with wooden walls, called the Brunswick and Perry's Docks. These it was determined to purchase and include in the proposed new docks, of which, however, they formed but a small part. Mr. Rennie and Mr. Ralph Walker were associated as engineers in carrying the works into execution, and they were finished and opened for business on the 4th of August, 1806. They consisted of an entrance lock into the Thames 210 feet long and 47 feet wide, with the cill laid 7 feet below low water of spring tides. This lock is connected with a triangular entrance basin, covering a space of 4½ acres, on the west side of which it communicates by a lock with a dock expressly provided for vessels outward bound, called the Export Dock, 760 feet long and 463 feet wide, covering a surface of 8⅓ acres. At the north end of the entrance basin is the Import Dock, 1410 feet long and 463 feet wide, covering a surface of 18⅔ acres. The depth of these basins is 22 feet below high water of ordinary spring tides. The total surface of dock room, including quays, sheds, and warehouses, is about 55 acres. The original capital of the East India Dock Company was 660,000*l*.; but

Mr. Rennie constructed and completed the Docks for a sum considerably within that amount. Eventually they were united to the West India Docks, under the joint directorate of the East and West India Dock

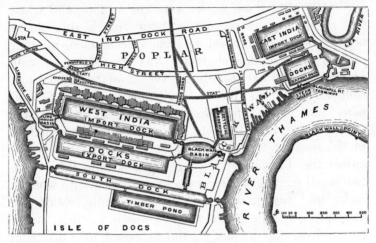

PLAN OF EAST AND WEST INDIA DOCKS.

Company.[1] Mr. Rennie also introduced into these Docks many improved methods of working; his machinery, invented by him for transporting immense blocks of mahogany by a system of railways and locomotive cranes, having, in the first six months, effected a saving in men's wages more than sufficient to defray their entire original cost, besides the increased expedition in the conduct of the whole Dock business.

---

[1] Among the improvements adopted by Mr. Rennie in these docks may be mentioned the employment of cast iron, then an altogether novel expedient, for the roofing of the sheds. One of these, erected by him in 1813, was 1300 feet long and 29 feet 6 inches in span, supported on cast iron columns $7\frac{1}{2}$ inches in diameter at bottom and $5\frac{3}{4}$ at top. Another, still more capacious, of 54 feet clear span between the supports, was erected by him over the mahogany warehouses in 1817. He also introduced an entirely new description of iron cranes, first employing wheel-work in connection with them, by which they worked much more easily and at a great increase of power. He entirely re-arranged the working of the mahogany sheds, greatly to the despatch of business and the economy of labour. His quick observation enabled him to point out new and improved methods of despatching work, even to those who were daily occupied in the docks, but whose eyes had probably become familiar with their hurry-scurry and confusion.

[Rennie's harbor works at other places were also of considerable size and significance. When advising projectors of harbor works he boldly anticipated increases in commerce and traffic. About 1793 he reported on the best means of improving the harbor of Wick, and on improving the navigation of the Clyde to accommodate the rapidly increasing trade of Glasgow. Another harbor on which Rennie was employed was Holyhead, a port of embarkation from Wales to Ireland. After the Act of Union was passed in 1801, the Government was determined to improve communications with Ireland. As Smiles observed, "everything had as yet been left to nature (at Holyhead), which had only provided plenty of deep water and many bold rocks."

Altogether, Rennie examined, reported upon, or improved more than forty harbors in England, Scotland, and Ireland between 1793 and 1821. In these harbors, as in all his engineering works, Rennie drew upon principles evolved by him over his years of experience. He would advise or countenance only works suited to the expanding economy and to withstanding the most rigorous tests of time. He would never in the name of economy reduce the strength of his piers and retaining walls below the limits he thought essential for stability. He took particular care to design harbors which would not only keep out the sea but also provide the easiest possible access to ships in stormy weather. This necessitated intelligent analyses of natural conditions and expenditures which some authorities were not willing to make. His reports and plans, therefore, were often not carried out.

The Napoleonic Wars caused the Government to heed the state of coastal defenses and the efficiency of naval installations. Because of his eminently practical abilities, Rennie was consulted. He planned a series of dams whose waters could have been released to inundate an army approaching London by the valley of the Lea (five of the dams were built), and he improved the defenses of coastal Kent by constructing the Hythe Military Canal.

Rennie also served on a commission of the House of Lords appointed to investigate the merits of a proposal made by Robert Fulton. The American, subsequently famous for his steamboat, proposed to blow up ships with his "torpedo." An old Danish brig was placed at his disposal for a test and "he succeeded, after an unresisted attack of two days . . . in blowing up the wretched carcass, and with it his own pretensions as an inventor." * Rennie had a "very mean opinion of Fulton, regarding him as a quack who traded upon the inventions of others."

Because some private manufacturing establishments had been organized more rationally than Admiralty shipyards, Rennie was also engaged by the Government — a Commission of Civil Officers of the Navy — to consider better means of turning out work with dispatch and economy. In his report of 14 May, 1807 he recommended the application of steam power for many purposes in the yards and that the yards be laid out to allow work of the same kind to be carried on by continuous operations. He planned and superintended improvements in facilities at the royal dockyard at Sheerness.

The Lords of the Admiralty called upon Rennie to report on improving the outer harbor of Plymouth. This haven lay open to the south and the fury of gales blowing from that quarter during the equinoxes. His report of 22 April, 1806 boldly recommended an extensive breakwater of a total length of 5,100 feet. Construction presented many problems. Rennie was appointed engineer-in-chief when the powers to proceed were given. When the work was completed in 1848 after his death, total rubble deposited was 3,670,444 tons, besides 22,149 cubic yards of masonry. This amount of material was at least equal to that in the Great Pyramid.

Another monument to Rennie's engineering genius is the Bell Rock Lighthouse, eleven miles off the mainland of Scotland near the entrance to the Firth of Forth. In 1805 Rennie reported on the best structure to erect on the ledge of rocks to the Commissioners of Northern Lights and their engineer, Robert Stevenson. In 1806 Rennie was named chief engineer for conducting the work, and Stevenson resident engineer. The edifice of stone, completed in 1810, resembled John Smeaton's renowned Eddystone Lighthouse, and Rennie acknowledged his debt to that great civil engineer. He believed, however, that some contemporary accounts of the project exaggerated the contribution of his assistant, Stevenson, in the construction of the Bell Rock.] *

CHAPTER VII.

On undertakings such as these, of great magnitude and importance, was Mr. Rennie engaged until the close of his useful and laborious life. There was scarcely a project of any large public work on which he was not consulted; sometimes furnishing the plans, and at other times revising the designs of others which were submitted to him. Numerous works of minor importance also occupied much of his attention, as is shown by the extent of his correspondence and the number of his reports, which contain an almost complete repository of engineering practice. Whilst he was engaged in designing and superintending the construction of his great London Bridges, the formation of Plymouth Breakwater, the building of the docks at Sheerness, the cutting of the Crinan Canal, and the drainage of the Fens by the completion of the Eau Brink Cut, he was at the same time consulted as to many important schemes for the supply of large towns with water. His report on the distribution of the water supplied by the York Buildings Company in the Strand—in which he proposed for the first time to appropriate a distinct service to the several quarters of the district supplied—was a masterpiece in its way; and the principles he then laid down have been generally followed by subsequent engineers. He also reported on the improved water-supply of Manchester, Edinburgh, Bristol, Leeds, Doncaster, Greenwich and Deptford, and many other large towns in England

and Scotland, as well as in the colonies and in foreign countries.[1]

In addition to the various mills and manufactories fitted up by him with new and improved machinery, we may mention that he advised the Bank of England on the subject of the more rapid manufacture of bank notes by the employment of the steam-engine; and he entirely re-arranged the Government machinery at Waltham for the better manufacture of gunpowder. He erected the anchor-forge at Woolwich Dockyard, considered to be the most splendid piece of machinery in its day; he supplied Baron Fagel (then Dutch minister in this country) with designs of dredging-engines for clearing the mud out of the rivers and canals of Holland; and he designed and constructed the celebrated machinery for making ropes according to Captain Huddart's patent.[2]

---

[1] In 1817, his fame having gone abroad as the most skilled water engineer of the day, Captain Dufour, of Geneva, came to England for the purpose of consulting him as to the extension and improvement of the waterworks of that city. Captain Dufour was introduced to Mr. Rennie by the mutual friend of both, the eminent Dr. Wollaston. Mr. Rennie made a careful and detailed report on the surveys and plans submitted to him, especially on the engine and pumping machinery of the proposed works; and his advice was followed, very much to the advantage of the citizens of Geneva.

[2] Captain Joseph Huddart, F.R.S., was a singularly estimable character. He was a man of great nautical experience, sound judgment, and excellent skill as a mechanic and engineer, and was often consulted by Mr. Rennie in reference to marine works of more than ordinary importance. His origin was humble, like that of so many of the early engineers; and, like them also, he was drawn to the pursuit by the force of his genius, rather than by the peculiar direction of his education. He was born at Allonby, in Cumber-

land, in 1740, the son of a shoemaker and small farmer. From his mother he inherited a determined spirit and a vigorous constitution, combined with sound moral principles, which he

CAPTAIN JOSEPH HUDDART, F.R.S.

nobly illustrated in his life. He received an ordinary share of education at the common school of his village, to which he added a knowledge of

In his capacity of advising engineer to the Admiralty, Mr. Rennie embraced every opportunity which his posi-

mathematics and astronomy, obtained from the son of his teacher, who had studied these branches at Glasgow University. He seems to have been an indefatigable learner, for he also acquired some knowledge of music from an itinerant music-teacher, whom he very shortly outstripped. His mechanical tastes early displayed themselves. Watching some mill-wrights employed in constructing a flour-mill, he copied the machinery which they erected, in a model which he finished as they completed their mill. He also made a model of a 74-gun ship, after the drawings given in Mungo Murray's 'Treatise on Navigation and Shipbuilding,' which he was so fortunate as to fall in with. At an early age he was employed in herding his father's cows on a hill-side overlooking the Solway Frith, and commanding a view of the coast of Scotland. There he took his books, with a desk of his own making, and while not forgetting the cattle, employed himself in reading, drawing, and mathematical studies. When a little older, his father set him on the cobbler's stool, and taught him shoe-making, though the boy's strong inclination was to be a sailor. But large shoals of herrings making their appearance about this time in the Solway Frith, a small fishing company was started by the Allonby people, in which his father had a share, and young Huddart was sent out with the boats, very much to his delight. He now began to study navigation, carrying on shoemaking in the winter and herring-fishing at the time the shoals were on the coast. On the death of his father, he succeeded to his share in the fishery, and took the command of a sloop employed in carrying the herrings to Ireland for sale. During his voyages he applied himself to chart-making, and his chart of St. George's Channel, which he afterwards published, is still one of the best. The herrings having left the frith, Huddart got the

command of a brig, his excellent character securing him the post, and he made a successful voyage in her to North America and back. His progress was steady and certain. A few years later we find him in command of an East Indiaman. After many successful voyages, in which he happily brought all his ships to port, and never met with any serious disaster, he retired from the service; having been in command of a ship of greater or less burden for a period of twenty-five years. He now published many of his charts, the results of the observations he had made during his numerous voyages. His eminent character, not less than his known scientific knowledge, secured his introduction to the Trinity House as an Elder Brother, and to the direction of the London and East India Docks, in which situations he was eminently useful. The lighting of the coast proceeded chiefly under his direction, and many new lighthouses were erected and floating-lights placed at various points at his recommendation. Among others, he superintended the construction of the lighthouse at Hurst Point. He also surveyed the harbours of White-haven, Boston, Hull, Swansea, St. Agnes, Leith, Holyhead, Woolwich Dockyard, and Sheerness; several of these in conjunction with his friend Mr. Rennie, who was always glad to have the benefit of his excellent judgment. He made many improvements in ship-building; but the invention for which the nation is principally indebted to him is his celebrated rope-making machinery, by which every part of a cable is made to bear an equal strain, greatly to the improvement of its strength and wearing qualities. This machinery, constructed for him by Mr. Rennie at Limehouse, was among the most perfect things of the kind ever put together. Captain Huddart died at his house in High-bury Terrace, London, in 1816, closing a life of unblemished integrity in the seventy-fifth year of his age.

tion afforded him of recommending the employment of
steam power in the Royal Navy. His advice met with
the usual reception from the inert official mind : first
indifference; next passive resistance; then active opposition when he pressed the matter further. Naval officers,
who had grown old in sailing tactics, could ill brook the
idea of navigating ships of war by a mechanical invention like the steam-engine, by which skill in seamanship,
of which the old salts were so proud, would be entirely
superseded. The navy had done well enough heretofore
without steam; why introduce it now? It was a smoky
innovation, and if permitted, would only render ships
liable to the constant risk of being blown up by boiler explosions. Lord Melville, however, listened to Mr. Rennie's
suggestions, and at length consented to the employment
of a small steam-vessel as a tug for a ship of war, by way
of experiment. Mr. Rennie accordingly hired the Mar-
gate steamboat, *Eclipse*, to tow the *Hastings*, 74, from
Woolwich to two miles below Gravesend, against a rising
tide. The experiment was made on the 4th of June,
1819, and proved so successful that the Admiralty were
induced to authorise a steamboat to be specially built at
Woolwich for similar service. This vessel was named
the *Comet*; it was built after the designs and under the
direction of the late Mr. Oliver Lang, assisted to a considerable extent by Mr. Rennie, who attended more particularly to the designing and fitting of the engines, which
were made by Boulton and Watt. The *Comet*, though a
small vessel, was the parent of other royal ships of vastly
greater dimensions. She was only 120 feet long between the
perpendiculars, and 22 feet 6 inches in extreme breadth;
the draught of water was about $6\frac{1}{2}$ feet, and the power of
her engines about 40 horses. The Admiralty had great
doubts as to the width of the paddle-boxes; but Mr.
Rennie encouraged them to make the experiment after
his design. " Steam-vessels," he observed, " are as yet
only in their infancy, and can scarcely be expected to

have arrived at anything approaching perfection. Much, however, will be learnt by experience; but unless some risk is run in the early application of the new power, no improvements are likely to be made." [1] For her size, the *Comet* proved a most efficient vessel—the best that had up to that time been constructed. She fully answered the purpose for which she was intended; and the result was so satisfactory, that vessels of increased size and power were from time to time built, until the prejudice amongst naval men against the employment of steam power having been got over, it was at length generally introduced in the Royal Navy. [2]

The last of Mr. Rennie's great designs was that of New London Bridge, which, however, he did not live to complete. The old bridge had been gradually falling into decay, and was felt to be an increasing obstacle to the navigation of the river. The starlings which protected the piers had of late years been seriously battered by the passing of hoys, barges, and lighters, on which they had inflicted equal injury in return; for vessels were constantly foundering on them, and many were sunk and their cargoes damaged or destroyed. Emptying stones into the river, to protect the decayed pile-work, had only

---

[1] Letter to the Admiralty, 22nd May, 1820.

[2] Mr. Rennie was engaged for many years in urging the introduction of steam power in the Royal Navy. In 1817 we find him writing to Lord Melville, Sir J. Yorke, Sir D. Milne, and others on the subject. It would appear that Lord Melville had declared that he was determined to employ steam-vessels as tugs, so soon as he could convince the Sea Lords of their advantages; on which Mr. Rennie compliments Sir D. Milne, saying that he is "glad to find that there is one admiral in the navy favourable to steamboats." In July, 1818, he laments that he cannot convince Sir G. Hope or Mr. Secretary Yorke of their utility, but

that he is persuaded their adoption *must* come at last. On the 30th May, 1820, he writes James Watt, of Birmingham, informing him that the Admiralty had at last decided upon having a steamboat, notwithstanding the strong resistance of the Navy Board. "My reasons," he says, "I understand, were satisfactory; but unless the Admiralty cram it down the throats of the Navy Board, nothing will be done; for of all the ignorant, obstinate, and stupid boards under the Crown, the Navy Board is the worst. I am so disgusted with them that, could I at the present moment with decency relinquish the works under them which I have in hand, I would do so at once."

had the effect of further obstructing the navigation, the scour of the current having formed two great banks of stone across the whole bed of the river : one about 100 feet below the bridge, and the other about the same distance above it, with two deep hollows between them and the piers, from 25 to 33 feet deep at low water. The piers and arches were also becoming decrepit. Though the top-hamper of houses had long been removed, and the piers patched and strengthened at various times, the bridge was every year becoming less and less adapted for accommodating the increasing traffic to and from the City. At last it was regarded as a standing nuisance, and generally condemned as a disgrace to the capital. To maintain the structure, inefficient and unsafe though it was, cost the City not less, on an average, than 3500*l.* a year, and the expense was likely to increase. The Corporation felt that they could no longer avoid dealing decisively with the subject. They then resolved to take the opinion of the best engineers and architects; and Mr. Daw, the architect of the Corporation, Mr. Chapman, the engineer, and Messrs. Alexander and Montague, two eminent City architects, were consulted as to the best steps to be taken under the circumstances. The result of their deliberations was a recommendation to the Corporation to remove eight of the arches and to substitute four larger ones, as well as to make extensive repairs in the remaining arches, piers, and superstructure. Their plan was referred to Mr. Rennie, Mr. Chapman, and Messrs. Montague, for further consideration; and, as was Mr. Rennie's custom before making his report, he proceeded to master the whole of the facts, on which alone a sound opinion could be formed. He had the tides and currents watched and noted, and the river carefully sounded above and below bridge, from Teddington Lock to the Hermitage entrance of the London Docks. He examined the piers down to their foundations, and explored the bottom of the river, making borings at various points between the one shore and the other. In the report which he made to the Corporation on the 12th of March, 1821, a great deal of new and accurate information was first brought to light respecting the flow of the tide through the arches, and the additional depth of water likely to be secured by their removal. Although it was pronounced quite practicable to carry out the alterations which had been recommended, and the erection of four new arches in lieu of the eight old ones, he was of opinion that the cost would be very considerable; and, after all, the old foundations would still present great defects, which could never be wholly cured. Mr. Rennie therefore suggested the propriety of building an entirely new bridge of five arches, with a lineal waterway of not less than 690 feet, in lieu of the then waterway of 231 feet, below the top level of the starlings, and 524 feet above them. Besides the greatly increased accommodation which would be provided by such a structure for the large traffic passing between London and Southwark, Mr. Rennie held that not the least advantage which it promised was the much greater facility which it would afford for the navigation of the river to and from the wharves above bridge; for coasters and even colliers, with striking masts, might then be enabled to navigate the whole extent of the City westwards. The increased waterway would also enable the waters descending from the interior to flow more readily away, floods often inflicting great damage along the shore, especially in the winter months, when the arches of the old bridge became choked up with ice.

The report was felt to be almost conclusive on the subject; and the more it was discussed the deeper grew the conviction in the minds of all that its recommendations ought to be adopted. The Corporation accordingly applied to Parliament, in the year 1821, for an Act enabling them to purchase the waterworks under the

arches of the old bridge, and to erect an entirely new
structure. The bill, of course, had its opponents; some
arguing that there was no necessity for a new bridge,
and that its erection would be only a useless waste of
money, whilst the old one could be repaired and made
fit for traffic at so much less outlay. The case in favour
of the new bridge was, however, too strong to be re-
sisted, and Mr. Rennie's evidence was considered so
clear and conclusive that committees of both Houses
unanimously approved the bill, and it duly received
the sanction of Parliament. Power was conferred by
the Act enabling the Treasury to advance from the
Consolidated Fund such sums as might be necessary
to supply any deficiency in the funds at the disposal
of the Corporation applicable to the erection of the
bridge; the Government regarding the work as one of
national importance, and consequently entitled to public
assistance.

During the progress of the bill through Parliament,
Mr. Rennie prepared the general outlines of a design of
the new structure. It consisted of five semi-elliptical
arches, the centre one 150 feet span, the two side arches
140 feet, and the two land arches 130 feet, making a
total lineal waterway of 690 feet; the height of the
soffit or under-side of the centre arch being 29 feet
6 inches above the level of Trinity high-watermark.
The general principle of this design was approved and
embodied in the bill. Very shortly after the Act had
passed, Mr. Rennie was seized by the illness which
carried him off, and it was accordingly left to others
to execute the great work which he had thus planned.
The Corporation of London then appealed to the
whole engineering and architectural world for com-
petitive designs, and at least thirty were prepared in
answer to their call. These were submitted to a Committee of the House of Commons in the year 1823, and
after long consideration the plan originally proposed

by Mr. Rennie was finally adopted; on which the Corporation of London selected his son, the present Sir John Rennie, engineer in chief, to carry it into effect; and the nomination having been approved by the Lords of the Treasury, in conformity with the pro-

NEW LONDON BRIDGE. [By Percival Skelton.]

vision of the Act, steps were forthwith taken to proceed with the work.* It is scarcely necessary to say how admirably Mr. Rennie's noble design has been executed. New London Bridge, in severe simplicity and unadorned elegance of design—in massive solidity, strength, and perfection of workmanship in all its parts—not less than as regards the capacious size of its arches and the breadth and width of its roadway and approaches—is perhaps the finest work of its kind in the world.[1]

It will be observed from the preceding chapters, that Mr. Rennie's life was one of constant employment, and that, apart from his great engineering works, his career contains but few elements of biographic interest. Indeed his works constitute his biography, overlaying, as they did, almost his entire life, and occupying nearly the whole of his available time. His personal wants were few; his habits regular; and his pleasures of the most moderate sort, consisting chiefly in reading and in the enjoyment of domestic life. At the age of twenty-nine he married Miss Mackintosh, an Inverness lady, who made his home happy; and he became the father of nine children, six of whom survived him. In the early part of his career in London he lived in the Great Surrey

---

[1] The new bridge was erected about thirty yards higher up the river than the old one, and involved the construction of new approaches on both sides. The first coffer-dam was put in on the Southwark side, and the first pile was driven on the 15th of March, 1824; the foundation stone was laid with great ceremony by H. R. H. the Duke of York, on the 15th of June, 1825, assisted by the Lord Mayor (Garrett), the Aldermen, and Common Council. The bridge was finally completed and formally opened by His Majesty King William the Fourth on the 1st of August, 1831— the time occupied in its construction having thus been seven years and three months. The total cost of the bridge and approaches was about two millions sterling. All the masonry below low water is composed of hard sandstone grit, from Bramley Fall, near Leeds; and the whole of the exterior masonry above low water is of the finest hard gray granite, from Aberdeen, Devonshire, and Cornwall. The actual width of the arches as executed is as follows: the centre arch is 152 feet 6 inches span; the two arches next the centre are 140 feet; and the two land arches 130 feet. The details of construction of the coffer-dams, piers, and floating and fixing the centres, were similar to those adopted by Mr. Rennie in building Waterloo and Southwark bridges. The total length of the bridge is 1005 feet; width from outside to outside, 56 feet; width of the footpaths, 88 feet; and of the carriageway, 35 feet. The total quantity of stone built into the bridge is 120,000 tons. The builders were Messrs. Joliffe and Banks, the greatest contractors of their day.

SECTION OF NEW LONDON BRIDGE.

Road, from which he afterwards removed to Stamford Street, not far from his works.

His close and often unremitting application early began to tell upon his health. In 1812, when arrived at the age of fifty-one, he was occasionally laid up by illness. While occupied one day in inspecting the works of Waterloo Bridge, he accidentally set his foot upon a loose plank, which tilted up, and he fell into the water, but happily escaped with only a damaged knee. Though unable for some time to stir abroad, he seized the opportunity of proceeding with the preparation of numerous reports, and of working up a long lee-way of correspondence. In the following year he was frequently confined to the house by a supposed liver-complaint; but his correspondence never flagged. He tried the effects of change of air at Cheltenham; but he had no time for repose, and after the lapse of only a week he was again in harness, giving evidence before a Committee of the House of Commons on Lough Erne drainage. He made another hurried visit to Cheltenham, but evidently took no rest; his absence from active business only affording him an opportunity for writing numerous letters to influential persons at the Admiralty on the subject of his grand scheme of the Northfleet Docks. To one of his correspondents we find him saying he was "better, though only half a man yet." In course of time, however, he partially recovered, and was forthwith immersed in business—engaged upon his docks, bridges, and breakwaters.

He very rarely "took play." In 1815 his venerable friend James Watt, of Birmingham, urged him to pay a visit with him to Paris, shortly after the battle of Waterloo. But Mr. Rennie was too full of work at the time to accept the invitation, and the visit was postponed until the following year, when he was accompanied by James Watt, jun., then of Aston Hall, near Birmingham. This journey was the first relaxation he had taken for a period of thirty years; yet it was not a mere holiday trip, but partly one of business, for it was his object to inspect with his own eyes the great dock and harbour works executed by Napoleon during the Continental war, of which he had heard so much, and to gather from the inspection such experience as might be of use to him in the improvement of the English dockyards on which he was then engaged. The two set out in September, 1816, passing by Dover to Calais and thence to Dunkirk, where Mr. Rennie carefully examined the jetties, arsenal, docks, and building-slips at that port. From thence they proceeded to Ostend, and afterwards to Antwerp, where our engineer admired the great skill and judgment with which the dock works there, still incomplete, had been laid out. From Antwerp they went to Paris, where they stayed only two days, and then to L'Orient and Brest, accompanied by Mr. Joliffe, the contractor. At both these ports Mr. Rennie took careful note of the depths, dimensions, and arrangements of the harbours in detail, receiving every attention from the authorities. At Cherbourg, in like manner, he examined the building-yards and docks, as well as the progress made with the famous Digue [1]—

---

[1] The Digue is of considerably greater extent than the breakwater at Plymouth, being above 2¼ English miles long. Up to the time of Mr. Rennie's visit, the work had been a series of attempts and failures, which, however, eventually produced experience, and led to success. Wooden cones filled with small stones were first tried; they were sunk so as to form a sea-rampart; but the cones were shattered to pieces by the force of the waves, and the stones were scattered about in the bottom of the sea. Then loose rubble-stones were tried; but the blocks were too small, and these, too, were driven asunder. Larger blocks were then used; but, for a time, the smaller stones beneath acted as rollers to the larger ones. At length, however, these found their bearing, and when Mr. Rennie visited the place, the slope formed by the sea-ridge of rubble was as much as 11 to 1. This greatly increased the contents of the breakwater, while its stability was not much to be depended on. Many accidents occurred to the work, and several extensive breaches were made through it by the force of the sea. At low water the height of the Digue was at some parts only three feet; at others, considerably more; whereas, in some places, the top of the work was from seven to eight feet below low water of spring tides. At length, after many years' labour and vast expense, the work has been brought to completion; and it now forms a very excellent defence for the fine war roadstead and arsenal of Cherbourg, greatly exceeding the

a rival to his own Breakwater at Plymouth. At Cherbourg he was joined by Mr. Whidbey, who had come over in a vessel of war to meet him. Mr. Rennie returned to England after less than a month's tour; and though he had made a labour of his pleasure-trip, the change of air and scene did him good, and he entered with zest upon the business of clearing off the formidable arrears of work which had accumulated during his absence. We may add, as an illustration of his habit of turning everything, even his pleasure, to account, that one of the first things he did on his return home was to make an elaborate report to Lord Melville, then First Lord of the Admiralty, of the results of his investigations of those foreign harbours which he had visited in the course of his journey.

After a few years' more devoted application, his health again began to give way. When consulted by Mr. Foljambe of Wakefield, in June, 1820, respecting a railway proposed to be laid down in that neighbourhood, he excused himself from entering upon the business because his hands were so full of work and his health was so delicate. Shortly after, we find him writing to a friend that the new works executed by him during the past year had cost about half a million, besides those in progress at Sheerness, which would cost a million. He was then busy with his investigations relative to New London Bridge, the report on which he prepared whilst laid up with gout. He persisted in going abroad as long as he could, and went to his doctor in a carriage for advice, instead of letting the latter come to him. But resistance, however brave, was useless against disease, and at length he was compelled to succumb. To the last he went on issuing instructions to his inspectors in different parts of the country relative to the works

humble dimensions which it presented when Mr. Rennie visited the place. The whole cost of the works amounted to upwards of seven millions sterling.

then in progress—the docks at Chatham and Sheerness, the harbours at Howth and Kingstown, the bridge at New Galloway, the Eau Brink Cut, the Aire and Calder Navigation, and the pumping-engines for Bottesham and Swaffham, in the Fens. He was especially anxious about the Eau Brink Cut, nearly ready for opening, urging his assistants to report from time to time, giving him full particulars of the progress made. In the midst of all this, he writes a letter to a harbour-master at Bridlington, giving him detailed instructions as to the arrangement of tide tables! His last business letter was written to the Navy Board respecting the proper kind of gates to be used for the dry dock at Pembroke : it was dated the 28th September, 1821. A little before this he had written to his friend Mr. Jerdan, the Edinburgh engineer, that he had completed all business connected with his preparation for the next session of Parliament, when he had many bills to carry through. But how often are the intentions of the bravest defeated! Day by day he grew weaker, struggling with the whole force of his will against the disease that was slowly mastering him. Although extremely ill, he insisted on rising from his bed, and tottered about, even taking an occasional airing in a carriage. In this state he continued until the 4th of October. On that day he did not get up. His mind had until then been as clear and vigorous as ever ; but now it began to wander. There was no resisting the hand of death, which was already upon him. He took no further heed of what passed around him, and about five in the evening a violent fit occurred, from which he never rallied. About an hour later he expired, in the sixty-first year of his age.

The portrait prefixed to this memoir expresses, so far as an accurate delineation of his features can do, the actual character of the man. It is grave and thoughtful, yet has an expression of mildness perfectly in unison with his gentle yet cheerful disposition. Raeburn painted his portrait, and Chantrey chiselled his bust; but the chalk drawing by Archibald Skirving,[1] after which our

---

[1] Archibald Skirving, like John Rennie, was the son of an East Lothian farmer. He was born in 1749, at Garleton, a farm belonging to the Earl of Wemyss. His father, Adam Skirving, was a well-known humorist and ballad-maker—one of his songs, ' Hey, Johnny Cope,' a description of the rout of the royal army at the battle of Prestonpans, being still popular in Scotland. Its publication gave great offence to some of Cope's officers, and one of them, Lieutenant Smith, went so far as to send Skirving a challenge, dated the George Inn, Haddington. When the messenger arrived with the missive, the farmer was in his yard, turning over manure. After reading the letter, he said, " Ye may gang back to Lieutenant Smith, and say to him, if he likes to come up-by here, I'll tak' a look at him; if I've a mind to fecht him, I'll fecht him; and if no, I'll do as he did—I'll rin awa' !" Many similar stories are told of the farmer's wit and humour, a considerable share of which was inherited by his son Archibald. In early life the latter went to Rome to study art, and remained in Italy nine years. He walked the whole way back from Rome, but, while passing through France, the revolutionary war broke out, and he was apprehended and thrown into prison, where he lay for nine months. He subsequently studied painting under David. Returned to Scotland, he pursued his art in a somewhat desultory manner, not being under the necessity of applying himself to it with that patient and continuous devotion which is essential to attaining high eminence in any profession. He painted when, where, and whom he pleased; and sometimes pursued a very eccentric course with his sitters. One gentleman's portrait he painted in such a manner as to give special prominence to a large wart upon his face. A lady who insisted on sitting to him, he put off with the ungallant remark that she " would ruin him for yellow."

Notwithstanding his eccentricity, Skirving was an extremely clever artist, and his crayon drawings have rarely been surpassed for vigour and brilliancy. He executed probably the best head of Burns, with whom he was intimate; and the portrait of John Rennie, which Mr. Holl has rendered with great skill, will give a good idea of Skirving's power as a delineator of character. Skirving and Rennie were intimate friends, although in most respects so unlike each other. Yet Skirving had as true a genius in him, and might have secured as great a reputation in his own walk, as his friend Rennie, had he worked as patiently and industriously. As he grew older, he became more eccentric and sarcastic. He dressed oddly, in a broad-brimmed white hat, without any neckcloth. He was at the Earl of Wemyss's house at Gosford one day, when the Countess was conversing with him as to the acquirements of her daughters in art. The young ladies were meanwhile occupied in making grimaces at the odd man behind his back, forgetting that they were standing opposite a mirror, in which he could see all their movements. " The young ladies," observed the painter, " may have studied art, but I never saw such ugly faces as those they make," pointing to the glass before him. Allan Cunningham relates the story of Skirving's calling on Chantrey while he was finishing the bust of Bird, the artist. " Well!— and who is that?" asked Skirving. " Bird, the eminent painter." " Painter ! — and what does he paint?" " Ludicrous subjects, Sir." " Ludicrous subjects ! — Have *you* sat?" " Yes—he has had one sitting; but when he heard that a gentleman with a white hat, who wore no neckcloth, had arrived from the North, he said, ' Go—go ; I know of a subject more ludicrous still: Mr. Skirving is come!'" This odd, but clever artist died at Inveresk, near Edinburgh, in 1819, at the advanced age of seventy.

engraving is made, is on the whole the most lifelike representation of the man as he lived. In person he was large, tall, and commanding; and strength was one of the attributes belonging to his family. But physical endurance has its limits, and we fear that Mr. Rennie taxed his powers beyond what they would fairly bear. He may be said to have died in harness, in the height of his fame, after threescore years, forty of which had been spent in hard work; still his death was premature, and in the case of a man of such useful gifts, was much to be lamented. But he himself held that life was made for work, and he could never bear to be idle. Work was with him not only a pleasure,—it was almost a passion. He sometimes made business appointments at as early an hour as five in the morning, and would continue incessantly occupied until late at night. It is clear that the most vigorous constitution could not long have borne up under such a tear and wear of vital energy as this.

He was very orderly, punctual, and systematic, and hence was enabled to get through a very large amount of business. No matter how numerous were the claims upon his time, nothing was neglected nor hurried. His reports were models of what such documents should be. They set forth all the facts bearing upon the topic under consideration in great detail; but with much plainness, force, and clearness. His harbour reports were especially masterly; in them he elaborately stated all the known facts as to the prevailing winds, currents, and tides, usually drawing very logical and conclusive inferences as to the particular plan which, under the circumstances, he considered it the most desirable to adopt. In his estimates he was careful to conceal nothing, stating the full sum which in his judgment the work under consideration would cost; nor would he understate the amount by one farthing in order to tempt projectors to begin any undertaking on which he was consulted.

He took the highest ground in his dealings with contractors. He held that the engineer was precluded by his position from mixing himself up with their business, and that if he dabbled in shares or contracts, either openly or underhand, half his moral influence was gone, and his character liable to be seriously compromised. Writing to Playfair at Edinburgh, in 1816, he said— " Engineers should be entirely independent of these connexions—not dabblers in shares—and free alike of contractors and contracts." By holding scrupulously to this course, Mr. Rennie established a reputation for truthfulness, honesty, and uprightness, not less honourable and exalted than his genius as an architect and engineer was illustrious.

He was a man of powerful and equally balanced mind— not so clever, as profound; not brilliant, but calm, serene, and solid, like one of his own structures.* While he lay on his deathbed, his last letters to his assistants urged upon them attention, punctuality, and despatch—qualities which he himself had illustrated so well in his own life. In his self-education he had overlooked no branch of science cultivated in his day; and in those which bore more especially upon his own calling, his knowledge was well-arranged, complete, and accurate.

Withal he was an exceedingly modest, unpretending, and retiring man. His great aim was to do the thing he was appointed to do in the best possible manner. He thought little of fame, but a great deal of character and duty. If his time was so entirely pre-occupied that he could not personally devote the requisite attention to any new undertaking brought before him, he would decline to enter upon it, and recommended the employment of some other leading engineer. He considered it his duty himself to go into the minutest details of every business on which he was consulted. He left as little as possible to subordinates, making his calculations and estimates himself; and he wrote and even copied his own reports;

deeming no point, however apparently unimportant, beneath his careful attention and study.

Hence great reliance was placed upon his judgment by those who consulted him ; and the accurate though comparatively reserved manner in which he expressed himself before Committees of Parliament gave all the greater weight to his evidence. " What I liked about Rennie," says one who knew him well, " was his severe truthfulness." When under examination on such occasions, he could always give a strong, clear reason, in support of any scheme he recommended, based upon his own careful preliminary study of the whole subject. But when asked any question outside the line of his actual knowledge, he had the honesty to say at once, " I do not know." He would not guess nor attempt to give ingenious answers to show his cleverness, nor act the special pleader in the witness-box, but confined himself solely to what he positively knew. *

In the course of his professional career, Mr. Rennie experienced the great advantage which he had derived from his early training as a millwright. His practical knowledge enabled him to select the best men to carry out his designs, and he took pride and pleasure in directing them how to do their work in the most efficient manner. His manufactory was indeed a school, in which some of the best mechanics of the day received a thorough training in machine work ; and many of his workmen, like himself, eventually raised themselves to the rank of large employers of skilled labour. Mr. Rennie was never ashamed to put his hand to any work where he could teach a lesson or facilitate despatch, and to the end of his career he continued as " handy " as he had been at the beginning.

A curious illustration of his expertness at smithwork occurred during a journey into Scotland, when on his way to visit the Earl of Eglinton at Eglinton Castle. He went by the stage-coach, in company

with some Ayrshire farmers and one or two rather important " Paisley boddies." [1] When travelling over a
very bad piece of the road, the jolting was such as to
break the axletree of the coach, and it came to a stand
on a solitary moor, with not a house in sight. Mr. Rennie
asked the coachman if there was any blacksmith near at
hand, and was told there was one a mile or two off.
" Well, then, help me to carry the parts of the axle there,
and I'll see to its being mended." The blacksmith, however, was not at home ; but Mr. Rennie forthwith lit the
forge fire, blew the bellows, and with the rather clumsy
assistance of one of his fellow-passengers, he very soon
welded the axle in a workmanlike manner, helped to
carry it back to the coach, and after the lapse of a few
hours the vehicle was again wheeling along the road
towards its journey's end. Mr. Rennie's fellow-passengers, who had been communicative and friendly during
the earlier part of the journey, now became very reserved,
and the " boddies" especially held themselves aloof from
" the blacksmith," who had so clearly revealed his calling
by the manner in which he had mended the broken axle.
Arrived at their journey's end for the day, the travellers
separated ; Mr. Rennie proceeding onwards to Eglinton
Castle. Next morning, when sitting at breakfast with his
noble host, a servant entered to say that a person outside
desired to have a word with the Earl. " Show him in."
The person entered. He proved to be one of Mr. Rennie's
fellow-travellers ; and great indeed were his surprise and
confusion at finding the identical " blacksmith " of the
preceding day breakfasting with my Lord ! The Earl
was much amused when Mr. Rennie afterwards described
to him the incident of the mending of the broken axle.

One of his few hobbies was for old books ; and if he
could secure a few minutes' leisure at any time, he would
wander amongst the old book-stalls in search of rare

---

[1] *Paisley Boddie*—a name applied in the West of Scotland to a person
belonging to Paisley.

volumes. Froissart's and Monstrelet's ' Chronicles' were amongst his favourites, and we find him on one occasion sending a present of duplicate copies to his friend Whidbey, accompanied with the wish that he might derive as much pleasure from their perusal as he himself always did from reading " honest John Froissart."   He also commissioned his friends, when travelling abroad, to pick up old books for him; and in 1820 we find him indulging his " extravagance," as he termed it, so far as to request Sir William Jolliffe to bring 300*l.* worth of old books for him from Paris.*

Although Mr. Rennie realized a competency by the practice of his profession, he did not accumulate a large fortune.   The engineer was then satisfied with a comparatively moderate rate of pay,[1] and Mr. Rennie's charge of seven guineas for an entire day's work was even objected to by General Brownrigg, the head of the Ordnance Department at the time.   " Why, this will never do," said the General, looking over the bill; "seven guineas a-day!   Why, it is equal to the pay of a Field Marshal!"   " Well," replied Mr. Rennie, " I am a Field Marshal in my profession; and if a Field Marshal in your line had answered your purpose, I suppose you would not have sent for me!"   " Then you refuse to make any abatement?"   " Not a penny," replied the engineer; and the bill was paid.

Mr. Rennie was blamed in his time for the costliness of his designs, and it was even alleged of him that he carried his love of durability to a fault.   But there is no doubt that the solidity of his structures proved the best economy in the long run.   Elevated by his genius and his conscientiousness above the thoughts of imme-

---

[1] We do not wonder to find Mr. Rennie complaining of the small remuneration of 350*l.* awarded to him by the Kennet and Avon Canal Company for constructing their canal works; and we are surprised to find his bill against the Manchester Waterworks Company, for his year and a-half's advice and service, amounting to only 159*l.* 7*s.*, his charge to them for a whole day's labour being only 6*l.* 6*s.*

diate personal gain, no consideration would induce him
to recommend or countenance in any way the construction of cheap or slight work. He held that the engineer
had not merely to consider the present but the future
in laying down and carrying out his plans. Hence his
designs of docks and harbours were usually framed so as
to be capable of future extension; and his bridges were
built not only for his own time, but with a view to the uses
of generations to come. In fine, Mr. Rennie was a great
and massive, yet a perfectly simple and modest man; and
though his engineering achievements may in some measure have been forgotten in the eulogies bestowed upon
more recent works, they have not yet been eclipsed, nor
indeed equalled; and his London bridges—not to mention his docks, harbours, breakwater, and drainage of
the Lincoln Fens—will long serve as the best exponents
of his genius.

The death of this eminently useful man was felt
to be a national loss, and his obsequies were honoured
by a public funeral. His remains were laid near those
of Sir Christopher Wren in St. Paul's Cathedral, the
dome of which overlooks his finest works. The same
motto might apply to him as to the great architect near
whose remains his lie—"Si monumentum quaeris, cir-
cumspice."

THOMAS TELFORD, F. R. S.

Engraved by W. Holl, after the portrait by Samuel Lane

LIFE OF THOMAS TELFORD.

CHAPTER I.

Eskdale.

Thomas Telford was born in one of the most solitary nooks of the narrow valley of the Esk, in the eastern part of the county of Dumfries, in Scotland. Eskdale runs north and south, its lower end having been in former times the western march of the Scottish border. Near the entrance to the dale is a pillar set upon a high hill, some eight miles to the eastward of the Gretna Green station of the Caledonian Railway,—which many travellers to and from Scotland may have observed,—a monument to the late Sir John Malcolm, Governor of Bombay, one of the distinguished natives of the district. It looks far over the English border-lands which stretch away towards the south, and marks the entrance to the mountainous parts of the valley which lie to the north. From that point upwards the dale gradually becomes narrower, the road winding along the river's banks, in some places high above the stream, dark-brown with peat water, which swiftly rushes over the rocky bed below. A few miles up from the lower end of Eskdale lies the little capital of the district, the town of Langholm; and there, in the market-place, stands another monument to the virtues of the Malcolm family in the statue erected to the memory of Admiral Sir Pulteney Malcolm, a distinguished naval officer. Above Langholm the country becomes more hilly and moorland. In many places only

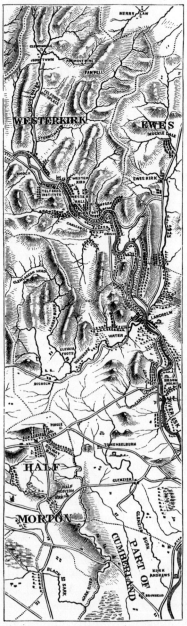

a narrow strip of haugh land by the river's side is left available for cultivation; until at length the dale contracts so much that the hills descend to the very road, and there are only to be seen their steep heathery sides sloping up towards the sky on either hand, and a narrow stream plashing and winding along the bottom of the valley among the rocks at their feet.

From this brief description of the character of Eskdale scenery, it may readily be supposed that the district is very thinly peopled, and that it never could have been capable of supporting a large number of inhabitants. Indeed, previous to the union of the crowns of England and Scotland, the principal branch of industry that existed in the Dale was of a lawless kind. The people living on the two sides of the border looked upon each other's cattle as their own, provided only they had the strength to "lift" them. They were, in truth, even during the time of peace, a kind of outcasts, against whom the united powers of England and Scotland were often employed. On the Scotch side of the Esk were the Johnstones and Armstrongs, and on the English the Graemes of Netherby; but both clans were alike wild and lawless. It was a popular border saying that "Elliots and Armstrongs ride thieves all;" and an old historian says of the Graemes that "they were all stark moss-troopers and arrant thieves; to England as well as Scotland outlawed." The neighbouring chiefs were no better: Scott of Buccleugh, from whom the modern Duke is descended, and Scott of Harden, the ancestor of the novelist, were both renowned freebooters.

There stand at this day on the banks of the Esk, only a few miles from the English border, the ruins of an old fortalice, called Gilnockie Tower, in a situation which in point of natural beauty is scarcely equalled even in Scotland. It was the stronghold of a chief popularly known in his day as Johnnie Armstrong.[1] He was a mighty freebooter in the time of James V., and the terror of his name is said to have extended as far as Newcastle-upon-Tyne, between which town and his castle on the Esk he was accustomed to levy black-mail, or "protection and forbearance money," as it was called. The King, however, determining to put down by the strong hand the depredations of the march men, made a sudden expedition along the borders; and Johnnie Armstrong having been so ill-advised as to make his appearance with his followers at a place called Carlenrig, in Etterick Forest, between Hawick and Langholm, James ordered him to instant execution. Had Johnnie Armstrong, like the Scotts and Kers and Johnstones of like calling, been imprisoned

---

[1] Sir Walter Scott, in his notes to the 'Minstrelsy of the Scottish Border,' says that the common people of the high parts of Liddlesdale and the country adjacent to this day hold the memory of Johnnie Armstrong in very high respect.

beforehand, he might possibly have survived to found
a British peerage; but as it was, the genius of the Armstrong dynasty was for a time extinguished, only, however, to reappear, after the lapse of a few centuries, in
the person of the eminent engineer of Newcastle-upon-
Tyne, the inventor of the Armstrong gun.

The two centuries and a half which have elapsed since
then have indeed effected extraordinary changes.[1] The
energy which the old borderers threw into their feuds
has not become extinct, but survives under more be-
nignant aspects, exhibiting itself in efforts to enlighten,
fertilize, and enrich the country which their wasteful
ardour before did so much to disturb and impoverish.
The heads of the Buccleugh and Elliot family sit in
the British House of Lords. The descendant of Scott of
Harden has achieved a world-wide reputation as a poet
and novelist; and the late representative of the Graemes
of Netherby—whose country seat now sits so peacefully
amidst its woods upon the English side of the border,
overlooking Lower Esk—was one of the most venerable
and respected of British statesmen. The border men,

---

[1] It was long before the Reformation flowed into the secluded valley
of the Esk; but when it did, the
energy of the Borderers displayed
itself in the extreme form of their
opposition to the old religion. The
Eskdale people became as resolute in
their covenanting as they had before
been in their freebooting; and the
moorland fastnesses of the moss-
troopers became the haunts of the
persecuted ministers in the reign of
the second James. A little above
Langholm is a hill known as "Peden's
View," and the well in the green hollow at its foot is still called "Peden's
Well"—that place having been the
haunt of Alexander Peden, the "pro-
phet." His hiding-place was among
the alder-bushes in the hollow, while
from the hill-top he could look up
the valley, and see whether the Johnstones of Wester Hall were coming.
Quite at the head of the same valley,

at a place called Craighaugh, on Eskdale Muir, one Hislop, a young cove-
nanter, was shot by Johnstone's men,
and buried where he fell; a gray
slabstone still marking the place of
his rest. Since that time, however,
quiet has reigned in Eskdale, and its
small population have gone about
their daily industry from one generation to another in peace. Yet, though
secluded and apparently shut out by
the surrounding hills from the outer
world, there is not a throb of the
nation's heart but pulsates along the
valley; and when the author visited
it, some two years since, he found
that a wave of the great Volunteer
movement had flowed into Eskdale;
and the "lads of Langholm" were
drilling and marching under their
chief, young Mr. Malcolm of the
Burnfoot, with even more zeal than
in the populous and far more exposed
towns and cities of the south.

who used to make such furious raids and forays—
have now come to regard each other, across the ima-
ginary line which divides them, as friends and neighbours ; and they meet as competitors for victory only at
agricultural meetings, where they strive for prizes for
the biggest turnips or the most effective reaping-ma-
chines ; whilst the men who followed their Johnstone
or Armstrong chiefs as prickers or hobilers to the fray
have, like Telford, crossed the border with powers
of road-making and bridge-building which have proved
a source of increased civilization and well-being to the
population of the entire United Kingdom.

The hamlet of Westerkirk, with its parish church and
school, lies in a narrow part of the valley, a few miles
above Langholm. Westerkirk parish is long and narrow,
its boundaries being the hill-tops on either side of the
dale. It is about seven miles long and two broad, with
a population of about 600 persons of all ages. Yet
this number is quite as much as the district is enabled
to support, as is proved by its remaining as nearly as
possible stationary from one generation to another.[1]
But what becomes of the natural increase of families ?
" Oh, they swarm off ! " was the explanation given to us

---

[1] The names of the families in the
valley remain very nearly the same
as they were three hundred years ago
—the Johnstones, Littles, Scotts, and
Beatties prevailing above Langholm ;
and the Armstrongs, Bells, Irwins,
and Graemes lower down towards
Canobie and Netherby. It is interesting to find that Sir David Lindesay, in
his curious drama published in ' Pin-
kerton's Scotish Poems ' (vol. ii., p.
156), gives these as among the names
of the borderers some three hundred
years since. One *Common Thift*,
when sentenced to condign punish-
ment, thus remembers his Border
friends in his dying speech :—

" Adew ! my bruther Annan theives,
    That holpit me in my mischeivis ;

Adew ! Grossars, Niksonis, and Bells,
Oft have we fairne owrthreuch the fells:
Adew ! Robsons, Howis, and Pylis,
That in our craft hes mony wilis:
Littlis, Trumbells, and Armestranges ;
Baileowes, Erewynis, and Elwandis,
Speedy of flicht, and slicht of handis ;
The Scotts of Eisdale, and the Gramis,
I haf na time to tell your nameis."

Telford, or Telfer, is an old name
in the same neighbourhood, com-
memorated in the well known border
ballad of ' Jamie Telfer of the fair
Dodhead.' Sir W. Scott says, in the
' Minstrelsy,' that " there is still a
family of Telfers, residing near Langholm, who pretend to derive their
descent from the Telfers of the Dodhead."

by a native of the valley. " If they remained at home,"
said he, " we should all be sunk in poverty, scrambling
with each other amongst these hills for a bare living.
But our peasantry have a spirit above that : they will
not consent to sink ; they look up ; and our parish schools
give them a power of making their way in the world,
each man for himself. So they swarm off—some to
America, some to Australia, some to India, and some,
like Telford, work their way across the border and up
to London, though he is the only one from this valley
who has yet reached Westminster Abbey."

One would scarcely have expected to find the birthplace of the builder of the Menai Bridge and other great
national works in so obscure a corner of the kingdom.
Possibly it may already have struck the reader with
surprise, that not only were all the engineers described in
the preceding pages self-taught in their professions, but
they were brought up mostly in remote country places,
far from the active life of great towns and cities. But
genius is of no locality, and springs alike from the farm-
house, the peasant's hut, or the herd's shieling. Strange
indeed it is that the men who have built our bridges,
docks, lighthouses, canals, and railways, should nearly
all have been country-bred boys : Edwards and Brindley
the sons of small farmers ; Smeaton, brought up in his
father's country house at Austhorpe ; Rennie, the son of
a farmer and freeholder ; and Stephenson, brought up in a
hamlet, an engine-tenter's son. But Telford, even more
than any of these, was a purely country-bred boy, and
was born and brought up in a valley so secluded that it
could not even boast of a cluster of houses of the dimensions of a village.

Telford's father was a herd on the sheep-farm of
Glendinning. The farm consists of green hills, lying
along the valley of the Meggat, a little burn, which
descends from the moorlands on the east, and falls into

TELFORD'S BIRTHPLACE. [1]

[By R. P. Leitch.]

the Esk near the hamlet of Westerkirk. John Telford's cottage was little better than a shieling, consisting of four mud walls, spanned by a thatched roof. It stood upon a knoll near the lower end of a deep gully worn in the hillside by the torrents of many winters. The ground stretches away from it in a long sweeping slope up to the sky, and is green to the top, except where the bare grey rocks in some places crop out to the day. From the knoll may be seen miles on miles of hills up and down the valley, winding in and out, sometimes branching off into smaller glens, each with its gurgling rivulet of peaty-brown water flowing down from the mosses above. Only a narrow strip of arable land is here and there visible along the

---

[1] The engraving represents the valley of the Meggat, showing the cottages of Glendinning in the distance.

bottom of the dale, all above being sheep-pasture, moors, and rocks. At Glendinning you seem to have got almost to the world's end. There the road ceases, and above it stretch trackless moors, the solitude of which is only broken by the wimpling sound of the burns on their way to the valley below, the hum of bees gathering honey among the heather, the whirr of a blackcock on the wing, the plaintive cry of the ewes at lambing-time, or the sharp bark of the shepherd's dog gathering the flock together for the fauld.

In this cottage on the knoll Thomas Telford was born on the 9th of August, 1757, and before the year was out he was already an orphan. The shepherd, his father, died in the month of November, and was buried in Westerkirk churchyard, leaving behind him his widow and her only child altogether unprovided for. We may here mention that one of the first things which that child did, when he had grown up to manhood and could "cut a headstone," was to erect one with the following inscription, hewn and lettered by himself, over his father's grave :—

" In Memory of John Telford, who, after living 33 Years an Unblameable Shepherd, died at Glendinning, November, 1757",—

a simple but poetical epitaph, which Wordsworth himself might have written.

The widow had a long and hard struggle with the world before her ; but she encountered it bravely. She had her boy to work for, and, destitute though she was, she had him to educate. She was helped, as the poor so often are, by those of her own condition, and there is no sense of degradation in receiving such help. One of the risks of benevolence is its tendency to lower the recipient to the condition of an alms-taker. Doles from poors'-boxes have this enfeebling effect; but a poor neighbour giving a destitute widow a help in her time of need is felt to be a friendly act, and is alike elevating to the character of both. Though misery such as is witnessed in large towns was quite unknown in the valley, there was poverty; but it was honest as well as hopeful, and no one felt ashamed of it. The farmers of the dale were very primitive [1] in their manners and habits, and being a warm-hearted, though by no means a demonstrative race, they were kind to the widow and her fatherless boy. They took him by turns to live with them at their houses, and gave his mother occasional employment. In spring-time she milked the ewes, in summer she made hay, and in harvest she went a-shearing; so that she not only contrived to live, but to be cheerful.

The house to which the widow and her son removed, at the Whitsuntide following the death of her husband, was at a place called The Crooks, about midway between

COTTAGE AT THE CROOKS. [By Percival Skelton.]

Glendinning and Westerkirk. It was a thatched cot-house, with two ends; in one of which lived Janet Telford (though more commonly known by her own name of Janet Jackson) and her son Tom, and in the other her neighbour Elliot; one door being common to both.

Young Telford grew up a healthy boy, and he was so full of fun and humour that he became known in the

---

[1] It may be mentioned as a curious fact that about the time of Telford's birth there were only two tea-kettles in the whole parish of Westerkirk, one of which was in the house of Sir James Johnstone of Wester Hall, and the other in that of Mr. Malcolm of the Burnfoot.

valley by the name of " Laughing Tam." When he was old enough to herd sheep he went to live with a relative, a shepherd like his father, and he spent most of his time with him in summer on the hill-side amidst the silence of nature. In winter he lived with one or other of the neighbouring farmers. He herded their cows or ran errands, receiving for recompense his meat, a pair of stockings, and five shillings a year for clogs. These were his first wages, and as he grew older they were gradually increased.

But Tom must now be put to school, and, happily, small though the parish of Westerkirk was, it possessed the advantage of that admirable institution the parish school. To the orphan boy the merely elementary teaching there provided was an immense boon. To master this was the first step of the ladder he was afterwards to mount; his own industry, energy, and ability must do the rest. To school accordingly he went, still working a-field or herding cattle during the summer months. Perhaps his own " penny fee " helped to pay the teacher's hire; but it is supposed that his uncle Jackson defrayed the principal part of the expense of his instruction. It was not much that he learnt; but in acquiring the arts of reading, writing, and figures, he learnt the beginnings of a great deal.

Apart from the question of learning, there was another manifest advantage to the poor boy in mixing freely at the parish school with the sons of the neighbouring farmers and proprietors. Such intercourse has an influence upon a youth's temper, manners, and tastes, which is quite as important in the education of character as the lessons of the master himself; and Telford often, in after-life, referred with pleasure to the benefits which he thus derived from his early school friendships. Amongst those to whom he was accustomed to look back with most pride, were the two elder brothers of the Malcolm family, both of whom rose to high rank in the service of their country; William Telford, a youth of

great promise, a naval surgeon, who died young; and
the brothers William and Andrew Little, the former of
whom settled down as a farmer in Eskdale, and the
latter, a surgeon, lost his eyesight when on service on the

WESTERKIRK CHURCH AND SCHOOL.

[By Percival Skelton, after his original Drawing.]

coast of Africa. Andrew Little afterwards established
himself as a teacher at Langholm, where he educated,
amongst others, General Sir Charles Pasley, Dr. Irving,
the Custodier of the Advocates' Library at Edinburgh,
and others known to fame beyond the bounds of their
native valley. Well might Telford say, when an old
man, full of years and honours, on sitting down to
write his autobiography, "I still recollect with pride
and pleasure my native parish of Westerkirk, on the
banks of the Esk, where I was born."

CHAPTER II.

THE time arrived when young Telford must be put to some regular calling. Was he to be a shepherd like his father and his uncle, or was he to be a farm-labourer, or put apprentice to a trade? There was not much choice; but at length it was determined to bind him to a stonemason. In Eskdale that trade was for the most part confined to the building of drystone walls, and there was very little more art employed in it than an ordinarily neat-handed labourer could manage. It was eventually determined to send the youth—and he was now a strong lad of about fifteen—to a mason at Lochmaben, a small town across the hills to the eastward, where a little more building and of a better sort—such as farm-houses, barns, and road-bridges—was carried on, than in his own immediate neighbourhood. There he remained only a few months; for his master using him badly, the high-spirited youth would not brook it, and ran away, taking refuge with his mother at The Crooks, very much to her dismay.

What was now to be done with Tam? He was willing to do anything or go anywhere rather than back to his Lochmaben master. In this emergency his cousin Thomas Jackson, the factor or land-steward at Wester Hall, offered to do what he could to induce Andrew Thomson, a small mason at Langholm, to take Telford for the remainder of his apprenticeship; and to him he went accordingly. The business carried on by his new master was of a very humble sort. Telford, in his autobiography, states that most of the farmers' houses in the district then consisted of " one storey of mud walls, or rubble stones bedded in clay, and thatched with straw, rushes, or heather ; the floors being of earth, and the fire in the middle, having a plastered creel chimney for the escape of the smoke ; and, instead of windows, small openings in the thick mud walls admitted a scanty light." The farm-buildings were of a similarly wretched description.

The principal owner of the landed property in the neighbourhood was the Duke of Buccleugh ; and shortly after the young Duke Henry succeeded to the title and estates in 1767, he introduced considerable improvements in the farmers' houses and farm-steadings, and the dwellings of the peasantry, as well as of the roads through Eskdale. In this way a demand sprang up for masons' labour, and Telford's master had no want of regular employment for his hands. Telford had the benefit of this increase in the building operations of the neighbourhood ; not only in raising rough walls and farm enclosures, but in erecting bridges across rivers wherever regular roads for wheel carriages were substituted for the horse-tracks formerly in use.

During the greater part of his apprenticeship Telford lived in the little town of Langholm, taking frequent opportunities of visiting his mother at The Crooks on Saturday evenings, and accompanying her to the parish-church of Westerkirk on Sundays. Langholm was then a very poor town, being no better in that respect than the district that surrounded it. It consisted chiefly of mud hovels, covered with thatch—the principal building in it being the Tolbooth, a stone and lime structure, the upper part of which was used as a justice-hall and the lower part as a gaol. There were, however, a few good houses in the little town occupied by people of the better class, and in one of these lived an

elderly lady, Miss Pasley,[1] one of the family of the
Pasleys of Craig. As the town was so small that
everybody in it knew everybody else, the ruddy-
cheeked, laughing mason's apprentice soon became generally known to all the townspeople, and amongst others
to Miss Pasley. When she heard that he was the
poor orphan boy from up the valley, the son of the hard-
working widow woman, Janet Jackson, so " eident " and
so industrious, her heart warmed to the mason's apprentice, and she sent for him to her house. That was a
proud day for Tam ; and when he called upon her, he
was not more pleased with Miss Pasley's kindness than
delighted at the sight of her little library of books,
which contained more volumes than he had ever before
seen.   He had by this time acquired a strong taste
for reading, and indeed exhausted all the little book
stores of his friends.   His joy may therefore be imagined when Miss Pasley volunteered to lend him some
books from her own library ! Of course the young
mason eagerly and thankfully availed himself of the
privilege ; and thus, while working as an apprentice
and afterwards as a journeyman, he gathered his first
stores of information in British literature, in which he
was accustomed to the close of his life to take such
pleasure.   He almost always had some book with
him, which he would snatch a few minutes to read
during the intervals of his work ; and in the winter
nights he occupied his spare time in poring over the
volumes that came in his way, usually with no better
light than what was afforded by the cottage fire.   On
one occasion Miss Pasley lent him ' Paradise Lost,' and
he took the book with him to the hill-side to read.
His delight was such that it fairly taxed his powers
of expression.   He could only say " I read and read,

---

[1] Aunt of Sir Charles Pasley, lately deceased.

and glowred; then read, and read again." He was also a great admirer of Burns, whose writings so inflamed his mind that at the age of twenty-two, when barely out of his apprenticeship, we find him breaking out in verse.[1]

By diligently reading all such books as he could borrow from friends and neighbours, Telford made considerable progress in his learning; and, what with his scribbling of "poetry" and various attempts at composition, he had become so good and legible a writer that he was often called upon by his less-educated fellows to pen letters for them to their distant friends. He was always willing to help them in this way; and, the other working people of the town making use of his services in the same manner, all the little domestic and family histories of the place soon became familiar to him. One evening a Langholm man asked Tom to write a letter for him to his son in England; and when the young scribe read over what had been written to the old man's dictation, the latter, at the end of almost every sentence, exclaimed, "Capital! capital!" and at the close he said, "Well! I say, Tam! Werricht himsel' couldna ha' written a better!"—the said Wright being a well-known lawyer or "writer" in Langholm.

His apprenticeship over, Telford went on working as a journeyman at Langholm, his wages at the time being only eighteenpence a-day. What was called the New Town was then in course of erection, and there are houses still pointed out in it, the walls of which Telford helped to put together. In the town are three arched door-heads of a more ornamental character than

---

[1] In his 'Epistle to Mr. Walter Ruddiman,' first published in 'Ruddiman's Weekly Magazine,' in 1779, occur the following lines addressed to Burns, in which Telford incidentally sketches himself at the time, and hints at his own subsequent meritorious career:—

"Nor pass the tentie curious lad,
Who o'er the ingle hangs his head,
And begs of neighbours books to read;
For hence arise
Thy country's sons, who far are spread,
Baith bold and wise."

the rest, of Telford's hewing; for he was already beginning to set up his pretensions as a craftsman, and took
pride in pointing to the superior handiwork which proceeded from his chisel. About the same time the bridge
connecting the Old with the New Town was built across
the Esk at Langholm, and upon that structure
he was also employed. Many of the stones in it
were hewn by his hand, and on several of the
blocks forming the land-breast his tool mark is
still to be seen.

Not long after the bridge was finished, an unusually
high flood or spate swept down the valley. The Esk
was "roaring red frae bank to brae," and it was generally feared that the new brig would be carried away.
Andrew Thomson, the master mason, was from home
at the time, and his wife, Tibby, knowing that he was
bound by his contract to maintain the fabric for a
period of seven years, was in a state of great alarm.
She ran from one person to another, wringing her
hands and sobbing, " Oh ! we'll be ruined—we'll all be
ruined ! " In her distress she thought of Telford, in
whom she had great confidence, and called out, " Oh !
where's Tammy Telfer—where's Tammy ? " He was
immediately sent for. It was evening, and he was soon
found at the house of Miss Pasley. When he came
running up, Tibby exclaimed, " Oh, Tammy ! they've
been on the brig, and they say it's shakin' ! It 'll be
doon ! " " Never you heed them, Tibby," said Telford,
clapping her on the shoulder, " there's nae fear o' the
brig. I like it a' the better that it shakes—it proves it's
weel put thegither." Tibby's fears, however, were not
so easily allayed; and insisting that she heard the brig
" rumlin," she ran up—so the neighbours afterwards used
to say of her—and set her back against the parapet as if
to hold it together. At this, it is said, " Tam hodged and
leuch ;" and Tibby, observing how easily he took it, at
length grew more calm. It soon became clear enough

that the bridge was sufficiently strong; for the flood sub-
sided without doing it any harm, and it has stood the
furious spates of nearly a century uninjured.

Telford acquired considerable general experience
about the same time as a house-builder, though the
structures on which he was engaged were of a humble
order, being chiefly small farm-houses on the Duke
of Buccleugh's estate, with the usual out-buildings.
Perhaps the most important of the jobs on which
he was employed was the manse of Westerkirk, where

VALLEY OF ESKDALE, WESTERKIRK IN THE DISTANCE.
[By Percival Skelton, after his original Drawing.]

he was comparatively at home. The hamlet stands
on a green hill-side at the entrance to the valley of
the Meggat. It consists of the kirk, the minister's
manse, the parish-school, and a few cottages, every
occupant of which was known to Telford. It is backed
by the purple moors, up which he loved to wander
in his leisure hours and read the poems of Fergusson
and Burns. The river Esk gurgles along its rocky
bed in the bottom of the dale, separated from the
kirkyard by a steep green field; whilst near at hand,

behind the manse, stretch the fine woods of Wester Hall, where Telford was often wont to roam. We can scarcely therefore wonder that, amidst such pastoral scenery, the descriptive poetic faculty of the country mason should have become so decidedly and strongly developed. It was while working at Westerkirk manse that Telford sketched the first draft of his descriptive poem, entitled 'Eskdale,' which was published in the 'Poetical Museum' [1] in 1784.

These early poetical efforts were at least useful as stimulating his self-education. For the practice of poetical composition, while it cultivates the sentiment of beauty in thought and feeling, is probably the best of all exercises in the art of writing correctly, grammatically, and expressively. By drawing a man out of his ordinary calling, too, it often furnishes him with a power of happy thinking which may in after life be a fountain of the purest pleasure; and this, we believe, proved to be the case with Telford, even though he ceased in later years to pursue the special cultivation of the art.

---

[1] The 'Poetical Museum,' Hawick, p. 267. 'Eskdale' was afterwards reprinted by Telford when living at Shrewsbury, when he added a few lines by way of conclusion. The poem describes very pleasantly the fine pastoral scenery of the district:—

"Deep 'mid the green sequester'd glens below,
Where murmuring streams among the alders flow,
Where flowery meadows down their margins spread,
And the brown hamlet lifts its humble head—
There, round his little fields, the peasant strays,
And sees his flock along the mountain graze;
And, while the gale breathes o'er his ripening grain,
And soft repeats his upland shepherd's strain,

And western suns with mellow radiance play,
And gild his straw-roof'd cottage with their ray,
Feels Nature's love his throbbing heart employ,
Nor envies towns their artificial joy."

The features of the valley are very fairly described. Its early history is then rapidly sketched; next its period of border strife, at length happily allayed by the union of the kingdoms, under which the Johnstones, Pasleys, and others, men of Eskdale, achieve honour and fame. Nor did he forget to mention Armstrong, the author of the 'Art of Preserving Health,' who seems to have been educated in the valley; and Mickle, the translator of the 'Lusiad,' whose father was minister of the parish of Langholm; both of whom Telford took a natural pride in as native poets of Eskdale.

Shortly after, when work became slack in the district, Telford undertook to do small jobs on his own account —such as the hewing of gravestones and ornamental doorheads. He prided himself especially upon his hewing, and from the specimens of his workmanship which are still to be seen in the churchyards of Langholm and Westerkirk, he had evidently attained considerable skill. On some of these pieces of masonry the year is carved—1779, or 1780. One of the most ornamental is that set into the wall of Westerkirk church, being a monumental slab, with an inscription and moulding, surmounted by a coat of arms, to the memory of James Pasley of Craig.

He had now learnt all that his native valley could teach him of the art of masonry; and, bent upon self-improvement and gaining a larger experience of life as well as knowledge of his trade, he determined to seek employment elsewhere. He accordingly left Eskdale for the first time, in 1780, and sought work in Edinburgh, where the New Town was then in course of erection on the elevated land, formerly green fields, extending along the north bank of the " Nor' Loch." A bridge had been thrown across the Loch in 1769, the stagnant pond or marsh in the hollow had been filled up, and Princes Street was rising as if by magic. Skilled masons were in great demand for the purpose of carrying out these and the numerous other architectural improvements which were in progress, and Telford had no difficulty in obtaining abundant employment. He remained at Edinburgh for about two years, during which he had the advantage of taking part in first-rate work and maintaining himself comfortably, whilst he devoted much of his spare time to drawing, in its application to architecture. He took the opportunity of visiting and carefully studying the fine specimens of ancient work at Holyrood House and Chapel, the Castle, Heriot's Hospital, and the numerous curious illustrations of middle age domestic architecture with which the Old Town abounds. He also made several journeys to the beautiful old chapel of Rosslyn, in the highly ornamented Gothic style, situated some miles to the south of Edinburgh, making careful drawings of the more important parts of that building.

When he had thus improved himself "and studied all that was to be seen in Edinburgh, in returning to the western border," he says, "I visited the justly celebrated Abbey of Melrose." There he was charmed by the delicate and perfect workmanship still visible even in the ruins of that fine old Abbey; and with his folio filled with sketches and drawings, he made his way back to Eskdale and the humble cottage at The Crooks. But not to remain there long. He merely wished to pay a parting visit to his mother and relations before starting upon a longer journey. "Having acquired," he says in his Autobiography, " the rudiments of my profession, I considered that my native country afforded few opportunities of exercising it to any extent, and therefore judged it advisable (like many of my countrymen) to proceed southward, where industry might find more employment and be better remunerated."

Before setting out he called upon all his old friends and acquaintances in the dale—the neighbouring farmers, who had befriended him and his mother when struggling with poverty—his schoolfellows, many of whom were preparing to migrate, like himself, from their native valley—and the many friends and acquaintances he had made whilst working as a mason in Langholm. Everybody knew that Tam was going south, and all wished him God speed. At length the leave-taking was over, and he set out for London in the year 1782, when twenty-five years of age. He had, like the little river Meggat, on the banks of which he was born, floated gradually on towards the outer world : first from the nook in the valley, to Westerkirk school ; then to Langholm and its little circle ; and now, like the Meggat, which flows with the Esk into the ocean, he was about to be borne away into the wide world. Tam, however, had confidence in himself, and no one had fears for him. As the neighbours said, wisely wagging their heads, "Ah, he's an auld-farran chap is Tam; he'll either mak a spoon or spoil a horn; any how, he's gatten a good trade at his fingers' ends."

Telford had made all his previous journeys on foot; but this one he rode on horseback. It happened that Sir James Johnstone, the laird of Wester Hall, had occasion to send a horse from Eskdale to a member of his family in London; but he had some difficulty in finding a person to take charge of it. It occurred to Mr. Jackson, the laird's factor, that this was a capital opportunity for his cousin Tom, the mason; and it was accordingly arranged that he should ride the horse to town. When a boy, he had learnt rough-riding sufficiently well for the purpose; and the better to fit him for the hardships of the road, Mr. Jackson lent him his buckskin breeches. Thus Tom set out from his native valley well mounted, with his little bundle of "traps" buckled behind him, and, after a prosperous journey, duly reached London, and delivered up the horse as he had been directed. Long after, Mr. Jackson used to tell the story of his cousin's first ride to London with great glee, and he always took care to wind up with—"but Tam forgot to send me back my breeks!"

LOWER VALLEY OF THE MEGGAT, THE CROOKS IN THE DISTANCE.
[By Percival Skelton.]

CHAPTER III.

A COMMON working man, whose sole property consisted in his mallet and chisels, his leathern apron and his industry, might not seem to amount to much in "the great world of London." But, as Telford afterwards used to say, very much depends on whether the man has got a head, with brains in it of the right sort, upon his shoulders. In London the weak man is simply a unit added to a vast floating crowd, and may be driven hither and thither, if he do not sink altogether; whilst the strong man will strike out, keep his head above water, and make a course for himself, as Telford did. There is indeed a wonderful impartiality about London. There the capable person usually finds his place. When work of importance is required, nobody cares to ask where the man who can do it best comes from, or what he has been; but what he is, and what he can do. Nor did it ever stand in Telford's way that his father had been a poor shepherd in Eskdale, and that he himself had begun his London career by working for weekly wages with his mallet and chisel.

After duly delivering up the horse, Telford proceeded to present a letter with which he had been charged by his friend Miss Pasley on leaving Langholm. It was addressed to her brother, Mr. John Pasley, an eminent London merchant — brother also of Sir Thomas Pasley, and uncle of the Malcolms. Miss Pasley requested his influence on behalf of the young mason from Eskdale, the bearer of the letter. Mr. Pasley received his countryman kindly, and furnished him with letters of introduction to Sir William Chambers, the architect of Somerset House, then in course of erection. It was the finest architectural work in progress in the metropolis, and Telford, desirous of improving himself by experience of the best kind, wished to be employed upon it. It did not, indeed, need any influence to obtain work there, for good hewers were in great demand; but our mason thought it well to make sure, and accordingly provided himself beforehand with the letter of introduction to the architect. He was employed immediately, and set to work amongst the hewers, receiving the usual wages for his work.

Mr. Pasley also furnished him with a letter to Mr. Robert Adam,[1] another distinguished architect of the time; and Telford seems to have been much gratified by the civility which he received from him. Sir William Chambers he found haughty and reserved, probably being too much occupied to bestow attention on the Somerset House hewer, whilst Adam he described as affable and communicative. " Although I derived no direct advantage from either," Telford says, " yet so powerful is manner, that the latter left the most favourable impression; while the interviews with both convinced me that my safest plan was to endeavour to advance, if by slower degrees, yet by independent conduct."

There was a good deal of fine hewer's work about Somerset House, and from the first Telford aimed at taking the highest place as an artist and tradesman in

---

[1] Robert and John Adam were architects of considerable repute in their day. Among their London erections were the Adelphi Buildings, in the Strand; Lansdowne House, in Berkeley Square; Caen Wood House, near Hampstead (Lord Mansfield's); Portland Place, Regent's Park; and numerous West End streets and mansions. The screen of the Admiralty and the ornaments of Drapers' Hall were also designed by them.

that line.[1]   Diligence, carefulness, and observation will always carry a man onward and upward; and before long we find that Telford had succeeded in advancing himself to the rank of a first-class mason. Judging by his letters written about this time to his friends in Eskdale, he seems to have been very cheerful and happy; and his greatest pleasure was in calling up recollections of his native valley. He was full of kind remembrances for everybody. " How is Andrew, and Sandy, and Aleck, and Davie?" he would say; and " remember me to all the folk of the nook." He seems to have made a round of the persons from Eskdale in or about London before he wrote, as his letters were full of messages from them to their friends at home; for in those days postage was dear, and as much as possible was necessarily packed within the compass of a working man's letter. In one, written after more than a year's absence, he says he envies the visit which a young surgeon of his acquaintance was about to pay to the valley; "for the meeting of long absent friends," he adds, "is a pleasure to be equalled by few other enjoyments here below."

He had now been more than a year in London, during which he had acquired much practical information both in the useful and ornamental branches of architecture. Was he to go on as a working mason? or what was to be his next move? He had been quietly making his observations upon his companions, and had come to the conclusion that they very much wanted spirit, and, more than all, fore-thought. He found very clever workmen about him with no idea whatever beyond their week's wages. For these they would

---

[1] Long after Telford had become famous, he was passing over Waterloo Bridge one day with a friend, when, pointing to some finely-cut stones in the corner nearest the bridge, he said: " You see those stones there; forty years since I hewed and laid them, when working on that building as a common mason."

make every effort : they would work hard, exert themselves to keep them up to the highest point, and very
readily "strike" to secure an advance; but as for the
next week, or the next year, he thought them exceedingly careless. On the Monday mornings they began
"clean;" and on Saturdays there was the week's earnings to spend. Thus they lived from one week to another
—their limited notion of "the week" seeming to bound
their existence.

Telford looked upon the week as only a stone in
one of the storeys of a building; and upon the succession of weeks, running on through years, he thought
that the complete life structure should be built up. He
thus describes one of the best of his fellow-workmen
at that time—the only individual he had formed an
intimacy with : "He has been six years at Somerset
House, and is esteemed the finest workman in London,
and consequently in England. He works equally in stone
and marble. He has excelled the professed carvers in
cutting Corinthian capitals and other ornaments about
this edifice, many of which will stand as a monument to
his honour. He understands drawing thoroughly, and
the master he works under looks on him as the principal
support of his business. This man, whose name is Mr.
Hatton, may be half a dozen years older than myself at
most. He is honesty and good nature itself, and is
adored by both his master and fellow-workmen. Notwithstanding his extraordinary skill and abilities, he has
been working all this time as a common journeyman,
contented with a few shillings a-week more than the
rest; but I believe your uneasy friend has kindled a
spark in his breast that he never felt before." [1]

In fact, Telford had formed the intention of inducing
this admirable fellow to join him in commencing business as builders on their own account. "There is nothing

---

[1] Letter to Mr. Andrew Little, Langholm, dated London, July, 1783.

done in stone or marble," he says, " that we cannot do in
the completest manner." Mr. Robert Adam, to whom the
scheme was mentioned, promised his support, and said
he would do all in his power to recommend them. But
the great difficulty was money, which neither of them
possessed; and Telford, with grief, admitting that this
was an "insuperable bar," the scheme went no further.

About this time Telford was consulted by Mr. Pulteney [1] respecting the alterations making in the mansion
at Wester Hall, and was often with him on this business.
We find him also writing down to Langholm for the
prices of roofing, masonry, and timber-work, with a
view to preparing estimates for a friend who was building a house in that neighbourhood. Although determined to reach the highest excellence as a manual
worker, it is clear that he was already aspiring to be
something more. Indeed, his steadiness, perseverance,
and general ability pointed him out as one well worthy
of promotion.

How he achieved his next step we are not informed;
but we find him, in July, 1784, engaged in superintending the erection of a house, after a design by Mr.
Samuel Wyatt, intended for the residence of the Commissioner (now occupied by the Port Admiral) at
Portsmouth Dockyard, together with a new chapel,

---

[1] Mr., afterwards Sir William Pulteney, was the second son of Sir
James Johnstone, of Wester Hall, and
assumed the name of Pulteney on his
marriage to Miss Pulteney, niece of
the Earl of Bath and of General Pulteney, by whom he succeeded to a
large fortune; he afterwards succeeded
to the baronetcy of his elder brother
James, who died without issue in
1797. Sir William Pulteney represented Cromarty, and afterwards
Shrewsbury, where he usually resided, in seven successive Parliaments.
He was a great patron of Telford's, as
we shall afterwards find. The story

is told in Eskdale that Miss Pulteney
had danced at a public ball with the
elder brother, Sir James, and afterwards expressed such admiration of
him that it was repeated to him, with
the hint that he might do worse than
offer his hand to the heiress. " There
is only one slight difficulty," said he,
" I have got a wife already; but," he
added, "if she would like a Johnstone, there's my brother Will—a
much better looking fellow than I
am." How it may have been brought
about is not stated, but Will did
make up to the heiress, and married
her, assuming the name of her family.

and several buildings connected with the Yard.   Telford took care to keep his eyes open to all the other works going forward in the neighbourhood, and he states that he had frequent opportunities of observing the various operations necessary in the foundation and construction of graving-docks, wharf-walls, and such like, which were amongst the principal occupations of his after-life.

The letters written by him from Portsmouth to his Eskdale correspondents about this time were cheerful and hopeful, like those he had sent from London. His principal grievance was that he received so few from home, but he supposed that opportunities for forwarding them by hand had not occurred, postage being so dear as scarcely then to be thought of. To tempt them to correspondence he sent copies of the poems which he still continued to compose in the leisure of his evenings : one of these was a 'Poem on Portsdown Hill.'   As for himself, he was doing very well.   The buildings were advancing satisfactorily ; but, " above all," said he, " my proceedings are entirely approved by the Commissioners and officers here—so much so that they would sooner go by my advice than my master's, which is a dangerous point, being difficult to keep their good graces as well as his. However, I will contrive to manage it." [1]

The following is his own account of the manner in which he was usually occupied during the winter months while at Portsmouth Dock :—" I rise in the morning at 7 (February 1st), and will get up earlier as the days lengthen until it come to 5 o'clock.   I immediately set to work to make out accounts, write on matters of business, or draw, until breakfast, which is at 9.   Then I go into the Yard about 10, see that all are at their posts, and am ready to advise about any matters that may require attention.   This, and going round the several

---

[1] Letter to Andrew Little, Langholm, dated Portsmouth, July 23rd, 1784.

works, occupies until about dinner-time, which is at 2 ; and after that I again go round and attend to what may be wanted. I draw till 5 ; then tea ; and after that I write, draw, or read until half after 9 ; then comes supper and bed. This is my ordinary round, unless when I dine or spend an evening with a friend ; but I do not make many friends, being very particular, nay, nice to a degree. My business requires a great deal of writing and drawing, and this work I always take care to keep under by reserving my time for it, and being in advance of my work rather than behind it. Then, as knowledge is my most ardent pursuit, a thousand things occur which call for investigation which would pass unnoticed by those who are content to trudge only in the beaten path. I am not contented unless I can give a reason for every particular method or practice which is pursued. Hence I am now very deep in chemistry. The mode of making mortar in the best way led me to inquire into the nature of lime. Having, in pursuit of this inquiry, looked into some books on chemistry, I perceived the field was boundless ; but that to assign satisfactory reasons for many mechanical processes required a general knowledge of that science. I have therefore borrowed a MS. copy of Dr. Black's Lectures. I have bought his ' Experiments on Magnesia and Quicklime,' and also Fourcroy's Lectures, translated from the French by one Mr. Elliot, of Edinburgh. And I am determined to study the subject with unwearied attention until I attain some accurate knowledge of chemistry, which is of no less use in the practice of the arts than it is in that of medicine." He adds, that he continues to receive the cordial approval of the Commissioners for the manner in which he performs his duties, and says, " I take care to be so far master of the business committed to me as that none shall be able to eclipse me in that respect." [1]   At

---

[1] Letter to Mr. Andrew Little, Langholm, dated Portsmouth Dockyard, Feb. 1, 1786.

the same time he states he is taking great delight in Freemasonry, and is about to have a lodge-room at the George Inn fitted up after his plans and under his direction. Nor does he forget to add that he has his hair powdered every day, and puts on a clean shirt three times a week. The Eskdale mason is evidently getting on, as he deserves. Yet he says that " he would rather have it said of him that he possessed one grain of good nature or good sense than shine the finest puppet in Christendom." " Let my mother know that I am well," he writes to Andrew Little, " and that I will print her a letter soon." [1]   For it was a practice of this good son, down to the period of his mother's death, no matter how much burdened he was with business, to set apart occasional times for the careful penning of a letter in *printed* characters, that she might be the more easily able to decipher it with her old and dimmed eyes by her cottage fireside at The Crooks. As a man's real disposition usually displays itself most strikingly in small matters—like light, which gleams the most brightly when seen through narrow chinks—it will probably be admitted that this trait, trifling though it may appear, was truly characteristic of the simple and affectionate nature of the hero of our story.

The buildings at Portsmouth were finished by the end of 1786, when Telford's duties at that place being at an end, and having no engagement beyond the termination of the contract, he prepared to leave, and began to look about him for other employment.

---

[1] Letter to Mr. Andrew Little, Langholm, dated Portsmouth Dockyard, Feb. 1, 1786.

CHAPTER IV.

MR. PULTENEY, member for Shrewsbury, was the owner of extensive estates in that neighbourhood by virtue of his marriage with the niece of the last Earl of Bath. Having resolved to fit up the Castle there as a residence, he bethought him of the young Eskdale mason, who had, some years before, advised him as to the repairs of the Johnstone mansion at Wester Hall. Telford was soon found, and engaged to go down to Shrewsbury to superintend the necessary alterations. Their execution occupied his attention for some time, and during their progress he was so fortunate as to obtain the appointment of Surveyor of Public Works for the county of Salop, most probably through the influence of his patron. Indeed, Telford was known to be so great a favourite with Mr. Pulteney that at Shrewsbury he usually went by the name of "Young Pulteney."[*]

Much of his attention was from this time occupied with the surveys and repairs of roads, bridges, and gaols, and the supervision of all public buildings under the control of the magistrates of the county. He was also frequently called upon by the corporation of the borough to furnish plans for the improvement of the streets and buildings of that fine old town ; and many alterations were carried out under his directions during the period of his residence there.

While the Castle repairs were in course of execution, he was called upon by the justices to superintend the erection of a new gaol, the plans for which had already been prepared and settled. The benevolent

Howard, who devoted himself with such zeal to gaol improvement, on hearing of the intentions of the magistrates, made a visit to Shrewsbury for the purpose of examining the plans; and the circumstance is thus adverted to by Telford in one of his letters to his Eskdale correspondent:—" About ten days ago I had a visit from the celebrated John Howard, Esq. I say *I*, for he was on his tour of gaols and infirmaries; and those of Shrewsbury being both under my direction, this was, of course, the cause of my being thus distinguished. I accompanied him through the infirmary and the gaol. I showed him the plans of the proposed new buildings, and had much conversation with him on both subjects. In consequence of his suggestions as to the former, I have revised and amended the plans, so as to carry out a thorough reformation; and my alterations having been approved by a general board, they have been referred to a committee to carry out. Mr. Howard also took objection to the plan of the proposed gaol, and requested me to inform the magistrates that, in his opinion, the interior courts were too small, and not sufficiently ventilated; and the magistrates, having approved his suggestions, ordered the plans to be amended accordingly. You may easily conceive how I enjoyed the conversation of this truly good man, and how much I would strive to possess his good opinion. I regard him as the guardian angel of the miserable. He travels into all parts of Europe with the sole object of doing good, merely for its own sake, and not for the sake of men's praise. To give an instance of his delicacy, and his desire to avoid public notice, I may mention that, being a Presbyterian, he attended the meeting-house of that denomination in Shrewsbury on Sunday morning, on which occasion I accompanied him; but in the afternoon he expressed a wish to attend another place of worship, his presence in the town having excited considerable curiosity, though his wish was to avoid public recognition. Nay, more,

he assures me that he hates travelling, and was born to
be a domestic man.   He never sees his country-house
but he says within himself, 'Oh! might I but rest here,
and never more travel three miles from home; then
should I be happy indeed!'   But he has become so
committed, and so pledged himself to his own conscience
to carry out his great work, that he says he is doubtful
whether he will ever be able to attain the desire of his
heart—life at home.   He never dines out, and scarcely
takes time to dine at all : he says he is growing old, and
has no time to lose.   His manner is simplicity itself.
Indeed, I have never yet met so noble a being.   He is
going abroad again shortly on one of his long tours of
mercy." [1]   The journey to which Telford here refers was
Howard's last.   In the following year he left England
to return no more; and the great and good man died
at Cherson, on the shores of the Black Sea, less than
two years after his interview with the young engineer
at Shrewsbury.

Telford writes to his Langholm friend at the same time,
that he is working very hard, and studying to improve
himself in branches of knowledge in which he feels himself deficient.   He is practising very temperate habits :
for half a year past he has taken to drinking water only,
avoiding all sweets, and eating no "nick-nacks."   He
has "sowens and milk" (oatmeal flummery) every night
for his supper.   His friend having asked his opinion of
politics, he says he really knows nothing about them ;
he had been so completely engrossed by his own business that he has not had time to read even a newspaper.
But, though an ignoramus in politics, he has been
studying *lime*, which is more to his purpose.   If his
friend can give him any information about that, he will
promise to read a newspaper now and then in the ensuing session of Parliament, for the purpose of forming

---

[1] Letter to Mr. Andrew Little, Langholm, dated Shrewsbury Castle, 21st
Feb., 1788.

some opinion of politics : he adds, however, "not if it interfere with my business—mind that!" His friend had told him that he proposed translating a system of chemistry. "Now you know," said he, "that I am chemistry mad; and if I were near you, I would make you promise to communicate any information on the subject that you thought would be of service to your friend, especially about calcareous matters and the mode of forming the best composition for building with, as well above as below water. But not to be confined to that alone, for you must know I have a book for the pocket,[1] which I always carry with me, into which I have extracted the essence of Fourcroy's Lectures, Black on Quicklime, Scheele's Essays, Watson's Essays, and various points from the letters of my respected friend Dr. Irving.[2] So much for chemistry. But I have also crammed into it facts relating to mechanics, hydrostatics, pneumatics, and all manner of stuff, to which I keep continually adding, and it will be a charity to me if you will kindly contribute your mite."[3] He says it has been, and will continue to be, his aim to endeavour to unite those "two frequently jarring pursuits, literature and business;" and he does not see why a man should be less efficient in the latter capacity because he has well informed, stored, and humanized his mind by the cultivation of letters. There was both good sense and sound practical wisdom in this view of Telford.

While the gaol was in course of erection, after the improved plans suggested by Howard, a variety of important matters occupied the county surveyor's attention. During

---

[1] This practice of noting down information, the result of reading and observation, was continued by Mr. Telford until the close of his life; his last pocket memorandum book, containing a large amount of valuable information on mechanical subjects— a sort of engineer's vade mecum—

being printed in the appendix to the 4to. 'Life of Telford' published by his executors in 1838. Pp. 663-90.

[2] A medical man, a native of Eskdale, of great promise, who died comparatively young.

[3] Letter to Mr. Andrew Little, Langholm.

the summer of 1788 he says he is very much occupied, having about ten different jobs on hand : roads, bridges, streets, drainage-works, gaol, and infirmary. Yet he had time to write verses, copies of which he forwarded to his Eskdale correspondent, inviting his criticism. Several of these were elegiac lines, somewhat exaggerated in their praise of the deceased, though doubtless sincere. One poem was in memory of George Johnstone, Esq., a member of the Wester Hall family, and another on the death of William Telford, an Eskdale farmer's son, an intimate friend and schoolfellow of our engineer.[1] These, however, were but the votive offerings of private friendship, persons more immediately about him knowing nothing of his stolen pleasures in versemaking. He continued to be shy of strangers, and was very " nice," as he calls it, as to those whom he admitted to his bosom.

Two circumstances of considerable interest occurred in the course of the same year (1788), which are worthy of passing notice. The one was the fall of the church of St. Chad's, at Shrewsbury ; the other was the discovery of the ruins of the Roman city of Uriconium, in the immediate neighbourhood. The church of St. Chad's was about four centuries old, and stood greatly

---

[1] It would occupy unnecessary space to cite these poems. The following, from the verses in memory of William Telford, relates to schoolboy days. After alluding to the lofty Fell Hills, which formed part of the sheep farm of his deceased friend's father, the poet goes on to say :—

" There, 'mongst those rocks I'll form a rural seat,
    And plant some ivy with its moss compleat ;
I'll benches form of fragments from the stone,
    Which, nicely pois'd, was by our hands o'erthrown,—

A simple frolic, but now dear to me,
Because, my Telford, 'twas performed with thee.
There, in the centre, sacred to his name,
I'll place an altar, where the lambent flame
Shall yearly rise, and every youth shall join
The willing voice, and sing the enraptured line.
But we, my friend, will often steal away
To this lone seat, and quiet pass the day ;
Here oft recall the pleasing scenes we knew
In early youth, when every scene was new,
When rural happiness our moments blest,
And joys untainted rose in every breast."

in need of repairs. The roof let in the rain upon the congregation, and the parish vestry met to settle the plans for mending it; but they could not agree about the mode of procedure. In this emergency Telford was sent for by the churchwardens, and requested to advise them what was best to be done. He accordingly examined the building, and found that not only the roof but the walls of the church were in a most decayed state. It appeared that, in consequence of graves having been dug in the loose soil close to the shallow foundation of the north-west pillar of the tower, it had sunk so as to endanger the whole structure. "I discovered," says he, "that there were large fractures in the walls, on tracing which I found that the old building was in a most shattered and decrepit condition, though until then it had been scarcely noticed. Upon this I declined giving any recommendation as to the repairs of the roof unless they would come to the resolution to secure the more essential parts, as the fabric appeared to me to be in a very alarming condition. I sent in a written report to the same effect." [1] The parish vestry again met, and the report was read; but the meeting exclaimed against so extensive a proposal, imputing mere motives of self-interest to the surveyor. "Popular clamour," says Telford, "overcame my report. 'These fractures,' exclaimed the vestrymen, 'have been there from time immemorial;' and there were some otherwise sensible persons, who remarked that professional men always wanted to carve out employment for themselves, and that the whole of the necessary repairs could be done at a comparatively small expense." [2] Telford at length left the meeting, advising that, if they wished to discuss anything besides the alarming state of the church, they had better adjourn to some other place, where there was no danger of its falling on their heads. The suggestion was received

---

[1] Letter to Mr. Andrew Little, Langholm, dated 16th July, 1788.   [2] Ibid.

with ridicule, and the vestry called in another person, a mason of the town, directing him to cut away the injured part of the pillar, in order to underbuild it. On the second evening after the commencement of these operations, the sexton was alarmed at the fall of lime-dust and mortar when he attempted to toll the great bell, on which he immediately desisted and left the church. Early next morning (on the 9th of July), while the workmen were waiting at the church door for the key, the bell struck four, and the vibration at once brought down the tower, which overwhelmed the nave, demolishing all the pillars along the north side, and shattering the rest. "The very parts I had pointed out," says Telford, " were those which gave way, and down tumbled the tower, forming a very remarkable ruin, which astonished and surprised the vestry, and roused them from their infatuation, though they have not yet recovered from the shock." [1]

The other circumstance to which we have above referred was the discovery of the Roman city of Uriconium, near Wroxeter, about five miles from Shrewsbury, in the year 1788. The situation of the place is extremely beautiful, the river Severn flowing along its western margin, and forming a barrier against what were once the hostile districts of West Britain. For many centuries the dead city had slept under the irregular mounds of earth which covered it, like those of Mossul and Nineveh. Farmers raised heavy crops of turnips and grain from the surface ; and they scarcely ever ploughed or harrowed the ground without turning up Roman coins or pieces of pottery. They also observed that in certain places the corn was more apt to be scorched in dry weather than in others— a sure sign to them that there were ruins underneath ; and their practice, when they wished to find stones for building, was to set a mark upon the scorched places

---

[1] Letter to Mr. Andrew Little, Langholm, dated 16th July, 1788.

when the corn was on the ground, and after harvest to dig down, sure of finding the store of stones which they wanted for walls, cottages, or farm-houses. In fact, the place came to be regarded in the light of a quarry, rich in ready-worked materials for building purposes. About this time a quantity of stone was wanted for the purpose of erecting a blacksmith's shop, and on digging down upon one of the marked places, the labourers came upon some ancient works of a more perfect appearance than usual. Curiosity was excited—antiquarians made their way to the spot—and lo! they pronounced the ruins to be neither more nor less than a Roman bath, in a remarkably perfect state of preservation. Mr. Telford was requested to apply to Mr. Pulteney, the lord of the manor, to prevent the destruction of these interesting remains, and also to permit the excavations to proceed, with a view to the buildings being completely explored. This was readily granted, and Mr. Pulteney authorised Telford himself to conduct the necessary excavations at his expense. This he promptly proceeded to do, and the result was, that an extensive hypocaust apartment was brought to light, with baths, sudatorium, dressing-room, and a number of tile pillars—all forming parts of a Roman floor—sufficiently perfect to show the manner in which they had been constructed and used.[1]

Among Telford's less agreeable duties about the same time was that of keeping the felons at work. He had to devise the ways and means of employing them without risk of their escaping, which gave him much trouble and anxiety. "Really," he says, "my felons are a very troublesome family. I have had a great deal of plague from them, and I have not yet got things quite in the train that I could wish. I have had a dress made for

[1] The discovery formed the subject of a paper read before the Society of Antiquaries in London on the 7th of May, 1789, published in the 'Archæologia,' together with a drawing of the remains supplied by Mr. Telford.

them of white and brown cloth, in such a way that they are pyebald. They have each a light chain about one leg. Their allowance in food is a penny loaf and a halfpenny worth of cheese for breakfast; a penny loaf, a quart of soup, and half a pound of meat for dinner; and a penny loaf and a halfpenny worth of cheese for supper; so that they have meat and clothes at all events. I employ them in removing earth, serving masons or bricklayers, or in any common labouring work on which they can be employed; during which time, of course, I have them strictly watched."

Much more pleasant was his first sight of Mrs. Jordan at the Shrewsbury theatre, where he seems to have been worked up to a pitch of rapturous enjoyment. She played for six nights there at the race time, during which there were various other entertainments. On the second day there was what was called an Infirmary Meeting, or an assemblage of the principal county gentlemen in the infirmary, at which, as county surveyor, Telford was present. They proceeded thence to church to hear a sermon preached for the occasion; after which there was a dinner, followed by a concert. He attended all. The sermon was preached in the new pulpit, which had just been finished after his designs, in the Gothic style; and he confidentially informed his Langholm correspondent that he believed the pulpit secured greater admiration than the sermon. With the concert he was completely disappointed, and he then became convinced that he could have no ear for music. Other people seemed very much pleased; but for the life of him he could make nothing of it. The only difference that he recognised between one tune and another was that there was a difference of noise. " It was all very fine," he says, " I have no doubt; but I would not give a song of Jock Stewart[1] for

---

the whole of them. The melody of sound is thrown away upon me. One look, one word of Mrs. Jordan, has more effect upon me than all the fiddlers in England. Yet I sat down and tried to be as attentive as any mortal could be. I endeavoured, if possible, to get up an interest in what was going on; but it was all of no use. I felt no emotion whatever, excepting only a strong inclination to go to sleep. It must be a defect; but it is a fact, and I cannot help it. I suppose my ignorance of the subject, and the want of musical experience in my youth, may be the cause of it." [1]

Telford's mother was still living in her old cottage at The Crooks. Since he had parted from her he had written many printed letters to keep her informed of his progress; and he never wrote to any of his friends in the dale without including some message or other to his mother. Like a good and dutiful son, he had taken care out of his means to provide for her comfort in her declining years. " She has been a good mother to me," he said, " and I will try and be a good son to her." In a letter written from Shrewsbury about this time, enclosing a ten pound note, seven pounds of which was to be given to his mother, he said, " I have from time to time written William Jackson [his cousin] and told him to furnish her with whatever she wants to make her comfortable ; but there may be many little things she may wish to have, and yet not like to ask him for. You will therefore agree with me that it is right she should have a little cash to dispose of in her own way. . . I am not rich yet ; but it will ease my mind to set my mother above the fear of want. That has always been my first object ; and next to that, to be the *somebody* which you have always encouraged me to believe I might aspire to become. Perhaps after all there may be something in it ! " [2]

---

[1] Letter to Mr. Andrew Little, Langholm, dated 3rd Sept., 1788.
[2] Letter to Mr. Andrew Little, Langholm, dated Shrewsbury, 8th October, 1789.

He now seems to have occupied much of his leisure hours in miscellaneous reading. Amongst the numerous books which he read, he expressed the highest admiration for Sheridan's 'Life of Swift.' But his Langholm friend, who was a great politician, having invited his attention to politics, Telford's reading gradually extended in that direction. Indeed the exciting events of the French Revolution then tended to make all men more or less politicians. The capture of the Bastille by the people of Paris in 1789 passed like an electric thrill through Europe. Then followed the Declaration of Rights; after which, in the course of six months, all the institutions which had before existed in France were swept away, and the reign of justice was fairly inaugurated upon earth! In the spring of 1791 the first part of Paine's 'Rights of Man' appeared, and Telford, like many others, read it, and was at once carried away by it. Only a short time before, he had admitted with truth that he knew nothing of politics; but no sooner had he read Paine than he felt completely enlightened. He now suddenly discovered how much reason he and everybody else in England had for being miserable. Whilst residing at Portsmouth he had quoted to his Langholm friend the lines from Cowper's 'Task,' then just published, beginning "Slaves cannot breathe in England;" but lo! Mr. Paine had filled his imagination with the idea that England was nothing but a nation of bondmen and aristocrats. To his natural mind the kingdom had appeared to be one in which a man had pretty fair play, could think and speak, and do the thing he would,—tolerably happy, tolerably prosperous, and enjoying many blessings. He himself had felt free to labour, to prosper, and to rise from manual to head work. No one had hindered him; his personal liberty had never been interfered with; and he had freely employed his earnings as he thought proper. But now the whole thing appeared a delusion. Those rosy-cheeked

old country gentlemen who came riding in to Shrewsbury to quarter sessions, and were so fond of their young
Scotch surveyor — occupying themselves in building
bridges, maintaining infirmaries, making roads, and regulating gaols—those county magistrates and members of
parliament, aristocrats all, were the very men who,
according to Paine, were carrying the country headlong
to ruin.

If Telford could not offer an opinion on politics before,
because he "knew nothing about them," he had now no
such difficulty. Had his advice been asked about the
foundations of a bridge, or the security of an arch, he
would have read and studied much before giving it; he
would have carefully inquired into the chemical qualities
of different kinds of lime—into the mechanical principles of weight and resistance, and such like; but he had
no such hesitation in giving an opinion about the foundations of a constitution more than a thousand years
old. Here, like other young politicians, with Paine's
book before him, he felt competent to pronounce a
decisive judgment at once. "I am convinced," said he,
writing to his Langholm friend, "that the situation of
Great Britain is such, that nothing short of some signal
revolution can prevent her from sinking into bankruptcy,
slavery, and insignificancy." He held that the national
expenditure was so enormous,[1] arising from the corrupt
administration of the country, that it was impossible the
"bloated mass" could hold together any longer; and as
he could not expect that "a hundred Pulteneys," such as
his employer, could be found to restore it to health, the
conclusion he arrived at was that ruin was "inevitable."[2]

---

[1] It was then under seventeen millions sterling, or about a fourth of
what it is now.

[2] Letter to Mr. Andrew Little,
Langholm, dated 28th July, 1791.
Notwithstanding the theoretical ruin
of England which pressed so heavy
on his mind at this time, we find

Telford strongly recommending his
correspondent to send any good wrights
he can find in his neighbbourhood to
Bath, where they would be enabled
to earn twenty shillings or a guinea
a week at piece-work—the wages paid
at Langholm for similar work being
only about half those amounts.

Fortunately for Telford, his intercourse with the towns-people of Shrewsbury was so small that his views on these subjects were never known; and we very shortly find him employed by the clergy themselves in building a new church in the town of Bridgenorth. His patron and employer, Mr. Pulteney, however, knew of his extreme views, and the knowledge came to him quite accidentally. He found that Telford had made use of his frank to send through the post a copy of Paine's 'Rights of Man' to his Langholm correspondent,[1] where the pamphlet excited as much fury in the minds of some of the people of that town as it had done in that of Telford himself. The "Langholm patriots" broke out into drinking revolutionary toasts at the Cross, and so disturbed the peace of the little town that some of them were confined for six weeks in the county gaol. Mr. Pulteney was very indignant at the liberty Telford had taken with his frank, and a rupture between them seemed likely to ensue; but the former was forgiving, and the matter went no further. It is only right to add, that as Telford grew older and wiser, he became more careful in jumping at conclusions on political topics. The events which shortly occurred in France tended in a great measure to heal his mental distresses as to the future of England. When the "liberty" won by the Parisians ran into riot, and the "Friends of Man" occupied themselves in taking off the heads of those who differed from them, he became wonderfully reconciled to the

---

[1] The writer of a memoir of Telford, in the 'Encyclopedia Britannica,' says :—"Andrew Little kept a private and very small school at Langholm. Telford did not neglect to send him a copy of Paine's 'Rights of Man ;' and, as he was totally blind, he employed one of his scholars to read it in the evenings. Mr. Little had received an academical education before he lost his sight; and, aided by a memory of uncommon powers, he taught the classics, and particularly Greek, with much higher reputation than any other schoolmaster within a pretty extensive circuit. Two of his pupils read all the Iliad, and all or the greater part of Sophocles. After hearing a long sentence of Greek or Latin distinctly recited, he could generally construe and translate it with little or no hesitation. He was always much gratified by Telford's visits, which were not infrequent, to his native district."

enjoyment of the substantial freedom which, after all, was secured to him by the English Constitution. At the same time he was so much occupied in carrying out his important works, that he found but little time to devote either to political speculation or to verse-making.

Whilst living at Shrewsbury, he had his poem of 'Eskdale' reprinted for private circulation. We have also seen several MS. verses by him, written about the same period, which do not appear ever to have been printed. One of these—the best—is entitled 'Verses to the Memory of James Thomson, author of "Liberty, a poem;"' another is a translation from Buchanan, 'On the Spheres;' and a third, written in April, 1792, is entitled 'To Robin Burns, being a postscript to some verses addressed to him on the establishment of an Agricultural Chair in Edinburgh.' It would unnecessarily occupy our space to print these effusions; and, to tell the truth, they exhibit few if any indications of poetic power. No amount of perseverance will make a poet of a man in whom the divine gift is not born. The true line of Telford's genius lay in building and engineering, in which direction we now propose to follow him.

SHREWSBURY CASTLE. [By Percival Skelton.]

[As county surveyor, Telford frequently advised the magistrates as to improvement and building of roads and bridges.
In 1792 he was engaged as an architect to prepare designs and
superintend construction of the new parish church at Bridgenorth. To improve his architectural style, he made a study tour
to London, Gloucester, Worcester, Oxford, and Bath. He was
gratified by his inspection of the fine buildings of Bath and by
his study of ancient architectural remains at the British
Museum. He proposed to proceed with his favorite study of
architecture, but his engineering work as county surveyor and
then his appointment in 1793 as engineer for the projected Ellesmere Canal interfered. Fully occupied by engineering, he wrote
to Andrew Little, his childhood friend who was teaching at
Langholm, that his present literary project would be retarded
for some time (Rennie was translating the Scottish humanist
George Buchanan from the Latin).

The Ellesmere Canal was projected to join the Mersey, the
Dee, and the Severn. Land owners of the region anticipated the
increased market and rents that had followed the completion
of Duke of Bridgewater's Canal. The Bridgewater Canal had
brought lime, coal, manure, and merchandise to the doors of the
land owners along the way and provided economical transportation for their produce to market. Competition with farmers
adjacent to urban centers was then feasible.

Telford assumed responsibility for the masonry work of
locks, embankments, cuttings, and aqueducts but, having had
comparatively little experience in earthwork, he often took
the advice of the eminent engineer William Jessop. The construction of two great masonry aqueducts — the Chirk, and the
Pont-Cysylltau — was a substantial engineering achievement.
Sir Walter Scott spoke of the 1007-foot Pont-Cysylltau ˙Aque-
duct as "the most impressive work of art he had ever seen."
The affairs of the Ellesmere Canal, which involved not only
construction but considerable attention to the creating and guid-
ing of traffic upon completed portions, occupied Telford for
some years.* He carried on, however, lesser projects simultane-
ously. In his few leisure hours, he read extensively in such works
as Robertson's *Disquisitions on Ancient India* and Alison's
Principles of Taste as well as "occasionally throwing off a bit
of poetry."

Telford had used a cast-iron trough to carry the canal across the Pont-Cysylltau Aqueduct. Working as he was in the neighborhood of the Black Country of which coal and iron were the principal products, his interest in the application of cast iron in construction grew. Well acquainted with the innovating iron master, John Wilkinson, and having seen the famous cast-iron bridge at Coalbrookdale, Telford decided in 1795 upon iron as the material for a bridge over the Severn at Buildwas. Subsequently Telford introduced iron in canal works and road bridges where before only timber or stone had been employed.

The boldest of his designs was a one-arch cast-iron bridge over the Thames at London. A committee of Parliament concluded that the proposed bridge was practical, and "preliminary works were actually begun," but abandoned primarily because of the expensive approaches necessary in the area of valuable river-front property.*

By 1800 Telford was involved in many projects in the region of England extending from Liverpool at the mouth of the Mersey to Bristol at the mouth of the Severn, and from Wales to Birmingham. He wrote:

> This is an extensive and rich district, abounding in coal, lime, iron, and lead. Agriculture too is improving, and manufactures are advancing in rapid strides towards perfection. Think of such a mass of population, industrious, intelligent, and energetic, in continual exertion. In short, I do not believe that any part of the world, of like dimensions, ever exceeded Great Britain, as it now is, in regard to the production of wealth and the practice of the useful arts.

To Andrew Little he also wrote, in the midst of his plans of docks, canals, and bridges, about Goethe's poems, Kotzebue's plays, and Bonaparte's campaign in Egypt. Telford thought of his own great exertions and diverse activities as engineer as not unlike the campaigns of the great Bonaparte.]

CHAPTER V.

IN our introduction to the Life of Rennie, we gave a rapid survey of the state of Scotland about the middle of last century. We found a country without roads, fields lying uncultivated, mines unexplored, and all branches of industry languishing, in the midst of an idle, miserable, and haggard population. Fifty years passed, and the state of the Lowlands had become completely changed. Roads had been made, canals dug, coal-mines opened up, iron works established; manufactures were extending in all directions; and Scotch agriculture, instead of being the worst, was admitted to be the best in the island.

"I have been perfectly astonished," wrote Romilly from Stirling, in 1793, "at the richness and high cultivation of all the tract of this calumniated country through which I have passed, and which extends quite from Edinburgh to the mountains where I now am. It is true, however, that almost everything which one sees to admire in the way of cultivation is due to modern improvements; and now and then one observes a few acres of brown moss, contrasting admirably with the corn-fields to which they are contiguous, and affording a specimen of the dreariness and desolation which, only half a century ago, overspread a country now highly cultivated and become a most copious source of human happiness." [1]

It must, however, be admitted, that the industrial progress to which we have above referred was confined

---

[1] Romilly's 'Autobiography.'

mainly to the Lowlands, and had scarcely penetrated the mountainous regions lying towards the north-west. The rugged nature of that part of the country interposed a formidable barrier to improvement, and the district had thus remained very imperfectly opened up. The only practicable roads were those which had been made by the soldiery after the rebellions of 1715 and '45, through counties which before had been inaccessible except by dangerous footpaths across high and rugged mountains. One was formed along the Great Glen of Scotland, in the line of the present Caledonian Canal, connected with the Lowlands by the road through Glencoe by Tyndrum down the western banks of Loch Lomond; another, more northerly, connected Fort Augustus with Dunkeld by Blair Athol; whilst a third, still further to the north and east, connected Fort George with Cupar-in-Angus by Badenoch and Braemar. These roads were about eight hundred miles in extent, and maintained at the public expense. But they were laid out for purposes of military occupation rather than for the convenience of the districts which they traversed. Hence they were comparatively little used, and the Highlanders, in passing from one place to another, for the most part continued to travel by the old cattle tracks along the mountain sides. But the population were so poor and so spiritless, and industry was as yet in so backward a state all over the Highlands, that the want of communication was little felt. The state of agriculture may be inferred from the fact that an instrument called the cas-chrom [1]—literally, the "crooked-foot"—

---

[1] The *cas-chrom* was a rude combination of a lever for the removal of rocks, a spade to cut the earth, and a foot-plough to turn it. We annex an illustration of this curious and now obsolete instrument. It weighed about eighteen pounds. In working it, the upper part of the handle, to which the left hand was applied, reached the workman's shoulder, and being slightly elevated, the point, shod with iron, was pushed into the ground horizontally; the soil being turned over by inclining the handle to the furrow side, at the same time making the heel act as a fulcrum to raise the point of the instrument. In turning up unbroken ground, it was first em-

the use of which had been forgotten for hundreds of
years in every other country in Europe, was almost the
only tool employed in tillage in those parts of the Highlands which were separated by impassable roads from
the rest of the United Kingdom. The country lying on
the west of the Great Glen was absolutely without a
road. The native population were by necessity peaceful.
Old feuds were restrained by the strong arm of the law,
if indeed the spirit of the clans had not been completely
broken by the upshot of the rebellion of Forty-five.
But the people had not yet learnt to bend their backs,
like the Sassenach, to the stubborn soil, and they sat
gloomily by their turf-fires at home, or wandered away
to settle in other lands beyond the seas. It even began
to be feared that the country would become entirely
depopulated ; and it became a matter of national concern
to devise methods of opening up the district so as to
develope its industry and afford improved means of suste-
nance for its population. The poverty of the inhabitants
rendered the attempt to construct roads—even had they
desired them—beyond their scanty means ; but the
ministry of the day entertained the opinion that by contributing a certain proportion of the necessary expense,
the proprietors of Highland estates might be induced to
advance the remainder ; and on this principle the construction of the new roads in those districts was undertaken.

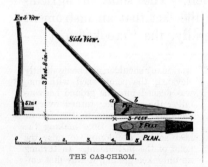

THE CAS-CHROM.

ployed with the heel uppermost, with
pushing strokes to cut the breadth of
the sward to be turned over ; after
which, it was used horizontally as
above described. We are indebted to
a Parliamentary Blue Book for our
representation of this interesting relic
of ancient agriculture. It is given in
the appendix to the 'Ninth Report of
the Commissioners for Highland Roads
and Bridges,' ordered by the House of
Commons to be printed, 19th April,
1821.

In 1802 Mr. Telford was called upon by the Government to make a survey of Scotland, and report as to the measures which were necessary for the improvement of the roads and bridges of that part of the kingdom, and also on the means of promoting the fisheries on the east and west coasts, with the object of better opening up the country and preventing further extensive emigration. Previous to this time he had been employed by the British Fisheries Society—of which his friend Sir William Pulteney was Governor—to inspect the harbours at their several stations, and to devise a plan for the establishment of a fishery on the coast of Caithness. He accordingly made an extensive tour of Scotland, examining, amongst other harbours, that of Annan; from which he proceeded northward by Aberdeen to Wick and Thurso, returning to Shrewsbury by Edinburgh and Dumfries.[1] He accumulated a large mass of data for his report, which was sent in to the Fishery Society, with charts and plans, in the course of the following year.

In July, 1802, he was requested by the Lords of the Treasury, most probably in consequence of the preceding report, to make a further survey of the interior of the Highlands, the result of which he communicated in his report presented to Parliament in the following year. Although full of important local business, "kept running," as he says, "from town to country, and from country to town, never when awake, and perhaps not always when asleep, have my Scotch surveys been absent from my mind." He had worked very hard at

---

[1] He was accompanied on this tour by Colonel Dixon, with whom he returned to his house at Mount Annan, in Dumfries. Telford says of him: "The Colonel seems to have roused the county of Dumfries from the lethargy in which it has slumbered for centuries. The map of the county, the mineralogical survey, the new roads, the opening of lime works, the competition of ploughing, the improving harbours, the building of bridges, are works which bespeak the exertions of no common man."—Letter to Mr. Andrew Little, dated Shrewsbury, 30th November, 1801.

his report, and hoped that it might be productive of some good.

The report was duly presented, printed,[1] and approved, and it formed the starting-point of a system of legislation with reference to the Highlands which extended over many years, and had the effect of completely opening up that romantic but rugged district of country, and extending to its inhabitants the advantages of improved intercourse with the other parts of the kingdom. Mr. Telford pointed out that the military roads which had been made there were altogether insufficient, and that the use of them was in many places very much circumscribed by the want of bridges over some of the principal rivers. For instance, the route from Edinburgh to Inverness, by the Central Highlands, was seriously interrupted at Dunkeld, where the Tay is broad and deep, and not always easy to be crossed by means of a boat. The route to the same place by the east coast was in like manner broken at Fochabers, where the rapid Spey could only be crossed by a dangerous ferry.

The difficulties encountered by the Bar, in travelling the north circuit about this time, are well described by Lord Cockburn in his 'Memorials.' "Those who are born to modern travelling," he says, "can scarcely be made to understand how the previous age got on. The state of the roads may be judged of from two or three facts. There was no bridge over the Tay at Dunkeld, or over the Spey at Fochabers, or over the Findhorn at Forres. Nothing but wretched pierless ferries, let to poor cottars, who rowed, or hauled, or pushed a crazy boat across, or more commonly got their wives to do it. There was no mail-coach north of Aberdeen till, I think, after the battle of Waterloo. What it must have been

---

[1] Ordered to be printed 5th of April, 1803.

a few years before my time may be judged of from Bozzy's ' Letter to Lord Braxfield,' published in 1780. He thinks that, besides a carriage and his own carriage-horses, every judge ought to have his sumpter-horse, and ought not to travel faster than the waggon which carried the baggage of the circuit. I understood from Hope that, after 1784, when he came to the Bar, he and Braxfield *rode* a whole north circuit ; and that, from the Findhorn being in a flood, they were obliged to go up its banks for about twenty-eight miles to the bridge of Dulsie before they could cross. I myself rode circuits when I was Advocate-Depute between 1807 and 1810. The fashion of every Depute carrying his own shell on his back, in the form of his own carriage, is a piece of very modern antiquity." [1]

North of Inverness, matters were still worse, if possible. There was no bridge over the Beauley or the Conan. The drovers coming south swam the rivers with their cattle. There being no roads, there was little use for carts. In the whole county of Caithness there was scarcely a farmer who owned a wheel-cart. Burdens were conveyed usually on the backs of ponies, but quite as often on the backs of women.[2] The interior of the county of Sutherland being almost inaccessible, the only track lay along the shore, amongst rocks and sand, covered by the sea at every tide. "The people lay scattered in inaccessible straths and spots among the mountains, where they lived in family with their pigs and kyloes (cattle), in turf cabins of the most miserable description ; they spoke only Gaelic, and spent the whole of their time in indolence and sloth. Thus they had gone on from father to son, with little change, except what the introduction of illicit distillation had wrought, and making little or no export from the country beyond

---

[1] ' Memorials of his Time,' by Henry Cockburn, pp. 341-3.
[2] ' Memoirs of the Life and Writings of Sir John Sinclair, Bart.,' vol. i., p. 339.

the few lean kyloes, which paid the rent and produced
wherewithal to pay for the oatmeal imported." [1]

Telford's first recommendation was, that a bridge
should be thrown across the Tay at Dunkeld, to connect
the improved lines of road proposed to be made on each
side of the river. He regarded this measure as of the
first importance to the Central Highlands; and as the
Duke of Athol was willing to pay one-half of the cost of
the erection, if the Government would defray the other—
the bridge to be free of toll after a certain period—it appeared to the engineer that this was a reasonable and just
mode of providing for the contingency. In the next
place, he recommended a bridge over the Spey, which
drained a great extent of mountainous country, and,
being liable to sudden inundations, was very dangerous
to cross. Yet this ferry formed the only link of communication between the whole of the northern counties.
The site pointed out for the proposed bridge was adjacent to the town of Fochabers, and here also the Duke
of Gordon and other county gentlemen were willing to
provide one-half the means for its erection.

Mr. Telford further described in detail the roads
necessary to be constructed in the north and west Highlands, with the object of opening up the western parts
of the counties of Inverness and Ross, and affording a
ready communication from the Clyde to the fishing
lochs in the neighbourhood of the Isle of Skye. As to
the means of executing these improvements, he suggested
that Government would be justified in dealing with the
Highland roads and bridges as exceptional and extraordinary works, and extending the public aid towards
carrying them into effect, as but for such assistance the
country must remain, perhaps for ages to come, imperfectly connected. His report further embraced certain
improvements in the harbours of Aberdeen and Wick,

---

[1] Extract of a letter from a gentleman residing in Sutherland, quoted in
'Life of Telford,' p. 465.

and a description of the country through which the
proposed line of the Caledonian Canal would necessarily
pass—a canal which had long been the subject of inquiry, but had not as yet emerged from a state of mere
speculation.

The new roads and bridges, and other improvements
suggested by the engineer, excited much interest in the
north.   The Highland Society voted him their thanks by
acclamation ; the counties of Inverness and Ross followed ;
and he had letters of thanks and congratulation from
many of the Highland chiefs.   " If they will persevere,"
says he, " with anything like their present zeal, they will
have the satisfaction of greatly improving a country that
has been too long neglected.   Things are greatly changed
now in the Highlands.   Even were the chiefs to quarrel,
de'il a Highlandman would stir for them.   The lairds
have transferred their affections from their people to
flocks of sheep, and the people have lost their veneration
for the lairds.   It seems to be the natural progress of
society ; but it is not an altogether satisfactory change.
There were some fine features in the former patriarchal
state of society ; but now clanship is gone, and chiefs and
people are hastening into the opposite extreme.   This
seems to me to be quite wrong." [1]

In the same year Telford was elected a member of the
Royal Society of Edinburgh, on which occasion he was
proposed and supported by three professors ; so that the
former Edinburgh mason was rising in the world and
receiving due honour in his own country.   The effect of
his report was such, that in the session of 1803 a Parliamentary Commission was appointed, under whose direction a series of practical improvements was commenced,
which issued in the construction of not less than 920
additional miles of roads and bridges throughout the
Highlands, one-half of the cost of which was defrayed by

---

[1] Letter to Mr. Andrew Little, Langholm, dated Salop, 18th February, 1803.

MAP OF TELFORD'S ROADS.

the Government and the other half by local assessment. But in addition to these main lines of communication, numberless county roads were formed by statute labour, under local road Acts and by other means; the land-owners of Sutherland alone having formed nearly 300 miles of district roads at their own cost.

By the end of the session of 1803 Telford received his instructions from Mr. Vansittart as to the survey he was forthwith to enter upon, with a view to commencing practical operations; and he again proceeded to the High-lands to lay out the roads and plan the bridges which were most urgently needed. The district of the Solway was, at his representation, included, with the object of improving the road from Carlisle to Portpatrick—the nearest point at which Great Britain meets the Irish coast, and where the sea passage forms only a sort of wide ferry.

It would occupy too much space, and indeed it is alto-gether unnecessary, to describe in detail the operations of the Commission and of their engineer in opening up the communications of the Highlands. Suffice it to say, that one of the first things taken in hand was the con-nection of the new lines of road by means of bridges at the more important points; such as at Dunkeld over the Tay, and near Dingwall over the Conan and Orrin. That at Dunkeld was the most important, as being the portal to the Central Highlands; and at the second meeting of the Commissioners Mr. Telford submitted his plan and estimates of the proposed bridge. In conse-quence of some difference with the Duke of Athol as to his share of the expense—which proved to be greater than he had taken into account—some delay occurred in the commencement of the work; but at length it was fairly begun, and after being three years in hand the structure was finished and opened for traffic in 1809.

The bridge is a handsome one of five river and two land arches. The span of the centre arch is 90 feet, of the two adjoining it 84 feet, and of the two side arches 74 feet; affording a clear waterway of 446 feet. The total breadth of the roadway and footpaths is 28 feet 6 inches. The cost of the structure was about 14,000*l.*, one-half of which was defrayed by the Duke of Athol. It forms a fine feature in a landscape not often surpassed, presenting within a comparatively small compass a great variety of character and beauty.

DUNKELD BRIDGE.

[By Percival Skelton, after a sketch by J. S. Smiles.]

The communication by road north of Inverness was also perfected by the construction of a bridge of five arches over the Beauley, and another of the same number over the Conan, the central arch being 65 feet span; and the formerly wretched bit of road between these points having been put in good repair, the town of Dingwall was thereupon rendered easily approachable from the south. At the same time a beginning was made with the construction of new roads through the districts most in need of them. The first contracted for was the Loch-na-Gaul road, from Fort William to Arasaig, on the western coast, nearly opposite the island of Egg. Another was begun from Loch Oich, on the line of the Caledonian Canal, across the middle of the Highlands, through Glengarry, to Loch Hourn on the western sea. Other roads were opened north and south; through Morvern to Loch Moidart; through Glen Morrison and Glen Sheil, and through the entire Isle of Skye; from Dingwall, eastward, to Lochcarron and Loch Torridon, quite through the county of Ross; and from Dingwall, northward, through the county of Sutherland as far as Tongue on the Pentland Frith; whilst another line, striking off at the head of the Dornoch Frith, proceeded along the coast in a north-easterly direction to Wick and Thurso, in the immediate neighbourhood of John o' Groats. There were numerous other subordinate lines, which it is unnecessary to specify in detail; but some idea may be formed of their extent, as well as of the rugged character of the country through which they were carried, when we state that they involved the construction of no fewer than twelve hundred bridges. Several important bridges were also erected at other points to connect existing roads, such as those at Ballater and Potarch over the Dee; at Alford over the Don; and at Craig-Ellachie over the Spey.

The last-named bridge is a remarkably elegant structure, thrown over the Spey at a point where the river, rushing obliquely against the lofty rock of Craig-Ellachie,[1] has formed for itself a deep channel not exceeding fifty yards in breadth. Only a few years before, there had not been any provision for crossing this river at its lower parts except the very dangerous ferry at Fochabers. The Duke of Gordon had, however, erected a suspension bridge at that town, and the inconvenience was in a great

---

[1] The names of Celtic places are highly descriptive. Thus Craig-Ellachie literally means, the rock of separation; Badenoch, bushy or woody; Cairngorm, the blue cairn; Lochinet, the lake of nests; Balknockan, the town of knolls; Dalnasealg, the hunting dale; All'n dater, the burn of the horn-blower; and so on.

measure removed. Its utility was so generally felt, that
the demand arose for a second bridge across the river;
for there was not another by which it could be crossed
for a distance of nearly fifty miles up Strath Spey. It
was a difficult stream to span by a bridge at any place,
in consequence of the violence with which the floods
descended at particular seasons. Sometimes, even in
summer, when not a drop of rain had fallen, the flood
would come down the Strath in great fury, sweeping
everything before it; this remarkable phenomenon being
accounted for by the prevalence of a strong south-westerly
wind, which blew the loch waters from their beds into the
Strath, and thus suddenly filled the valley of the Spey.[1]
The same phenomenon, similarly caused, is also frequently
observed in the neighbouring river, the Findhorn, cooped
up in its deep rocky bed, where the water sometimes
comes down in a wave six feet high, like a liquid wall,
sweeping everything before it. To meet such a contingency, it was deemed necessary to provide abundant
waterway, and to build a bridge offering as little resistance as possible to the passage of the Highland floods.
Telford accordingly designed for the passage of the river
at Craig-Ellachie a light cast iron arch of 150 feet span,
with a rise of 20 feet, the arch being composed of four
ribs, each consisting of two concentric arcs forming
panels, which are filled in with diagonal bars. The
roadway is 15 feet wide, and is formed of another arc of
greater radius, attached to which is the iron railing; the
spandrels being filled by diagonal ties, forming trellis-
work. Mr. Robert Stephenson took objection to the two
dissimilar arches, as liable to subject the structure, from
variations of temperature, to very unequal strains. Nevertheless this bridge, as well as many others constructed
by Mr. Telford after a similar plan, has stood perfectly

---

[1] Sir Thomas Dick Lauder has
vividly described the destructive character of the Spey-side inundations in
his capital book on the 'Morayshire
Floods.'

well, and to this day remains a very serviceable structure.
Its appearance is highly picturesque. The scattered pines
and beech trees on the side of the impending mountain, the meadows along the valley of the Spey, and the
western approach road to the bridge cut deeply into the
face of the rock, combine, with the slender appearance
of the iron arch, in rendering this spot one of the most
remarkable in Scotland.[1]

CRAIG-ELLACHIE BRIDGE. [By Percival Skelton]

An iron bridge of a similar span to that at Craig-
Ellachie had previously been constructed across the
head of the Dornoch Frith at Bonar, near the point
where the waters of the Shin join the sea. The very
severe trial which this structure sustained from the
tremendous blow of an irregular mass of fir-tree logs,

---

[1] 'Report of the Commissioners on Highland Roads and Bridges.' Appendix to ' Life of Telford,' p. 400.

consolidated by ice, as well as, shortly after, from the blow of a schooner which drifted against it on the opposite side, and had her two masts knocked off by the collision, gave him every confidence in the strength of this form of construction, and he accordingly repeated it in several of his subsequent bridges, though none of them are comparable in beauty with that of Craig-Ellachie.

Thus, in the course of eighteen years, 920 miles of capital roads, connected together by no fewer than 1200 bridges, were added to the road communications of the Highlands, at an expense defrayed partly by the localities immediately benefited, and partly by the nation. The effects of these twenty years' operations were such as follow the making of roads everywhere—development of industry and increase of civilization. In no districts were the benefits derived from them more marked than in the remote northern counties of Sutherland and Caithness. The first stage-coaches that ran northward from Perth to Inverness were tried in 1806, and became regularly established in 1811; and by the year 1820 no fewer than forty arrived at the latter town in the course of every week, and the same number departed from it. Others were established in various directions through the Highlands, which became as accessible as any English county.

Agriculture made rapid progress. The use of carts became practicable, and manure was no longer carried to the field on women's backs. Sloth and idleness disappeared before the energy, activity, and industry which were called into life by the improved communications; better built cottages took the place of the old mud biggins with a hole in the roof to let out the smoke; the pigs and cattle were treated to a separate table; the dunghill was turned to the outside of the house; tartan tatters gave place to the produce of Manchester and Glasgow looms; and very soon few young persons were to be found who could not both

read and write English. But not less remarkable were
the effects of the road-making upon the industrial habits
of the people. Before Telford went into the Highlands,
they did not know how to work, having never been
accustomed to labour continuously and systematically.
Telford himself thus describes the moral influences of his
Highland contracts :—" In these works," says he, " and
in the Caledonian Canal, about three thousand two
hundred men have been annually employed. At first,
they could scarcely work at all ; they were totally unac-
quainted with labour ; they could not use the tools.
They have since become excellent labourers, and of
the above number we consider about one-fourth left us
annually, taught to work. These undertakings may,
indeed, be regarded in the light of a working academy,
from which eight hundred men have annually gone
forth improved workmen. They have either returned
to their native districts with the advantage of having
used the most perfect sort of tools and utensils (which
alone cannot be estimated at less than ten per cent. on
any sort of labour), or they have been usefully distributed through the other parts of the country. Since
these roads were made accessible, wheelwrights and
cartwrights have been established, the plough has been
introduced, and improved tools and utensils are generally used. The plough was not previously employed ;
in the interior and mountainous parts they used crooked
sticks, with iron on them, drawn or pushed along. The
moral habits of the great masses of the working classes
are changed ; they see that they may depend on their
own exertions for support : this goes on silently, and is
scarcely perceived until apparent by the results. I consider these improvements among the greatest blessings
ever conferred on any country. About two hundred
thousand pounds has been granted in fifteen years. It
has been the means of advancing the country at least a
century."

[Funds were also available to improve harbors of the Scottish coast. These would complement the roads and bridges on which progress was being made. The Parliamentary Commission adopted Telford's plan for the further improvement of Wick Harbor, and work began in 1808. The improved harbor helped a little poverty-stricken village become one of the world's great fishing stations. Telford specified alterations and improvements in numerous other small harbors, but his principal works in Scotland were those of Aberdeen and Dundee. Improvements at Aberdeen by Telford and other engineers contributed to the enlargement and beautification of the city, the rapid progress of the shipbuilding industry, and the development of thriving commerce in wool, cotton, flax, and iron. The tonnage entering the port increased from 50,000 tons in 1800 to 300,000 in 1860.]

CHAPTER VI.

THE formation of a navigable highway through the chain of lochs occupying the Great Glen of the Highlands, extending diagonally across Scotland from the Atlantic to the North Sea, had long been regarded as a work of national importance. As early as 1773, James Watt, when following the business of a land-surveyor at Glasgow, made a survey of the country at the instance of the Commissioners of Forfeited Estates. He pronounced the canal practicable, and pointed out how it could best be constructed. There was certainly no want of water, for Watt was drenched with rain during most of his survey, and had difficulty in preserving even his journal book. "On my way home," he says, "I passed through the wildest country I ever saw, and over the worst conducted roads."

Twenty years later, in 1793, Mr. Rennie was consulted as to the canal, and he also prepared a scheme; but nothing was done. The project was, however, revived in 1801, during the war with Napoleon, when various inland ship canals—such as those from London to Portsmouth, and from Bristol to the English Channel—were under consideration, with the view of protecting British shipping against French privateers. But there was another reason for urging the formation of the canal through the Great Glen of Scotland, which was regarded as of importance before the introduction of steam enabled vessels to set the winds and tides at comparative defiance. It was this: vessels sailing from the eastern ports to America had to beat up the Pentland Frith often against adverse winds and stormy seas, which rendered the navigation both tedious and dangerous. Thus it was cited by Sir Edward Parry, in his evidence before Parliament in favour of completing the Caledonian Canal, that of two vessels despatched from Newcastle on the same day—one bound for Liverpool by the north of Scotland, and the other for Bombay by the English Channel and the Cape of Good Hope—the latter reached its destination first! Another case may be mentioned, of an Inverness vessel, which sailed for Liverpool on a Christmas Day, reached Stromness Harbour, in Orkney, on the 1st of January, and lay there windbound, with a fleet of other traders, until the middle of April following! In fact the Pentland Frith, which is the throat connecting the Atlantic and German Oceans, and through which the former rolls its long majestic waves with tremendous force, was the dread of mariners, and it was considered an object of national importance to mitigate the dangers of the passage towards the western seas.

As the lochs occupying the chief part of the bottom of the Great Glen were of sufficient depth to be navigable by large vessels, it was thought that if they could be connected by a ship canal, so as to render the line of navigation continuous, it would be used by shipping to a large extent, and prove of great public service. Five hundred miles of dangerous navigation by the Orkneys and Cape Wrath would thus be saved, and ships of war, were this track open to them, might reach the north of Ireland in two days from Fort George, near Inverness. When the scheme of the proposed canal was revived in 1801, Mr. Telford was requested to make a survey and send in his report on the subject. He immediately wrote to his friend James Watt, saying, " I have so long accustomed myself to look with a degree of reverence to your work, that I am particularly anxious to learn what occurred to you in this business while the whole was fresh in your mind. The object appears to me so great and so desirable, that I am convinced you will feel a pleasure in bringing it again under investigation, and I am very desirous that the thing should be fully and fairly explained, so that the public may be made aware of its extensive utility. If I can accomplish this, I shall have done my duty ; and if the project is not executed now, some future period will see it done, and I shall have the satisfaction of having followed you and promoted its success." [1] We may here state that Telford's survey agreed with Watt's in the most important particulars, and that he largely cited his descriptions of the intended scheme in his own report.

Mr. Telford's first inspection of the district was made in 1801, and his report was sent in to the Treasury in the course of the following year. Lord Bexley, then Secretary to the Treasury, took a warm personal interest in the project, and lost no opportunity of actively promoting it. A board of commissioners was eventually appointed to carry out the formation of the canal. Mr. Telford, on being appointed principal engineer of the undertaking, was requested at once to proceed to Scotland and prepare the necessary working survey. He was accompanied on the occasion by Mr. Jessop as consulting engineer.* Twenty thousand pounds were granted under the provisions of the 43 Geo. III. (chap. cii.), and the works were commenced, in the beginning of 1804, by the formation of a dock or basin adjoining the intended tide-lock at Corpach, near Bannavie.

The basin at Corpach formed the southernmost point of the intended canal. It is situated at the head of Loch Eil, amidst some of the grandest scenery of the Highlands. Across the Loch is the little town of Fort William, one of the forts established at the end of the

---

[1] 'The Origin and Progress of the Mechanical Inventions of James Watt.' By J. P. Muirhead, Esq., M.A. Vol. i., p. cxxvi.

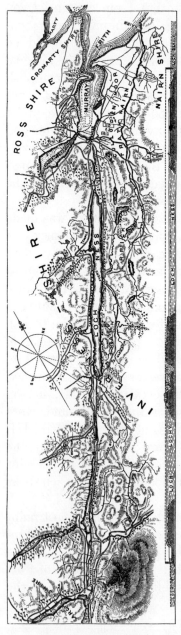

MAP OF CALEDONIAN CANAL.

seventeenth century to keep the wild Highlanders in subjection. Above it, hills over hills arise, of all forms and sizes, and of all hues, from grass-green below to heather-brown and purple above, capped with heights of weather-beaten grey; whilst towering over all stands the rugged mass of Ben Nevis —a mountain almost unsurpassed for picturesque grandeur. Along the western foot of the range, which extends for some six or eight miles, lies a long extent of brown bog, on the verge of which, by the river Lochy, stand the ruins of Inverlochy Castle.

The works at Corpach involved great labour, and extended over a long series of years. The difference between the level of Loch Eil and Loch Lochy is ninety feet, whilst the distance between them was less than eight miles, and it was therefore necessary to climb up the side of the hill by a flight of eight gigantic locks, clustered together, and which Telford named Neptune's Staircase. The ground passed over was in some places very difficult, requiring large masses of embankment, the slips of which in the course of the work frequently occasioned serious embarrassment. The basin on Loch Eil, on the other hand, was constructed amidst rocks, and considerable difficulty was experienced in getting in the necessary coffer-dam for the construction of the entrance to the sea-lock, the entrance-sill of which was laid upon the rock itself, so that there was a depth of 21 feet of water upon it at high water of neap tides.[1]

At the same time that the works at Corpach were commenced, the dock or basin at the north-eastern extremity of the canal, situated at Clachnagarry, on the shore of Loch Beauly, was also laid out, and the excavations and embankments were carried on with considerable activity. This dock was constructed about 967 yards long, and upwards of 162 yards in breadth, giving an area of about 32 acres,—forming, in fact, a harbour for the vessels using the canal. The dimensions of the artificial waterway were of unusual size, as the intention was to adapt it throughout for the passage of a 32-gun frigate of that day, fully equipped and laden with stores. The canal, as originally resolved upon, was designed to be 110 feet wide at the surface, and 50 feet at the bottom, with a depth in the middle of 20 feet; though these dimensions were somewhat modified in the execution of the work. The locks were of corresponding large dimensions, each being from 170 to 180 feet long, 40 broad, and 20 deep.

Between these two extremities of the canal—Corpach on the south-west and Clachnagarry on the north-east—extends the chain of fresh-water lochs: Loch Lochy on the south; next Loch Oich; then Loch Ness; and lastly, furthest north, the small Loch of Doughfour. The whole length of the navigation is 60 miles 40 chains, of which the navigable lochs constitute about 40 miles,

---

[1] For professional details of this work, see 'Life of Telford,' p. 305.

leaving only about 20 miles of canal to be constructed, but of unusually large dimensions and through a very difficult country.

The summit loch of the whole is Loch Oich, the surface of which is exactly a hundred feet above high water-mark, both at Inverness and Fort William; and to this sheet of water the navigation climbs up by a series of locks from both the eastern and western seas. The whole number of these is twenty-eight: the entrance-lock at Clachnagarry, constructed on piles, at the

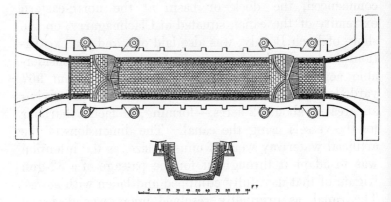

0 10 20 30 40 50 60 70 80 90 100 FT

LOCK, CALEDONIAN CANAL.

end of huge embankments, forced out into deep water, at Loch Beauly;[1] another at the entrance to the capacious artificial harbour above mentioned at Muirtown; four connected locks at the southern end of this basin; a

---

[1] In the 'Sixteenth Report of the Commissioners of the Caledonian Canal,' the following reference is made to this important work, which was finished in 1812:—" The depth of the mud on which it may be said to be artificially seated is not less than 60 feet; so that it cannot be deemed superfluous, at the end of seven years, to state that no subsidence is discoverable; and we presume that the entire lock, as well as every part of it, may now be deemed as immovable, and as little liable to destruction, as any other large mass of masonry. This was the most remarkable work performed under the immediate care of Mr. Matthew Davidson, our superintendent at Clachnagarry, from 1804 till the time of his decease. He was a man perfectly qualified for the employment by inflexible integrity, unwearied industry, and zeal to a degree of anxiety, in all the operations committed to his care."|*

regulating lock a little to the north of Loch Doughfour; five contiguous locks at Fort Augustus, at the south end of Loch Ness; another, called the Kytra Lock, about midway between Fort Augustus and Loch Oich; a regulating lock at the north-east end of Loch Oich; two contiguous locks between Lochs Oich and Lochy; a regulating lock at the south-west end of Loch Lochy; next, the grand series of locks, eight in number, called " Neptune's Staircase," at Bannavie, within a mile and a quarter of the sea; two locks, descending to Corpach basin; and lastly, the great entrance or sea-lock at Corpach.

As may naturally be supposed, the execution of these great works involved vast labour and anxiety. They were designed with much skill, and executed with equal ability. There were lock-gates to be constructed, prin- cipally of cast iron, sheathed with pine planking; public road bridges crossing the line of the canal, of which there were eight, constructed of cast iron and swung horizontally. There were many mountain streams, swollen to torrents in winter, crossing under the canal, for which abundant accommodation had to be provided, involving the construction of numerous culverts, tunnels, and under-bridges of large dimensions. There were also powerful sluices provided to let off the excess of water sent down from the adjacent mountains into the canal during winter. Three of these, of large dimen- sions, high above the river Lochy, are constructed at a point where the canal is cut through the solid rock; and the sight of the mass of waters rushing down into the valley beneath, gives an impression of power which, once seen, is never forgotten.

These great works were only brought to a completion after the labours of many years, during which the diffi- culties encountered in their construction had swelled the cost of the canal far beyond the original estimate. The rapid advances which had taken place in the interval

in the prices of labour and materials also tended greatly
to increase the expenses, and, after all, the canal, when
completed and opened, was comparatively little used.
This was doubtless owing, in a great measure, to the
rapid changes which occurred in the system of navigation shortly after the projection of the undertaking.
For these Telford was not responsible. He was called
upon to make the canal, and he did so in the best
manner. Engineers are not required to speculate as
to the commercial value of the works they are required
to construct; and there were circumstances connected
with the scheme of the Caledonian Canal which removed
it from the category of mere commercial adventures.
It was a Government project, and it proved a failure.
Hence it formed a prominent topic for discussion in
the journals of the day; but the attacks made upon the
Government because of their expenditure on the hapless
undertaking were perhaps more felt by Telford, who
was its engineer, than by all the ministers of state con-
joined. *

"The unfortunate issue of this great work," writes
the present engineer of the canal, to whom we are indebted for many of the preceding facts, "was a grievous
disappointment to Mr. Telford, and was in fact the one
great bitter in his otherwise unalloyed cup of happiness
and prosperity. The undertaking was maligned by
thousands who knew nothing of its character. It became 'a dog with a bad name,' and all the proverbial
consequences followed. The most absurd errors and
misconceptions were propagated respecting it from year
to year, and it was impossible during Telford's lifetime
to stem the torrent of popular prejudice and objurga-
tion. It must, however, be admitted, after a long
experience, that Telford was greatly over-sanguine in
his expectations as to the national uses of the canal, and
he was doomed to suffer acutely in his personal feelings,
little though he may have been personally to blame,

the consequences of what in this commercial country is regarded as so much worse than a crime, namely, a financial mistake." [1]

Mr. Telford's great sensitiveness made him feel the ill success of this enterprise far more than most other men would have done. He was accustomed to throw himself into the projects on which he was employed with an enthusiasm almost poetic. He regarded them not merely as so much engineering, but as works which were to be instrumental in opening up the communications of the country and extending its civilization. Viewed in this light, his canals, roads, bridges, and harbours were unquestionably of great national importance, though their commercial results might not in all cases justify the estimates of their projectors. To refer to like instances—no one can doubt the immense value and public uses of Mr. Rennie's Waterloo Bridge or Mr. Robert Stephenson's Britannia and Victoria Bridges, though every one knows that, commercially, they have been failures. But it is probable that neither of these eminent engineers gave himself anything like the anxious concern that Telford did about the financial issue of his undertaking. Were railway engineers to fret and vex themselves about the commercial value of the schemes in which they have been engaged, there

---

[1] The misfortunes of the Caledonian Canal did not end with the life of Telford. The first vessel passed through it from sea to sea in October, 1822, by which time it had cost about a million sterling, or double the original estimate. Notwithstanding this large outlay, it appears that the canal was opened before the works had been properly completed ; and the consequence was that they very shortly fell into decay. It even began to be considered whether the canal ought not to be abandoned. In 1838, Mr. James Walker, C.E., an engineer of the highest eminence, examined it, and reported fully on its then state, strongly recommending its completion as well as its improvement. His advice was eventually adopted, and the canal was finished accordingly, at an additional cost of about 200,000*l*., and the whole line was re-opened in 1847, since which time it has continued in useful operation. The passage from sea to sea at all times can now be depended on, and it can usually be made in forty-eight hours. As the trade of the North increases, the uses of the canal will probably become much more decided than they have heretofore proved.

are few of them but would be so haunted by the ghosts of wrecked speculations that they could scarcely lay their heads upon their pillows for a single night in peace.

While the Caledonian Canal was in progress, Mr. Telford was occupied in various works of a similar kind in England and Scotland, and also one in Sweden. In 1804, while on one of his journeys to the North, he was requested by the Earl of Eglinton and others to examine a project for making a canal from Glasgow to Saltcoats and Ardrossan, on the north-western coast of the county of Ayr, passing near the important manufacturing town of Paisley. A new survey of the line was made, and the works were carried on during several successive years until a very fine capacious canal was completed, on the same level, as far as Paisley and Johnstown. But the funds of the company falling short, the works were stopped, and the canal was carried no further. Besides, the measures so actively employed by the Clyde Trustees to deepen the bed of that river and enable ships of large burden to pass up as high as Glasgow, had proved so successful that the ultimate extension of the canal to Ardrossan, so as to avoid the shoals of the Clyde, was no longer necessary, and the prosecution of the work was accordingly abandoned. But as Mr. Telford has observed, no person suspected, when the canal was laid out in 1805, " that steamboats would not only monopolise the trade of the Clyde, but penetrate into every creek where there is water to float them, in the British Isles and the continent of Europe, and be seen in every quarter of the world."

Another of the navigations on which Mr. Telford was long employed was that of the river Weaver in Cheshire. It was only twenty-four miles in extent, but of considerable importance to the country through which it passed, accommodating the salt-manufacturing districts, of which the towns of Nantwich, Northwich, and

Frodsham are the centres. The channel of the river was extremely crooked and much obstructed by shoals, when Telford took the navigation in hand in the year 1807, and a number of essential improvements were made in it, by means of new locks, weirs, and side cuts, which had the effect of greatly improving the communications of these important districts.

In the following year we find our engineer consulted, at the instance of the King of Sweden, on the best mode of constructing the Gotha Canal, between Lake Wenern and the Baltic, to complete the communication with the North Sea. In 1808, at the invitation of Count Platen,*Mr. Telford visited Sweden and made a careful survey of the district. The service occupied him and his assistants two months, after which he prepared and sent in a series of detailed plans and sections, together with an elaborate report on the subject. His plans having been adopted, he again visited Sweden in 1810, to inspect the excavations which had already been begun, when he supplied the drawings for the locks and bridges. With the sanction of the British Government, he at the same time furnished the Swedish contractors with patterns of the most improved tools used in canal making, and took with him a number of experienced lock-makers and navvies for the purpose of instructing the native workmen. The construction of the Gotha Canal was an undertaking of great magnitude and difficulty, similar in many respects to the Caledonian Canal, though much more extensive. The length of artificial canal was 55 miles, and of the whole navigation, including the lakes, 120 miles. The locks are 120 feet long and 24 feet broad; the width of the canal at bottom being 42 feet, and the depth of water 10 feet. The results, so far as the engineer was concerned, were much more satisfactory than in the case of the Caledonian Canal. Whilst in the one case he had much obloquy to suffer for the services he had given,

in the other he was honoured and fêted as a public
benefactor, the King conferring upon him the Swedish
order of knighthood, and presenting him with his portrait set in diamonds.

[Among the various other canals in England which Telford
was employed to construct or improve upon were the Gloucester
and Berkeley Canal, in 1818; the Grand Trunk Canal, in 1822;
the Harecastle Tunnel, in 1824–27; the Birmingham Canal, in
1824; and the Birmingham and Liverpool Junction Canals, in
1825. Smiles comments that "Telford was justly proud of his
canals . . . the finest works of their kind that had yet been
executed in England."]

CHAPTER VII.

Mr. Telford's extensive practice as a bridge-builder
led his friend Southey to designate him " Pontifex
Maximus." Besides the numerous bridges erected by
him in the West of England, we have found him furnishing designs for about twelve hundred in the Highlands, of various dimensions, some of stone and others
of iron. His practice in bridge-building had, therefore, been of an unusually extensive character, and
Southey's sobriquet was not ill applied. But besides
being a great bridge-builder, Telford was also a great
road-maker. With the progress of industry and trade,
the easy and rapid transit of persons and goods had
come to be regarded as an increasing object of public
interest. Fast coaches now ran regularly between all
the principal towns of England; every effort being
made, by straightening and shortening the roads, cutting down hills, and carrying embankments across valleys
and viaducts over rivers, to render travelling by the
main routes as easy and expeditious as possible.

Attention was especially turned to the improvement
of the longer routes, and to perfecting the connection of London with the chief towns of Scotland and
Ireland. Telford was early called upon to advise as to
the repairs of the road between Carlisle and Glasgow,
which had been allowed to fall into a wretched state;
as well as the formation of a new line from Carlisle,
across the counties of Dumfries, Kirkcudbright, and
Wigton, to Port Patrick, for the purpose of ensuring a
more rapid communication with Belfast and the northern

parts of Ireland. Although Glasgow had become a place of considerable wealth and importance, the roads to it, north of Carlisle, continued in a very unsatisfactory state. It was only in July, 1788, that the first mail-coach from London had driven into Glasgow by that route, when it was welcomed by a procession of the citizens on horseback, who went out several miles to meet it. But the road had been shockingly made, and before long had become almost impassable. Robert Owen states that, in 1795, it took him two days and three nights' incessant travelling to get from Manchester to Glasgow, and he mentions that the coach had to cross a well-known dangerous mountain at midnight, called Trickstone Bar, which was then always passed with fear and trembling.[1]

As late as the year 1814 we find a Parliamentary Committee declaring the road between Carlisle and Glasgow to be in so ruinous a state as often seriously to delay the mail and endanger the lives of travellers. The bridge over Evan Water was so much decayed, that one day the coach and horses fell through it into the river, when " one passenger was killed, the coachman survived only a few days, and several other persons were dreadfully maimed ; two of the horses being also killed." [2] The remaining part of the bridge continued for some time unrepaired, just space enough being left for a single carriage to pass. The road trustees seemed to be helpless, and did nothing ; a local subscription was tried and failed, the district passed through being very poor ; but as the road was absolutely required for more than merely local purposes, it was eventually determined to undertake its reconstruction as a work of national importance, and 50,000l. was granted by Parliament with this object, under the provisions of the Act passed in 1816. The

---

[1] 'Life of Robert Owen,' by himself.

[2] 'Report from the Select Committee on the Carlisle and Glasgow Road,' 28th June, 1815.

works were placed under Mr. Telford's charge; and
an admirable road was very shortly under construction
between Carlisle and Glasgow. That part of it between Hamilton and Glasgow, eleven miles in length,
was however left in the hands of local trustees, as was
the diversion of thirteen miles at the boundary of the
counties of Lanark and Dumfries, for which a previous
Act had been obtained.

The length of new line constructed by Mr. Telford
was sixty-nine miles, and it was probably the finest piece
of road which up to that time had been made. The
engineer paid especial attention to two points : first, to
lay it out as nearly as possible upon a level, so as to
reduce the draught to horses dragging heavy vehicles,—one in thirty being about the severest gradient
at any part of the road. The next point was to make
the working, or middle portion of the road, as firm and
substantial as possible, so as to bear, without shrinking,
the heaviest weight likely to be brought over it. With
this object he specified that the metal bed was to be formed
in two layers, rising about four inches towards the centre
—the bottom course being of stones (whinstone, limestone, or hard freestone), seven inches in depth. These
were to be carefully set by hand, with the broadest ends
downwards, all crossbonded or jointed, no stone being
more than three inches wide on the top. The spaces between them were then to be filled up with smaller stones,
packed by hand, so as to bring the whole to an even and
firm surface. Over this a top course was to be laid,
seven inches in depth, consisting of properly broken hard
whinstones, none exceeding six ounces in weight, and
each to be able to pass through a circular ring, two inches
and a half in diameter; a binding of gravel, about an
inch in thickness, being placed over all. A drain crossed
under the bed of the bottom layer to the outside ditch in
every hundred yards. The result was an admirably easy,
firm, and dry road, capable of being travelled upon in

all weathers, and standing in comparatively small need of repairs.[1]

Owing to the mountainous nature of the country through which this road passes, the bridges are unusually numerous and of large dimensions. Thus, the

---

[1] A similar practice was introduced in England about the same time by Mr. Macadam; and, though his method was not so thorough as that of Telford, it was usefully employed on nearly all the leading high roads of the kingdom. Mr. Macadam's atten-

J. L. MACADAM.

tion was first directed to the subject while acting as one of the trustees of a road in Ayrshire. Afterwards, while employed as Government agent for victualling the navy in the western parts of England, he continued the study of road-making, keeping in view the essential conditions of a compact and durable substance and a smooth surface. At that time road legislation was principally directed to the breadth of the wheels of vehicles; whilst Macadam was of opinion that the main point was to attend to the nature of the roads on which they were to travel. Most roads were then made with gravel, or flints tumbled upon them in their natural state, and so rounded that they had no points of contact, and rarely consolidated. When a heavy vehicle of

any sort passed over them, their loose structure presented no resistance; the roads were thus constantly standing in need of repair, and they were bad even at the best. He pointed out that the defect did not arise from the want of materials, which were not worn out by the traffic, but merely displaced. The practice he urged was this: to break the stones into angular fragments, so that a bed several inches in depth should be formed, the best adapted for the purpose being fragments of granite, greenstone, or basalt; to watch the repairs of the road carefully during the process of consolidation, filling up the inequalities caused by the traffic passing over it, until a hard and level surface had been obtained, when the road would last for years without further attention. In 1815 Mr. Macadam devoted himself with great enthusiasm to roadmaking as a profession, and being appointed surveyor-general of the Bristol roads, he had full opportunities of exemplifying his system. It proved so successful that the example set by him was quickly followed over the entire kingdom. Even the streets of many large towns were *Macadamised*. In carrying out his improvements, however, Mr. Macadam spent several thousand pounds from his own resources, and in 1825, having proved this expenditure before a committee of the House of Commons, the amount was reimbursed to him, together with an honorary tribute of two thousand pounds. Mr. Macadam died poor, but, as he himself said, "at least an honest man." By his indefatigable exertions and his success as a roadmaker, by greatly saving animal labour, facilitating commercial intercourse, and rendering travelling easy and expeditious, he entitled himself to the reputation of a public benefactor.

CARTLAND CRAGS BRIDGE.

[By R. P. Leitch, after a Drawing by J. S. Smiles.]

Fiddler's Burn Bridge is of three arches, one of 150 and two of 105 feet span each. There are fourteen other bridges, presenting from one to three arches, of from 20 to 90 feet span. But the most picturesque and remarkable bridge constructed by Telford in that district was upon another line of road subsequently carried out by him, in the upper part of the county of Lanark, and crossing the main line of the Carlisle and Glasgow road almost at right angles. Its northern and eastern part formed a direct line of communication between the great cattle-markets of Falkirk, Crief, and Doune, and Carlisle and the West of England. It was carried over deep ravines by several lofty bridges, the most formidable of which was that across the Mouse Water at Cartland Crags, about a mile to the westward of Lanark. The stream here flows through a deep rocky chasm, the sides of which are in some places about four hundred feet high. At a point where the height of the rocks is considerably less, but still most formidable, Telford spanned the ravine with the beautiful bridge represented in the engraving on the preceding page, its parapet being 129 feet above the surface of the water beneath.

The reconstruction of the western road from Carlisle to Glasgow, which Telford had thus satisfactorily carried out, shortly led to similar demands from the population on the eastern side of the kingdom. The spirit of road reform was now fairly on foot. Fast·coaches and wheel-carriages of all kinds had become greatly improved, so that the usual rate of travelling had advanced from five or six to nine or ten miles an hour. The desire for the rapid communication of political and commercial intelligence was found to increase with the facilities for supplying it; and, urged by the public wants, the Post-Office authorities were stimulated to unusual efforts in this direction. Numerous surveys were made and roads laid out, so as to improve the main line of communication between London and Edinburgh and the intermediate towns. The first part of this road taken in hand was the worst—that lying to the north of Catterick Bridge, in Yorkshire. A new line was surveyed by West Auckland to Hexham, passing over Carter Fell to Jedburgh, and thence to Edinburgh; but rejected as too crooked and uneven. Another was tried by Aldstone Moor and Bewcastle, and rejected for the same reason. The third line proposed was eventually adopted as the best, passing from Morpeth, by Wooler and Coldstream, to Edinburgh;

saving rather more than fourteen miles between the two points, and securing a line of road of much more favourable gradients.

The principal bridge on this new highway was at Pathhead, over the Tyne, about eleven miles south of Edinburgh. To maintain the level, so as to avoid the winding of the road down a steep descent on one side of the valley and up an equally steep ascent on the other, Telford ran out a lofty embankment from both sides, connecting their ends by means of a spacious bridge. The structure at Pathhead is of five arches, each 50 feet span, with 25 feet rise from their springing, 49 feet above the bed of the river. Bridges of a similar character were also thrown over the deep ravines of Cranston Dean and Cotty Burn, in the same neighbourhood. At the same time a useful bridge was built on the same line of road at Morpeth, in Northumberland, over the river Wansbeck. It consisted of three arches, of which the centre one was 50 feet span, and two side-arches 40 feet each; the breadth between the parapets being 30 feet.

The advantages derived from the construction of these new roads were found to be so great, that it was proposed to do the like for the remainder of the line between London and Edinburgh; and at the instance of the Post-Office authorities, with the sanction of the Treasury, Mr. Telford proceeded to make detailed surveys of an entire new post-road between London and Morpeth. In laying it out, the main points which he endeavoured to secure were directness and flatness; and 100 miles of the proposed new Great North Road, south of York, were laid out in a perfectly straight line. This survey, which was begun in 1824, extended over several years; and all the requisite arrangements had been made for beginning the works, when the result of the locomotive competition at Rainhill, in 1829, had the effect of directing attention to that new method of travelling, fortunately in time to prevent what would have proved, for the most part, an unnecessary expenditure, on works soon to be superseded by a totally different order of things.

The most important road-improvements actually carried out under Mr. Telford's immediate superintendence were those on the western side of the island, with the object of shortening the distance and facilitating the communication between London and Dublin by way of Holyhead, as well as between London and Liverpool. At the time of the Union the mode of transit between the capital of Ireland and the metropolis of the United Kingdom was tedious, difficult, and full of peril. In crossing the Irish Sea to Liverpool, the packets were frequently tossed about for days together. On the Irish side there was scarcely the pretence of a port, the landing-place being within the bar of the river Liffey, inconvenient at all times, and in rough weather extremely dangerous. To avoid the long voyage to Liverpool, the passage began to be made from Dublin to Holyhead, the nearest point of the Welsh coast. Arrived there, the passengers were landed upon rugged, unprotected rocks, without a pier or landing conveniences of any kind.[1] But the traveller's perils were not at an end,—comparatively speaking they were only begun. From Holyhead, across the island of Anglesea, there was no made road, but only a miserable track, circuitous and craggy, full of terrible jolts, round bogs and over rocks, for a distance of twenty-four miles. Having reached the Menai Strait, the passengers had

---

[1] A diary is preserved of a journey to Dublin from Grosvenor Square, London, 12th June, 1787, in a coach and four, accompanied by a postchaise and pair, and five outriders. The party reached Holyhead in four days, at a cost of 75l. 11s. 3d. The state of intercourse between this country and the sister island at this part of the account is strikingly set forth in the following entries: "Ferry at Bangor, 1l. 10s.; expenses of the yacht hired to carry the party across the channel, 28l. 7s. 9d.; duty on the coach, 7l. 13s. 4d.; boats on shore, 1l. 1s.: total, 114l. 3s. 4d."—Roberts's 'Social History of the Southern Counties,' p. 504.

again to take to an open ferry-boat before they could gain the main land. The tide ran with great rapidity through the Strait, and, when the wind blew strong, the boat was liable to be driven far up or down the channel, and was sometimes swamped altogether. The perils of the Welsh roads had next to be encountered, and these were in as bad a condition at the beginning of the present century as those of the Highlands above described. Through North Wales they were rough, narrow, steep, and unprotected, mostly unfenced, and in winter almost impassable. The whole traffic on the road between Shrewsbury and Bangor was conveyed by a small cart, which passed between the two places once a week in summer. As an illustration of the state of the roads in South Wales, which were quite as bad as those in the North, we may state that, in 1803, when the late Lord Sudeley took home his bride from the neighbourhood of Welshpool to his residence only thirteen miles distant, the carriage in which the newly married pair rode stuck in a quagmire, and the occupants, having extricated themselves from their perilous situation, performed the rest of their journey on foot.

The first step taken was to improve the landing-places on both the Irish and Welsh sides of St. George's Channel, and for this purpose Mr. Rennie was employed in 1801. The result was, that Howth on the one coast, and Holyhead on the other, were fixed upon as the most eligible sites for packet stations. Improvements, however, proceeded slowly, and it was not until 1810 that a sum of 10,000l. was granted by Parliament to enable the necessary works to be begun. Attention was then turned to the state of the roads, and here Mr. Telford's services were called into requisition. As early as 1808 it had been determined by the Post-Office authorities to put on a mail-coach between Shrewsbury and Holyhead; but it was pointed out that the roads in North Wales were so rude and dangerous that it was

doubtful whether the service could be conducted in
safety. Attempts were made to enforce the law with
reference to their repairs, and no less than twenty-one
townships were indicted by the Postmaster-General.
The route was found too perilous even for a riding
post, the legs of three horses having been broken in one
week.[1] The road across Anglesea was quite as bad.
Sir Henry Parnell mentioned, in 1819, that the coach
had been overturned beyond Gwynder, going down one
of the hills, when a friend of his was thrown a considerable distance from the roof into a pool of water.
Near the post-office of Gwynder, the coachman had
been thrown from his seat by a violent jolt, and broken
his leg. The post-coach, and also the mail, had been
overturned at the bottom of Penmyndd Hill; and the
route was so dangerous that the London coachmen, who
had been brought down to "work" the country, refused
to continue the duty because of its excessive dangers.
Of course, anything like a regular mail-service through
such a district was altogether impracticable.

The indictments of the townships proved of no
use; the localities were too poor to provide the means
required to construct a line of road sufficient for the conveyance of mails and passengers between England and
Ireland. The work was really a national one, to be
carried out at the national cost. How was this best
to be done? Telford recommended that the old road
between Shrewsbury and Holyhead (109 miles long)
should be shortened by about four miles, and made as
nearly as possible on a level; the new line proceeding
from Shrewsbury by Llangollen, Corwen, Bettws-y-
Coed, Capel-Curig, and Bangor, to Holyhead. Mr.
Telford also proposed to cross the Menai Strait by
means of a cast iron bridge, hereafter to be described.

---

[1] 'Second Report from Committee on Holyhead Roads and Harbours,' 1810.
(Parliamentary paper.)

Although a complete survey was made in 1811, nothing was done for several years. The mail-coaches continued to be overturned, and stage-coaches, in the tourist season, to break down as before.[1] The Irish mail-coach took forty-one hours to reach Holyhead from the time of its setting out from St. Martin's-le-Grand; the journey was performed at the rate of only 6¾ miles an hour, the mail arriving in Dublin on the third day. The Irish members made many complaints of the delay and dangers to which they were exposed in travelling up to town. But, although there was much discussion, there was no money voted until the year 1815, when Sir Henry Parnell vigorously took the question in hand and successfully carried it through. A Board of Parliamentary Commissioners was appointed, of which he was chairman, and, under their direction, the new Shrewsbury and Holyhead road was at length commenced and carried to completion, the works extending over a period of about fifteen years. The same Commissioners exercised an authority over the roads between London and Shrewsbury; and numerous improvements were also made in the main line at various points, with the object of facilitating communication between London and Liverpool, as well as between London and Dublin.

---

[1] Many parts of the road are extremely dangerous for a coach to travel upon. At several places between Bangor and Capel-Curig there are a number of dangerous precipices without fences, exclusive of various hills that want taking down. At Ogwen Pool there is a very dangerous place where the water runs over the road, extremely difficult to pass at flooded times. Then there is Dinas Hill, that needs a side fence against a deep precipice. The width of the road is not above twelve feet in the steepest part of the hill, and two carriages cannot pass without the greatest danger. Between this hill and Rhyddlanfair there are a number of dangerous precipices, steep hills, and difficult narrow turnings. From Corwen to Llangollen the road is very narrow, long, and steep; has no side fence, except about a foot and a half of mould or dirt, which is thrown up to prevent carriages falling down three or four hundred feet into the river Dee. Stage-coaches have been frequently overturned and broken down from the badness of the road, and the mails have been overturned; but I wonder that more and worse accidents have not happened, the roads are so bad. — Evidence of Mr. William Akers, of the Post-Office, before Committee of the House of Commons, 1st June, 1815.

The rugged nature of the country through which the new road passed, along the slopes of rocky precipices and across inlets of the sea, rendered it necessary to build many bridges, to form many embankments, and cut away long stretches of rock, in order to secure an easy and commodious route. The line of the valley of the Dee, to the west of Llangollen, was selected, the road proceeding along the scarped sides of the mountains, crossing from point to point by lofty embankments where necessary; and, taking into account the character of the country, it must be acknowledged that a wonderfully level road was secured. Whilst the gradients on the old road had in some cases been as steep as 1 in 6½, passing along the edge of unprotected precipices, the new one was so laid out as to be no more than 1 in 20 at any part, while it was wide and well protected along its whole extent. Mr. Telford pursued the same system that he had adopted in the formation of the Carlisle and Glasgow road, as regards metalling, cross-draining, and fence-walling; for the latter purpose using schistus, or slate rubble-work, instead of sandstone. The largest bridges were of iron; that at Bettws-y-Coed, over the Conway—called the Waterloo Bridge, constructed in 1815—being a very fine specimen of Telford's iron bridge-work.

Those parts of the road which had been the most dangerous were taken in hand first, and, by the year 1819, the route had been rendered comparatively commodious and safe. Angles were cut off, the sides of hills were blasted away, and several heavy embankments run out across formidable arms of the sea. Thus, at Stanley Sands, near Holyhead, an embankment was formed 1300 yards long and 16 feet high, with a width of 34 feet at the top, along which the road was laid. Its breadth at the base was 114 feet, and both sides were coated with rubble stones, as a protection against storms. By the adoption of this expedient, a mile and a half was saved in a distance of six miles. Heavy embankments were also run out, where bridges were thrown across chasms and ravines, to maintain the general level. From Ty-Gwynn to Lake Ogwen, the road along the face of the rugged hill and across the river Ogwen was entirely new made, of a uniform width of 28 feet between the parapets, with an inclination of only 1 in 22 in the steepest place. A bridge was thrown over the deep chasm forming the channel of the Ogwen, the embankment being carried forward from the rock-cutting, protected by high breastworks. From Capel - Curig to near the great waterfall over the river Lugwy, about a mile of new road was cut; and a still greater length from Bettws across the river Conway and along the face of Dinas Hill to Rhyddlanfair, a distance of 3 miles, its steepest descent being 1 in 22, diminishing to 1 in 45. By

ROAD DESCENT NEAR BETTWS-Y-COED

[By Percival Skelton]

this improvement, the most difficult and dangerous pass along the route through North Wales was rendered safe and commodious. Another point of almost equal difficulty occurred near Ty-Nant, through the rocky pass of Glynn Duffrws, where the road was confined between

steep rocks and rugged precipices : there the way was
widened and flattened by blasting, and thus reduced to

ROAD ABOVE NANT FFRANCON, NORTH WALES
[By Percival Skelton, after his original Drawing.]

the general level; and so on eastward to Llangollen
and Chirk, where the main Shrewsbury road to London
was joined.[1]

---

[1] The Select Committee of the
House of Commons, in reporting as to
the manner in which these works
were carried out, stated as follows :—
"The professional execution of the
new works upon this road greatly sur-
passes anything of the same kind in
these countries. The science which
has been displayed in giving the
general line of the road a proper inclination through a country whose
whole surface consists of a succession
of rocks, bogs, ravines, rivers, and
precipices, reflects the greatest credit
upon the engineer who has planned
them; but perhaps a still greater degree of professional skill has been
shown in the construction, or rather
the building, of the road itself. The
great attention which Mr. Telford
has bestowed to give to the surface of
the road one uniform and moderately
convex shape, free from the smallest
inequality throughout its whole
breadth; the numerous land drains,
and, when necessary, shores and tunnels of substantial masonry, with
which all the water arising from
springs or falling in rain is instantly
carried off; the great care with which
a sufficient foundation is established
for the road, and the quality, solidity,
and disposition of the materials that
are put upon it, are matters quite new
in the system of roadmaking in these
countries."—'Report from the Select
Committee on the Road from London
to Holyhead in the year 1819.'

By means of these admirable roads the traffic of North Wales continues to be mainly carried on to this day. Although railways have superseded coach-roads in the more level districts, the hilly· nature of Wales precludes their formation in that quarter to any considerable extent; and even in the event of railways being constructed, a large part of the traffic of every country must necessarily continue to pass over the old high roads. Without them even railways would be of comparatively little value; for a railway station is of use chiefly because of its easy accessibility, and thus, both for passengers and merchandise, the common roads of the country are as useful as ever they were, though the main post-roads have in a great measure ceased to be employed for the purpose for which they were originally designed.

The excellence of the roads constructed by Mr. Telford through the formerly inaccessible counties of North Wales was the theme of general praise; and their superiority, compared with those of the richer and more level districts in the midland and western English counties, becoming the subject of public comment, he was called upon to execute like improvements upon that part of the postroad which extended between Shrewsbury and the metropolis. A careful survey was made of the several routes from London northward by Shrewsbury as far as Liverpool; and the short line by Coventry, being 153 miles from London to Shrewsbury, was selected as the one to be improved to the utmost. Down to 1819 the road between London and Coventry was in a very bad state, being so laid as to become a heavy slough in wet weather. There were also many steep hills to be cut down, in some parts of deep clay, in others deep sand. A mail-coach had been tried to Banbury; but the road below Aylesbury was so bad, that the Post-Office authorities were obliged to give it up. The twelve miles from Towcester to Daventry

were still worse. The line of way was covered with
banks of dirt; in winter it was a puddle of from four to
six inches deep—quite as bad as it had been in Arthur
Young's time; and when horses passed along the road,
they came out of it a mass of mud and dirt.[1] There were
also several steep and dangerous hills to be crossed.
The loss of horses by fatigue in travelling by that
route was represented at the time to be very great.
Even the roads in the immediate neighbourhood of the
metropolis were little better, those under the Highgate
and Hampstead trust being pronounced in a wretched
state. They were badly formed, on a clay bottom, and
being undrained, were almost always wet and sloppy.
The gravel was usually tumbled on and spread unbroken,
so that the materials, instead of becoming consolidated,
were only rolled about by the wheels of the carriages
passing over them. Mr. Telford applied the same
methods in the reconstruction of these roads that he had
already adopted in Scotland and Wales, and the same
improvement was shortly experienced in the more easy
passage of vehicles of all sorts and the great acceleration
of the mail service.

In addition to the reconstruction of these roads, that
along the coast from Bangor, by Conway, Abergele,
St. Asaph, and Holywell, to Chester, was greatly improved. It formed the mail road from Dublin to Liverpool, and it was considered of importance to render it as
safe and level as possible. The principal new cuts on
this line were those along the rugged skirts of the huge
Penmaen-Mawr; around the base of Penmaen-Bach to
the town of Conway; and between St. Asaph and Holywell, to ease the ascent of Rhyall Hill. But more
important than all, as a means of completing the main

---

[1] Evidence of William Waterhouse before the Select Committee, 10th March,
1819.

line of communication, there were the great bridges over the Conway and the Menai Straits to be constructed. These dangerous ferries had still to be crossed in open boats, sometimes in the night, when the luggage and mails were exposed to great risks, and passengers occasionally lost with them. It was therefore determined, after long consideration, to erect bridges over these formidable straits, and Mr. Telford was employed to execute the works,—in what manner we propose to describe in the succeeding chapter.

CHAPTER VIII.

The Menai and Conway Bridges.

THE erection of a bridge over the Straits of Menai had long been matter of speculation amongst engineers. As early as 1776 Mr. Golborne proposed his plan of an em-

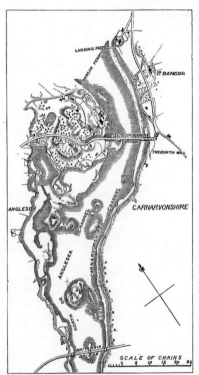

MAP OF MENAI STRAIT. [Ordnance Survey]

bankment, with a bridge in the middle of it; and a few years later, in 1785, Mr. Nichols proposed a wooden viaduct, furnished with drawbridges, at Cadnant Island. Later still, Mr. Rennie proposed his design of a cast iron bridge. But none of these plans were carried out, and the whole subject remained in abeyance until the year 1810, when a commission was appointed to inquire and report as to the state of the roads between Shrewsbury, Chester, and Holyhead. The result was, that Mr. Telford was called upon to report as to the most effectual method of bridging the Menai Strait, and thus completing the communication with the port of embarkation for Ireland.

Mr. Telford submitted alternative plans for a bridge

over the Strait: one at the Swilly Rock, consisting of
three cast iron arches of 260 feet span, with a stone
arch of 100 feet span between each two iron ones, to
resist their lateral thrust; and another at Ynys-y-moch,
to which he himself attached the preference, consisting
of a single cast iron arch of 500 feet span, the crown
of the arch to be 100 feet above high water of spring
tides, and the breadth of the roadway to be 40 feet.

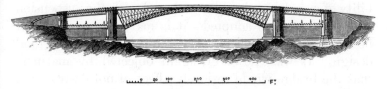

TELFORD'S PROPOSED CAST IRON BRIDGE.

The principal objection taken to this plan by engineers
generally, was the supposed difficulty of erecting a proper
centering to support the arch during construction; and
the mode by which Mr. Telford proposed to overcome
this may be cited in illustration of his ready ingenuity
under such circumstances. He proposed to suspend the
centering from above instead of supporting it from below
in the usual manner—a contrivance afterwards revived
by another very skilful engineer, the late Mr. Brunel.
Frames, fifty feet high, were to be erected on the top of
the abutments, and on these strong blocks, or rollers and
chains, were to be fixed, by means of which, and by the
aid of windlasses and other mechanical powers, each
separate piece of centering was to be raised into, and
suspended in, its proper place. Mr. Telford regarded

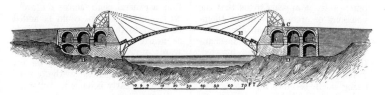

PROPOSED PLAN OF SUSPENDED CENTERING.

this method of constructing centres as applicable to stone as well as to iron arches; and indeed it is applicable, as Mr. Brunel held, to the building of the arch itself.[1] Mr. Telford anticipated that, if the method recommended by him were successfully adopted on the large scale proposed at Menai, all difficulties with regard to carrying bridges over deep ravines would be done away with, and a new era in bridge-building begun. For this and other reasons—but chiefly because of the much greater durability of a cast iron bridge compared with the suspension bridge afterwards adopted—it is matter of regret that he was not permitted to carry out this novel and grand design. It was, however, again objected by mariners that the bridge would seriously affect, if not destroy, the navigation of the Strait; and this plan, like Mr. Rennie's, was eventually rejected.

Several years passed, and during the interval Mr. Telford was consulted as to the construction of a bridge over Runcorn Gap on the Mersey, above Liverpool. As the river was there about twelve hundred feet wide, and much used for purposes of navigation, a bridge of the ordinary construction was found inapplicable. But as he was required to furnish a plan of the most suitable structure, he proceeded to consider how the difficulties

---

[1] In an article in the 'Edinburgh Review,' No. cxli., from the pen of Sir David Brewster, the writer observes:—"Mr. Telford's principle of suspending and laying down from above the centering of stone and iron bridges is, we think, a much more fertile one than even he himself supposed. With modifications, by no means considerable, and certainly practicable, it appears to us that the voussoirs or arch-stones might themselves be laid down from above, and suspended by an appropriate mechanism till the keystone was inserted. If we suppose the centering in Mr. Telford's plan to be of iron, this centering itself becomes an iron bridge, each rib of which is composed of ten pieces of fifty feet each; and by increasing the number of suspending chains, these separate pieces or voussoirs having been previously joined together, either temporarily or permanently, by cement or by clamps, might be laid into their place, and kept there by a single chain till the road was completed. The voussoirs, when united, might be suspended from a general chain across the archway, and a platform could be added to facilitate the operations." This is as nearly as possible the plan afterwards revived by Mr. Brunel, and for the originality of which, we believe, he has generally the credit, though it clearly belongs to Telford.

of the case were best to be met.   The only practicable
plan, he thought, was a bridge constructed on the principle of suspension.   Expedients of this kind had long
been employed in India and America, where wide rivers
were crossed by means of bridges formed of ropes and
chains; and even in this country a suspension bridge,
though of a very rude kind, had long been in use near
Middleton on the Tees, where, by means of two common
chains stretched across the river, upon which a footway
of boards was laid, the colliers were enabled to pass
from their cottages to the colliery on the opposite bank.
Captain (afterwards Sir Samuel Brown) took out a patent
for forming suspension bridges in 1817; but it appears
that Telford's attention had been directed to the subject
before this time, as he was first consulted respecting the
Runcorn Bridge in the year 1814, when he proceeded to

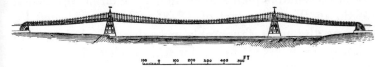

OUTLINE OF TELFORD'S PROPOSED BRIDGE AT RUNCORN

make an elaborate series of experiments on the tenacity
of wrought iron bars, with the object of employing this
material in his proposed structure.   After he had made
upwards of two hundred tests of malleable iron of various
qualities, he proceeded to prepare his design of a bridge,
which consisted of a central opening of 1000 feet span,
and two side openings of 500 feet each, supported by pyramids of masonry placed near the low water lines.   The
roadway was to be 30 feet wide, divided into one central
footway and two distinct carriageways of 12 feet each.
At the same time he prepared and submitted a model of
the central opening, which satisfactorily stood the various
strains which were applied to it.   This Runcorn design
of 1814 was of a very magnificent character, perhaps
superior even to that of the Menai Suspension Bridge,

afterwards erected; but unhappily the means were not
forthcoming to carry it into effect.* The publication of
his plan and report had, however, the effect of directing
public attention to the construction of bridges on the
suspension principle; and many were shortly after
designed and erected by Telford and other engineers in
different parts of the kingdom.

Mr. Telford continued to be consulted by the Commissioners of the Holyhead Roads as to the completion of
the last and most important link in the line of communication between London and Holyhead, by bridging the
Straits of Menai; and at one of their meetings in 1815,
shortly after the publication of his Runcorn design, the
inquiry was made whether a bridge upon the same principle was not applicable in this particular case. The
engineer was instructed again to examine the Straits and
submit a suitable plan and estimate, which he proceeded
to do in the early part of 1818. The site selected by
him as the most favourable was that which had previously been fixed upon for the projected cast iron
bridge, namely at Ynys-y-moch—the shores there being
bold and rocky, affording easy access and excellent foundations, whilst by spanning the entire channel between
the low water lines, and the roadway being kept uni-
formly 100 feet above the highest water at spring tide,
the whole of the navigable waterway would be left
entirely uninterrupted. The distance between the centres
of the supporting pyramids was proposed to be of the
then unprecedented width of 550 feet, and the height of
the pyramids 53 feet above the level of the roadway.
The main chains were to be sixteen in number, with a
deflection of 37 feet, each composed of thirty-six bars of

MENAI BRIDGE.

half-inch-square iron, so placed as to give a square of
six on each side, making the whole chain about four inches
in diameter, welded together for their whole length,
secured by bucklings, and braced round with iron wire;
whilst the ends of these great chains wére to be secured
by a mass of masonry, built over stone arches between
each end of the supporting piers and the adjoining shore.
Four of these arches were to be on the Anglesea, and
three on the Caernarvonshire side, each of them fifty-two
feet six inches span.   The roadway was to be divided, as
in the Runcorn design—a carriageway twelve feet wide
on each side, and a footpath of four feet in the middle.
Mr. Telford's plan was supported by Mr. Rennie and
other engineers of eminence; and the Select Committee
of the House of Commons, being satisfied as to its practicability, recommended Parliament to pass a Bill and
make a grant of money to enable the work to be carried
into effect.

The necessary Act passed in the session of 1819,
and Mr. Telford immediately proceeded to Bangor to
make preparations for commencing the works.   The
first proceeding was to blast off the inequalities from
the surface of the rock, called Ynys-y-moch, situated on
the western or Holyhead side of the Strait, at that time
only accessible at low water.   The object was to form an
even surface upon it for the foundation of the west main
pier.   It used to be at this point, where the Strait
was narrowest, that horned cattle were driven down,
preparatory to swimming them across the channel to
the Caernarvon side, when the tide was weak and at
its lowest ebb.   Many cattle were, nevertheless, often
carried away when the current was too strong for the
animals to contend against.

At the same time a landing-quay was erected on
Ynys-y-moch, which was connected with the shore by
an embankment carrying lines of railway.   Along this
horses drew the sledges laden with stone required for the

work; the material being brought in barges from the quarries opened at Penmon Point, on the north-eastern extremity of the Isle of Anglesea, a little to the westward of the northern opening of the Strait. When the surface of the rock had been levelled and the causeway completed, the first stone of the main pier was laid by Mr. W. A. Provis,* the resident engineer, on the 10th of August, 1819; but not the slightest ceremony was observed on the occasion.

Later in the autumn preparations were made for proceeding with the foundations of the eastern main pier on the Bangor side of the Strait. After excavating the beach to a depth of seven feet, a solid mass of rock was reached, which served the purpose of an immoveable foundation for the pier. At the same time workshops were erected; builders, artisans, and labourers were brought together from distant quarters; vessels and barges were purchased or built for the special purpose of the work; a quay was constructed at Penmon Point for loading the stones for the piers; and all the requisite preliminary arrangements were made for proceeding with the building operations in the ensuing season.

A careful specification of the masonry work was drawn up, and the contract first let to Messrs. Stapleton and Hall; but as they did not proceed satisfactorily, and desired to be released from the contract, it was re-let on the same terms to Mr. John Wilson, one of Mr. Telford's principal contractors for mason work on the Caledonian Canal. The building operations were begun with great vigour early in 1820. The three arches on the Caernarvonshire side and the four on the Anglesea side were first proceeded with. They are of immense magnitude, and occupied four years in construction, having been finished late in the autumn of 1824. These piers are 65 feet in height from high water line to the springing of the arches, the span of each being 52 feet 6 inches. The work of the main piers also made satisfactory

progress, and the masonry proceeded so rapidly that
stones could scarcely be got from the quarries in sufficient quantity to keep the builders at work. By the
end of June about three hundred men were employed.

The two main piers, each 153 feet in height, upon
which the main chains of the bridge were to be suspended,
were built with great care
and under rigorous inspection. In these, as indeed in
most of the masonry of the
bridge, Mr. Telford adopted
the same practice which he
had employed in his pre-
vious bridge structures, that
of leaving large void spaces,
commencing above high
water mark and continuing
them up perpendicularly
nearly to the level of the
roadway. " I have else-
where expressed my con-
viction," he says, when
referring to the mode of
constructing these piers,
" that one of the most
important improvements
which I have been able to
introduce into masonry consists in the preference of
cross-walls to rubble, in the
structure of a pier, or any
other edifice requiring strength. Every stone and joint
in such walls is open to inspection in the progress of the
work, and even afterwards, if necessary ; but a solid
filling of rubble conceals itself, and may be little better
than a heap of rubbish confined by side walls." [1] The

SECTION OF MAIN PIER

HIGH   WATER
LOW   WATER

0   10   20   30   40   50   60   70   FT
                                        80

---

[1] ' Life of Telford,' p. 221.

walls of these main piers were built from within as well as from without all the way up, and the inside was as carefully and closely cemented with mortar as the external face. Thus the whole pier was bound firmly together, and the utmost strength given, while the weight of the superstructure upon the lower parts of the work was reduced to its minimum.

Over the main piers the small arches intended for the roadways were constructed, each being 15 feet to the springing of the arch, and 9 feet wide. Upon these arches the masonry was carried upwards, in a tapering form, to a height of 53 feet above the level of the road. As these piers were to carry the immense weight of the suspension chains, great pains were taken with their construction, and each stone, from top to bottom, was firmly bound together with iron dowels to prevent the possibility of their being separated or bulged by the immense pressure they had to withstand.[1]

The most important point in the execution of the details of the bridge, where the engineer had no past experience to direct him, was in the designing and fixing of the wrought iron work. Mr. Telford had continued his experiments as to the tenacity of bar iron, until he had obtained several hundred distinct tests; and at length, after the most mature deliberation, the patterns and dimensions were finally arranged by him, and the contract for the manufacture of the whole was let to Mr. Hazeldean, of Shrewsbury, in the year 1820. The iron was to be of the best Shropshire, drawn at Upton forge,

---

[1] To guard against the effects of lateral pressure resulting from this heavy mass of masonry bearing on the arches forming the roadway, six strong wrought iron ties, four inches wide and two inches thick, firmly bolted at each end, were introduced horizontally at the springing of the arches over the carriage ways, and thus any deficiency of strength at that point was effectually provided for. Strong cast iron blocks or saddles were placed upon the top of the piers to bear the suspension chains, and they were fitted with wrought iron self-acting rollers and brass bushes, for the purpose of regulating the contraction and expansion of the iron, by moving themselves either way according to the temperature of the atmosphere, without the slightest derangement to any part of the work.

and finished and proved at his works, under the inspection of a person appointed by the engineer.[1]

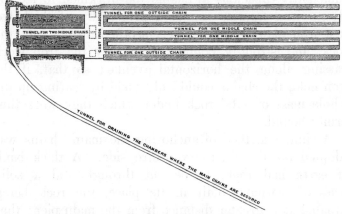

CUT SHOWING FIXING OF THE CHAINS IN THE ROCK.

The mode by which the land ends of those enormous suspension chains were rooted to the solid ground on either side of the Strait, was remarkably ingenious and effective. Three oblique tunnels were made by blasting the rock on the Anglesea side; they were each about six feet in diameter, the excavations being carried down an inclined plane to the depth of about twenty yards. A considerable width of rock lay be-

---

[1] The cross sections of the bars forming the main chains were 3¼ inches square; they were proved to be capable, according to Mr. Telford's experiments, of bearing a strain of not less than 87¾ tons before fracture; but in order not to strain the iron unduly, the proof was limited to 35 tons, or about 11 tons to every square inch of cross section. Every piece of iron introduced in the work was submitted to this test by means of a very accurate and powerful proving machine constructed for the purpose; while under the strain, it was frequently struck with a hammer, and after being proved, every separate piece was well cleansed, put into a stove, and when brought to a gentle heat was immersed in a trough containing linseed oil. It was then taken out after a short time, again put into the stove, and, when dried, appeared as if covered with varnish. A coat of linseed oil paint was finally put over all, and the iron bar was in this state sent to the bridge for use. "No precautions were spared," writes Mr. Telford, "to render every part perfectly true, and therefore secure; for as any variation in the length of the numerous bars would produce unequal bearings, each was subjected to a fresh adjustment by means of a steel model, upon which they were bored when cold, so that a cross-bolt passed through a certain number, in most cases through eight bars, so as to form four chains, thus accurately attached to each other."

tween each tunnel, but at the bottom they were all united by a connecting horizontal avenue or cavern, sufficiently capacious to enable the workmen to fix the strong iron frames, composed principally of thick flat cast iron plates, which were engrafted deeply into the rock, and strongly bound together by the iron work passing along the horizontal avenue ; so that, if the iron held, the chains could only yield by tearing up the whole mass of solid rock under which they were thus firmly bound.

A similar method of anchoring the main chains was adopted on the Caernarvonshire side. A thick bank of earth had there to be cut through, and a solid mass of masonry built in its place, the rock being situated at a greater distance from the main pier ; thus involving a greater length of suspending chain, and a disproportion in the catenary or chord line on that side of the bridge. The excavation and masonry thus rendered necessary proved a work of vast labour, and its execution occupied a considerable time ; but by the beginning of the year 1825 the suspension pyramids, the land piers and arches, and the rock tunnels, had all been completed, and the main chains firmly secured in them ; the work being sufficiently advanced to enable the suspending of the chains to be proceeded with. This was by far the most difficult and anxious part of the undertaking.

With the same careful forethought and provision for every contingency which had distinguished the engineer's procedure in the course of the work, he had made frequent experiments to ascertain the actual power which would be required to raise the main chains to their proper curvature. A valley lay convenient for the purpose, a little to the west of the bridge on the Anglesea side. Fifty-seven of the intended vertical suspending rods, each nearly ten feet long and an inch square, having been fastened together, a piece of chain was attached to one end to make the chord line 570 feet in length; and experiments having been made and comparisons drawn, Mr. Telford ascertained that the absolute weight of one of the main chains of the bridge between the points of suspension was 23½ tons, requiring a strain of 39½ tons to raise it to its proper curvature. On this calculation the necessary apparatus required for the hoisting was prepared. The mode of action finally determined on for lifting the main chains, and fixing them into their places, was to build the central portion of each upon a raft 450 feet long and 6 feet wide, then to float it to the site of the bridge, and lift it into its place by capstans and proper tackle.

At length all was ready for hoisting the first great chain, and about the middle of April, 1825, Mr. Telford left London for Bangor to superintend the operations. An immense assemblage collected to witness the sight; greater in number than any that had been collected in the same place since the men of Anglesea, in their war-paint, rushing down to the beach, had shrieked defiance across the Straits at their Roman invaders on the Caernarvon shore. Numerous boats arrayed in gay colours glided along the waters; the day—the 26th of April—being bright, calm, and in every way propitious. At half-past two, about an hour before high water, the raft bearing the main chain was cast off from near Treborth Mill, on the Caernarvon side. Towed by four boats, it began gradually to move from the shore, and with the assistance of the tide, which caught it at its further end, it swung slowly and majestically round to its position between the main piers, where it was moored. One end of the chain was then bolted to that which hung down the face of the Caernarvon pier; whilst the other was attached to ropes connected with strong capstans fixed upon the Anglesea side, the ropes passing by means of blocks over the top of the pyramid of the Anglesea pier. The capstans for hauling in the ropes bearing the main chain, were two in number, manned by about 150 labourers.   When all

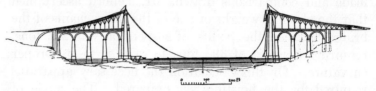

CUT OF BRIDGE, SHOWING STATE OF SUSPENSION CHAIN.

was ready, the signal was given to "go along!"   A band of fifers struck up a lively tune; the capstans were instantly in motion, and the men stepped round in a steady trot.   All went well.   The ropes gradually coiled in.   As the strain increased, the pace slackened a little; but "Heave away! now she comes!" was sung out. Round went the men, and steadily and safely rose the ponderous chain.   The tide had by this time turned, and bearing upon the side of the raft, now getting freer of its load, the current floated it away from under the middle of the chain still resting on it, and it swung easily off into the water.   Until this moment a breathless silence pervaded the watching multitude; and nothing was heard amongst the working party on the Anglesea side but the steady tramp of the men at the capstans, the shrill music of the fife, and the occasional order to "Hold on!" or "Go along!"   But no sooner was the raft seen floating away, and the great chain safely swinging in the air, than a tremendous cheer burst forth along both sides of the Straits.   The rest of the work was only a matter of time.   The most anxious moment had passed.   In an hour and thirty-five minutes after the commencement of the hoisting, the chain was raised to its proper curvature, and fastened to the land portion of it which had been previously placed over the top of the Anglesea pyramid.   Mr. Telford ascended to the point of fastening, and satisfied himself that a continuous and safe connection had been formed from the Caernarvon fastening on the rock to that on Anglesea. The announcement of the fact was followed by loud and prolonged cheering from the workmen, echoed by the spectators, and extending along the Straits on both sides, until it seemed to die away along the shores in the distance. Three foolhardy workmen, excited by the day's proceedings, had the temerity to scramble along the upper surface of the chain—which was only nine inches wide and formed a curvature of 590 feet—from one side of the Strait to the other!

Far different were the feelings of the engineer who had planned this magnificent work. Its failure had been predicted; and, like Brindley's Barton Viaduct, it had been freely spoken of as a "castle in the air." Telford had, it is true, most carefully tested every point by repeated experiment, and so conclusively proved the sufficiency of the iron chains to bear the immense weight they would have to support, that he was thoroughly convinced as to the soundness of his principles of construction; and satisfied that, if rightly manufactured and properly put together, the chains would hold together and the piers would sustain them. Still there was necessarily an element of uncertainty in the undertaking. It was the largest structure of the kind that had ever been attempted. There was the contingency of a flaw in the iron; some possible scamping in its manufacture; some little point which, in the multiplicity of details to be attended to, he might have overlooked, or which his subordinates might have neglected. It was, indeed, impossible but that he should feel intensely anxious as to the result of the day's operations. Mr. Telford afterwards stated to a friend, only a few months before his death, that for some time previous to the opening of the bridge, his anxiety was so extreme that he could scarcely sleep; and that a continuance of that condition must have very soon completely undermined his health. We are not, therefore, surprised to learn that when his friends rushed to congratulate him on the result of the first day's experiment, which decisively proved the strength and solidity of the bridge, they should have found the engineer upon his knees engaged in prayer. A vast load had been taken off his mind; the perilous enterprise of the day had been accomplished without loss of life; and his spontaneous act was thankfulness and gratitude.

MENAI BRIDGE.
[By Percival Skelton, after his original Drawing.]

The suspension of the remaining fifteen chains was accomplished without difficulty. The last was raised and fixed on the 9th of July, 1825, when the entire line was completed. On fixing the final bolt, a band of music descended from the top of the suspension pier on the Anglesea side to a scaffolding erected over the centre of the curved part of the chains, and played the National Anthem amidst the cheering of many thousand persons assembled along the shores of the Strait; whilst the workmen marched in procession along the bridge, upon which a temporary platform had been placed, and the *St. David* steam-packet of Chester passed under the chains towards the Smithy Rocks and back again, thus reopening the navigation of the Strait. In August the road platform was commenced, and in September the trussed bearing bars were all suspended. The road was constructed of timber in a substantial manner, the planking being spiked together, with layers of patent felt between the planks, and the carriage-way being protected by oak guards placed seven feet and a half apart. Side railings were added and toll-houses and approach-roads completed by the end of the year; and the bridge was opened for public traffic on Monday, the 30th of January, 1826, when the London and Holyhead mail-coach passed over it for the first time, followed by the Commissioners of the Holyhead roads, the engineer, several stage-coaches, and a multitude of private persons too numerous to mention.

We may briefly add a few facts as to the quantities of materials used, and the dimensions of this remarkable structure. The total weight of iron was 2187 tons, in 33,265 pieces. The total length of the bridge is 1710 feet, or nearly a third of a mile; the distance between the points of suspension of the main bridge being 579 feet. The total sum expended by Government in its erection, including the embankment and about half a mile of new line of road on the Caernarvon side, together with the toll-houses, was 120,000*l*.

Shortly after the Menai Bridge was commenced, it was determined by the Commissioners of the Holyhead road that a bridge of similar design should be built over the estuary of the Conway, immediately opposite the old castle at that place, and which had formerly been crossed by an open ferry boat. The first stone was laid on the 3rd of April, 1822, and, the works having proceeded satisfactorily, the bridge and embankment approaching it were completed by the summer of 1826. But the

operations being of the same kind as those connected
with the larger structure above described, though of a
much less difficult character, it is unnecessary to enter
into any details as to the several stages of its construction.   In this bridge the width between the centres of
the supporting towers is 327 feet, and the height of the
under side of the roadway above high water of spring
tides only 15 feet.   The heaviest work was an embankment at its eastern approach, 2015 feet in length and
about 300 feet in width at its highest part.

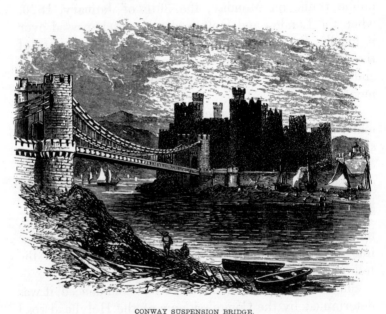

CONWAY SUSPENSION BRIDGE.

[By Percival Skelton, after his original Drawing.]

• • • • •

It will have been observed, from the Lives of those
Engineers which we have thus far been enabled to lay
before the reader, how much has been done by skill and
industry to open up and develope the material resources
of the kingdom. The stages of improvement which we

have recorded exhibit a measure of the vital energy which has from time to time existed in the nation. In the earlier periods the war was with nature; the sea was held back by embankments; the Thames, instead of being allowed to overspread the wide marshes on either bank, was confined within limited bounds, by which the navigable depth of its channel was increased at the same time that a wide extent of land was rendered available for agriculture.

In those early days the great object was to render the land more habitable, comfortable, and productive. Marshes were reclaimed, and wastes subdued. But so long as the country remained comparatively closed, and intercourse was restricted by the want of bridges and roads, improvement was extremely slow. Whilst roads are the consequence of civilization, they are also among its most influential causes. We have seen even the blind Metcalf acting as an effective instrument of progress in the northern counties by the formation of long lines of road. Brindley and the Duke of Bridgewater carried on the work in the same districts, and conferred upon the north and north-west of England the blessings of cheap and effective water communication. Smeaton followed and carried out similar undertakings in still remoter places, joining the east and west coasts of Scotland by the Forth and Clyde Canal, and building bridges in the far north. Rennie made harbours, built bridges, and hewed out docks for shipping, the increase in which had kept pace with the growth of our home and foreign trade. He was followed by Telford, whose long and busy life, as we have seen, was occupied in building bridges and making roads in all directions, in districts of the country formerly inaccessible, and therefore comparatively barbarous. At length the wildest districts of the Highlands and the most rugged mountain valleys of North Wales were rendered as easy of access as the comparatively level counties in the immediate neighbourhood of the metropolis.

During all this while the wealth and industry of the country had been advancing with rapid strides. London had grown in population and importance. Many improvements had been effected in the river, but the dock accommodation was still found insufficient; and, as the recognised head of his profession, Mr. Telford, though now grown old and fast becoming infirm, was called upon to supply the requisite plans. He had. been engaged upon great works for upwards of thirty years, previous to which he had led the life of a working mason. But he had been a steady, temperate man all his life; and though nearly seventy, when consulted as to the proposed new docks, his mind was as able to deal with the subject in all its bearings as it had ever been; and he undertook the work.

[In 1824 a company was formed to provide a dock between the Tower and the London Docks and nearer the heart of the City than existing ones. There were problems which Telford's plans had to solve but, when complete, St. Katherine Docks were regarded as a masterpiece of harbor construction. Other designs and works of Telford executed towards the close of his career were bridges at Tewkesbury, Gloucester, Edinburgh, and Glasgow. He also, like previous outstanding engineers, planned drainage projects, and he was especially concerned with the Fen drainage of the North Level. Toward the close of his life, Telford spoke with pride of the economic and social development of the region following the execution of this and other drainage projects.]

CHAPTER IX.

WHEN Mr. Telford had occasion to visit London on business during the early period of his career, his quarters were at the Salopian Coffee House, now the Ship Hotel, at Charing Cross. It is probable that his Shropshire connections led him in the first instance to the 'Salopian;' but the situation being near to the Houses of Parliament, and in many respects convenient for the purposes of his business, he continued to live there for no less a period than twenty-one years. During that time the Salopian became a favourite resort for engineers; and not only Telford's provincial associates, but numerous visitors from abroad (where his works attracted even more attention than they did in England) took up their quarters there. Several apartments were specially reserved for Telford's exclusive use, and he could always readily command any additional accommodation for purposes of business or hospitality. The successive landlords of the Salopian at length came to regard the engineer as a fixture, and even bought and sold him from time to time with the goodwill of the business. When he finally resolved, on the persuasion of his friends, to take a house of his own, and gave notice of his intention of leaving, the landlord, who had but recently entered into possession, almost stood aghast. "What! leave the house!" said he; "Why, Sir, I have just paid 750*l.* for you!" On explanation it appeared that this price had actually been paid by him to the outgoing landlord, on the assumption that Mr. Telford was a fixture of the hotel; the previous tenant having paid

450*l.* for him; the increase in the price marking very
significantly the growing importance of the engineer's
position.   There was, however, no help for the discon-
solate landlord, and Telford left the Salopian to take
possession of his new house at 24, Abingdon Street.
Labelye, the engineer of Westminster Bridge, had formerly occupied the dwelling; and, at a subsequent period,
Sir William Chambers, the architect of Somerset House.
Telford used to take much pleasure in pointing out to
his visitors the painting of Westminster Bridge, impa-
nelled in the wall over the parlour mantelpiece, made
for Labelye by an Italian artist whilst the bridge works
were in progress.   In that house Telford continued to
live until the close of his life.

One of the subjects in which he took much interest
during his later years was the establishment of the Institute of Civil Engineers.   In 1818 a society had been
formed, consisting principally of young men educated
to civil and mechanical engineering, who occasionally
met to discuss matters of interest relating to their
profession.   As early as the time of Smeaton, a social
meeting of engineers was occasionally held at an inn
in Holborn, which was discontinued, in 1792, in consequence of some personal differences amongst the
members.   It was revived in the following year, under
the auspices of Mr. Jessop, Mr. Naylor, Mr. Rennie,
and Mr. Whitworth, and joined by other gentlemen of
scientific distinction.   They were accustomed to dine
together every fortnight at the Crown and Anchor in
the Strand, spending the evening in conversation on
engineering subjects.   But as the numbers and importance of the profession increased, the desire began to be
felt, especially amongst the junior members, for an institution of a more enlarged character.   Hence the movement among the younger men to which we have alluded,
and which led to an invitation to Mr. Telford to accept
the office of President of their proposed Engineers' In-

stitute. To this he consented, and entered upon the duties of the office on the 21st of March, 1820.

During the remainder of his life, Mr. Telford continued to watch over the progress of the society, which gradually grew in importance and usefulness. He supplied it with the nucleus of a reference library, now become of great value to its members. He established the practice of recording the proceedings,[1] minutes of discussions, and substance of the papers read, which has led to the accumulation, in the printed records of the Institute, of a vast body of information as to engineering practice. In 1828 he exerted himself strenuously and successfully in obtaining a Charter of Incorporation for the society; and finally, at his death, he left the Institute their first bequest of 2000*l*., together with many valuable books, and a large collection of documents which had been subservient to his own professional labours.

In the distinguished position which he occupied, it was natural that Mr. Telford should be called upon, as he often was, towards the close of his life, to give his opinion and advice as to projects of public importance. Where strongly conflicting opinions were entertained on any subject, his help was occasionally found most valuable; for he possessed great tact and suavity of manner, which often enabled him to reconcile opposing interests when they stood in the way of important enterprises.

---

[1] We are informed by Joseph Mitchell, Esq., C.E., of the origin of this practice. Mr. Mitchell was a pupil of Mr. Telford's, living with him in his house at 24, Abingdon Street. It was the engineer's custom to have a dinner-party every Tuesday, after which his engineering friends were invited to accompany him to the Institution, the meetings of which were then held on Tuesday evenings in a house in Buckingham Street, Strand. The meetings did not usually consist of more than from twenty to thirty persons. Mr. Mitchell took notes of the conversations which followed the reading of the papers. Mr. Telford afterwards found his pupil extending the notes, on which he asked permission to read them, and was so much pleased that he took them to the next meeting, and read them to the members. Mr. Mitchell was then formally appointed reporter of conversations to the Institute; and the custom having been continued, a large mass of valuable practical information has thus been placed on record.

In 1828 he was appointed one of the commissioners to investigate the subject of the supply of water to the metropolis, in conjunction with Dr. Roget and Professor Brande, and the result was the very able report published in that year. Only a few months before his death, in 1834, he prepared and sent in an elaborate separate report, containing many excellent practical suggestions, which had the effect of strongly stimulating the water companies, and eventually led to great improvements.

On the subject of roads he was the very highest authority, his friend Southey jocularly styling him the " Colossus of Roads " as well as " Pontifex Maximus." The Russian Government frequently consulted him with reference to the new roads with which that great empire was being opened up. The Polish road from Warsaw to Briesc on the Russian frontier, 120 miles in length, was constructed after his plans, and it remains, we believe, the finest road in the Russian dominions to this day.

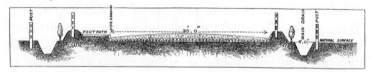

SECTION OF POLISH ROAD.

He was consulted by the Austrian Government on the subject of bridges as well as roads. Count Széchenyi recounts the very agreeable and instructive interview which he had with Telford when he called to consult him as to the bridge proposed to be erected across the Danube, between the towns of Buda and Pesth. On a suspension bridge being suggested by the English engineer, the Count, with surprise, asked if such an erection was *possible* under the circumstances he had described ? " We do not consider anything to be impossible," replied Telford ; " impossibilities exist chiefly in the prejudices of mankind, to which some are slaves, and from which few are able to emancipate themselves and enter on the path of truth." But supposing a suspension bridge were not deemed advisable under the circumstances, and it were considered necessary altogether to avoid motion, "then," said he, "I should recommend you to erect a cast iron bridge of three spans, each 400 feet; such a bridge will have no motion, and though half the world lay a wreck, it would still stand."[1] The suspension bridge was eventually resolved upon; it was constructed by one of Mr. Telford's ablest pupils, Mr. Tierney Clark, between the years 1839 and 1850, and is one of the greatest triumphs of English engineering to be found in Europe, the Buda-Pesth people proudly declaring it to be "the eighth wonder of the world."

At a time when speculation was very rife—in the year 1825—Mr. Telford was consulted respecting a grand scheme for cutting a canal across the Isthmus of Darien; and about the same time he was employed to resurvey the line for a ship canal—which had before occupied the attention of Whitworth and Rennie—between Bristol and the English Channel. But although he gave great attention to this latter project, and prepared numerous plans and reports upon it, and although an Act was actually passed enabling it to be carried out, the scheme was eventually abandoned, like the preceding ones with the same object, for want of the requisite funds.

Our engineer had a perfect detestation of speculative jobbing in all its forms, though on one occasion he could not help being used as an instrument by schemers. A public company was got up at Liverpool, in 1827, to form a broad and deep ship canal, of about seven miles in length, from opposite Liverpool to near Helbre Isle, in the estuary of the Dee : its object was to enable shipping to avoid the variable shoals and sand-banks which obstruct the entrance to the Mersey. Mr. Telford entered on the project with great zeal, and his name was widely quoted

---

[1] Supplement to Weale's 'Bridges,' Count Széchenyi's Report, p. 18.

in its support. It appeared, however, that one of its
principal promoters, who had secured the right of pre-
emption of the land on which the only possible entrance
to the canal could be formed on the northern side, suddenly closed with the corporation of Liverpool, who were
opposed to the plan, and "sold" his partners as well as
the engineer for a large sum of money. Telford, disgusted at being made the instrument of an apparent
fraud upon the public, destroyed all the documents
relating to the scheme, and never afterwards spoke of it
except in terms of extreme indignation.

About the same time the formation of locomotive
railways was extensively discussed, and schemes were set
on foot to construct them between several of the larger
towns. But Mr. Telford was now about seventy years
old; and, desirous of limiting the range of his business
rather than extending it, he declined to enter upon
this new branch of engineering. Yet, in his younger
days, he had surveyed numerous lines of railway—
amongst others, one as early as the year 1805, from
Glasgow to Berwick, down the vale of the Tweed. A
line from Newcastle-on-Tyne to Carlisle was also surveyed and reported on by him some years later; and the
Stratford and Moreton Railway was actually constructed
under his direction. He made use of railways in all his
large works of masonry, for the purpose of facilitating
the haulage of materials to the points at which they were
required to be deposited or used. There is a paper of
his on the Inland Navigation of the County of Salop,
contained in 'The Agricultural Survey of Shropshire,'
in which he speaks of the judicious use of railways, and
recommends that in all future surveys "it be an instruction to the engineers that they do examine the county
with a view of introducing iron railways wherever difficulties may occur with regard to the making of navigable
canals." When the project of the Liverpool and Manchester Railway was started, we are informed that he
was offered the appointment of engineer; but he declined,

partly because of his advanced age, but also out of a feeling of duty to his employers, the Canal Companies, stating that he could not lend his name to a scheme which, if carried out, must so materially affect their interests.*

Towards the close of his life he was afflicted by deafness, which made him feel exceedingly uncomfortable in mixed society. Thanks to a healthy constitution, unimpaired by excess and invigorated by active occupation, his working powers had lasted longer than those of most men. He was still cheerful, clear-headed, and skilful in the arts of his profession, and felt the same pleasure in useful work that he had ever done. It was, therefore, with difficulty that he could reconcile himself to the idea of retiring from the field of honourable labour, which he had so long occupied, into a state of comparative inactivity. But he was not a man who could be idle, and he determined, like his great predecessor Smeaton, to occupy the remaining years of his life in arranging his engineering papers for publication. Vigorous though he had been, he felt that the time was shortly approaching when the wheels of life must stand still altogether. Writing to a friend at Langholm, he said, "Having now been occupied for about seventy-five years in incessant exertion, I have for some time past arranged to decline the contest; but the numerous works in which I am engaged have hitherto prevented my succeeding. In the mean time I occasionally amuse myself with setting down in what manner a long life has been laboriously, and I hope usefully, employed." And again, a little later, he writes: "During the last twelve months I have had several rubs; at seventy-seven they tell more seriously than formerly, and call for less exertion and require greater precautions. I fancy that few of my age belonging to the valley of the Esk remain in the land of the living." [1]

One of the last works on which Mr. Telford was pro-

---

[1] Letter to Mrs. Little, Post Office, Langholm, 28th August, 1833.

fessionally consulted was at the instance of the Duke of Wellington—not many years younger than himself, but of equally vigorous intellectual powers—as to the improvement of Dover Harbour, then falling rapidly to decay. The long-continued south-westerly gales of 1833-4 had the effect of rolling an immense quantity of shingle up Channel towards that port, at the entrance to which it became deposited in unusual quantities, so as to render it at times altogether inaccessible. The Duke, as a military man, took a more than ordinary interest in the improvement of Dover, as the military and naval station nearest to the French coast; and it fell to him as Lord Warden of the Cinque Ports to watch over the preservation of the harbour, situated at a point in the English Channel which he regarded as of great strategic importance in the event of a continental war. He therefore desired Mr. Telford to visit the place and give his opinion as to the most advisable mode of procedure with a view to improving the harbour. The result was a report, in which the engineer recommended a plan of sluicing, similar to that adopted by Mr. Smeaton at Ramsgate, and which was afterwards carried out with considerable success by Mr. James Walker, C.E.

This was his last piece of professional work. A few months later he was laid up by bilious derangement of a serious character, which recurred with increased violence towards the close of the year; and on the 2nd of September, 1834, Thomas Telford closed his useful and honoured career, at the advanced age of seventy-seven. With that absence of ostentation which characterised him through life, he directed that his remains should be laid, without ceremony, in the burialground of the parish church of St. Margaret's, Westminster. But the members of the Institute of Civil Engineers, who justly deemed him their benefactor and chief ornament, urged upon his executors the propriety of interring him in Westminster Abbey. He was buried there accordingly, near the middle of the nave; where the letters, "Thomas Telford, 1834," mark the place beneath which he lies.[1] The adjoining stone bears the inscription, "Robert Stephenson, 1859," that engineer having during his life expressed the wish that his body should be laid near that of Telford; and the son of the Killingworth engineman thus sleeps by the side of the son of the Eskdale shepherd.

TELFORD'S BURIAL PLACE IN WESTMINSTER ABBEY.

[By Percival Skelton.]

It was a long, a successful, and a useful life which thus ended. Every step in his upward career, from the

---

[1] A statue of him, by Bailey, has since been placed in the east aisle of the north transept, known as the Islip Chapel. It is considered a fine work, but its effect is quite lost in consequence of the crowded state of the aisle, which has very much the look of a sculptor's workshop. The subscription raised for the purpose of erecting the statue was 1000*l*., of which 200*l*. was paid to the Dean for permission to place it within the Abbey.

poor peasant's hut in Eskdale to Westminster Abbey, was nobly and valorously won. The man was diligent and conscientious; whether as a working mason hewing stone blocks at Somerset House, as a foreman of builders at Portsmouth, as a road surveyor at Shrewsbury, or as an engineer of bridges, canals, docks, and harbours. The success which followed his efforts was thoroughly well deserved. He was laborious, pains-taking, and skilful; but, what was better, he was honest and upright. He was a most reliable man; and hence he came to be extensively trusted. Whatever he undertook, he endeavoured to excel in. He would be a first-rate hewer, and he became so. He was himself accustomed to attribute much of his success to the thorough way in which he had mastered the humble beginnings of this trade. He was even of opinion that the course of manual training he had undergone, and the drudgery, as some would call it, of daily labour—first as an apprentice, and afterwards as a journeyman mason—had been of greater service to him than if he had passed through the curriculum of a University. Writing to his friend, Miss Malcolm, respecting a young man who desired to enter the engineering profession, he in the first place endeavoured to dissuade the lady from encouraging the ambition of her *protégé*, the profession being overstocked, and offering very few prizes in proportion to the large number of blanks. " But," he added, " if civil engineering, notwithstanding these discouragements, is still preferred, I may point out that the way in which both Mr. Rennie and myself proceeded, was to serve a regular apprenticeship to some practical employment—he to a millwright, and I to a general housebuilder. In this way we secured the means, by hard labour, of earning a subsistence; and, in time, we obtained by good conduct the confidence of our employers and the public; eventually rising into the rank of what is called Civil Engineering. This is the true way of acquiring practical skill, a thorough knowledge of the materials employed in construction, and last, but not least, a perfect knowledge of the habits and dispositions of the workmen who carry out our designs. This course, although forbidding to many a young person, who believes it possible to find a short and rapid path to distinction, is proved to be otherwise by the two examples I have cited. For my own part, I may truly aver that ' steep is the ascent, and slippery is the way.' " [1]

That Mr. Telford was enabled to continue to so advanced an age employed on laborious and anxious work, was no doubt attributable in a great measure to the cheerfulness of his nature. He was, indeed, a most happy-minded man. It will be remembered that, when a boy, he had been known in his valley as " Laughing Tam." The same disposition continued to characterize him even in his old age. He was playful and jocular, and rejoiced in the society of children and young people, especially when well-informed and modest. But when they pretended to acquirements they did not possess, he was quick to detect and see through them. One day a youth expatiated to him in very large terms about a friend of his, who had done this and that, and made so and so, and could do all manner of wonderful things. Telford listened with great attention, and when the youth had done, he quietly asked, with a twinkle in his eye, " Pray, can your friend lay eggs ? "

When in society he gave himself up to it, and thoroughly enjoyed it. He did not sit apart, a moody and abstracted " lion ; " nor desire to be regarded as " the great engineer," pondering new Menai Bridges ; but he appeared in his natural character of a simple, intelligent, cheerful companion ; as ready to laugh at his own jokes as at other people's ; and he was as communicative to a child as to any philosopher of the party.

---

[1] Letter to Miss Malcolm, Burnfoot, Langholm, dated 7th Oct., 1830.

Robert Southey, than whom there was no better judge of a loveable man, said of him, " I would go a long way for the sake of seeing Telford and spending a few days in his company." Southey had the best opportunity of knowing him well ; for he performed a long tour with him through Scotland, in 1819. And a journey in company, extending over many weeks, is, probably, better than anything else, calculated to bring out the weak as well as the strong points of a friend : indeed, many friendships have completely broken down under the severe test of a single week's tour. But Southey on that occasion firmly cemented a friendship which lasted until Telford's death. On one occasion the latter called at the poet's house, in company with Sir Henry Parnell, when engaged upon the survey of one of his northern roads. Unhappily Southey was absent at the time ; and, writing about the circumstance to a correspondent, he said, " This was a great mortification to me, inasmuch as I owe Telford every kind of friendly attention, and like him heartily." [1]

Campbell, the poet, was another early friend of our engineer ; and the attachment seems to have been mutual. Writing to Dr. Currie, of Liverpool, in 1802, Campbell says : " I have become acquainted with Telford the engineer, ' a fellow of infinite humour,' and of strong enterprising mind. He has almost made me a bridge-builder already ; at least he has inspired me with new sensations of interest in the improvement and ornament of our country. Have you seen his plan of London Bridge ? or his scheme for a new canal in the North Highlands, which will unite, if put in effect, our Eastern and Atlantic commerce, and render Scotland the very emporium of navigation ? Telford is a most useful cicerone in London. He is so universally acquainted, and so popular in his manners, that he can

---

[1] 'Selections from the Letters of Robert Southey.' Edited by J. W. Warter, B.D. Vol. iii., p. 326.

introduce one to all kinds of novelty, and all descriptions of interesting society."[1] Shortly after, Campbell named his first son after Telford, who stood godfather for the boy. Indeed, for many years, Telford played the part of Mentor to the young and impulsive poet, advising him about his course in life, trying to keep him steady, and holding him aloof as much as possible from the seductive allurements of the capital. But it was a difficult task, and Telford's numerous engagements necessarily left the poet at many seasons very much to himself. It appears that they were living together at the Salopian when Campbell composed the first draft of his poem of Hohenlinden; and several important emendations made in it by Telford were adopted by Campbell. Although the two friends pursued different roads in life, and for many years saw little of each other, they often met again, especially after Telford took up his abode at his house in Abingdon Street, where Campbell was a frequent and always a welcome guest.

When engaged upon his surveys, our engineer was the same simple, cheerful, laborious man. While at work, he gave his whole mind to the subject in hand, thinking of nothing else for the time; dismissing it at the close of each day's work, but ready to take it up afresh with the next day's duties. This was a great advantage to him as respected the prolongation of his working faculty. He did not take his anxieties to bed with him, as many do, and rise up with them in the morning; but he laid down the load at the end of each day, and resumed it all the more cheerfully when refreshed and invigorated by natural rest. It was only while the engrossing anxieties connected with the suspension of the Menai Bridge were weighing heavily upon his mind, that he could not sleep; and then, age

---

[1] Beattie's 'Life and Letters of Thomas Campbell,' vol. i., p. 451.

having stolen upon him, he felt the strain almost more than he could bear. But that great anxiety over, his spirits speedily resumed their wonted elasticity.

When engaged upon the construction of the Carlisle and Glasgow road, he was very fond of getting a few of the "navvy men," as he called them, to join him at an ordinary at the Hamilton Arms Hotel, Lanarkshire, each paying his own expenses; and though Telford told them he could not drink, yet he would carve and draw corks for them. One of the rules he laid down was, that no business was to be introduced from the moment they sat down to dinner. All at once, from being the plodding, hard-working engineer, with responsibility and thought in every feature, Telford unbended and relaxed, and became the merriest and drollest of the party. He possessed a great fund of anecdote available for such occasions, had an extraordinary memory for facts relating to persons and families, and the wonder to many of his auditors was, how in all the world a man living in London should know so much better about their locality and many of its oddities than they did themselves.

In his leisure hours at home, which were but few, he occupied himself a good deal in the perusal of miscellaneous literature, never losing his taste for poetry. He continued to indulge in the occasional composition of verses until a comparatively late period of his life; one of his most successful efforts being a translation of the 'Ode to May,' from Buchanan's Latin poems, executed in a very tender and graceful manner. That he might be enabled to peruse engineering works in French and German, he prosecuted the study of those languages, and with such success that he was shortly able to read them with comparative ease. He occasionally occupied himself in literary composition on subjects connected with his profession. Thus he wrote for the Edinburgh Encyclopedia, conducted by his friend Sir David (then Dr.) Brewster, the elaborate and able articles on Architecture, Bridge-building, and Canal making. Besides his contributions to that work, he advanced a considerable sum of money to aid in its publication, which remained a debt due to his estate at the period of his death.

Although occupied as a leading engineer for nearly forty years—having certified contractors' bills during that time amounting to many millions sterling—he died in comparatively moderate circumstances. Eminent constructive ability was not very highly remunerated in Telford's time, and his average income did not amount to more than is paid to the resident engineer of any modern railway. But Telford's charges were perhaps unusually low—so much so that a deputation of members of the profession on one occasion formally expostulated with him on the subject.

Although he could not be said to have an indifference for money, he yet estimated it as a thing worth infinitely less than character. His wants were few, and his household expenses small; and though he entertained many visitors and friends, it was in a quiet way and on a moderate scale. The small regard he had for personal dignity may be inferred from the fact, that to the last he continued the practice, which he had learnt when a working mason, of darning his own stockings.[1] But he had nevertheless the highest idea of the dignity of his profession; not, however, because of the money it would

---

[1] Mr. Mitchell says: "He lived at the rate of about 1200l. a year. He kept a carriage, but no horses, and used his carriage principally for making his journeys through the country on business. I once accompanied him to Bath and Cornwall, when he made me keep an accurate journal of all I saw. He used to lecture us on being independent, even in little matters, and not ask servants to do for us what we might easily do for ourselves. He carried in his pocket a small book containing needles, thread, and buttons, and on an emergency was always ready to put in a stitch. A curious habit he had of mending his stockings, which I suppose he acquired when a working mason. He would not permit his housekeeper to touch them, but after his work at night, about nine or half-past, he would go upstairs, and take down a lot, and sit mending them with great apparent delight in his own room till bed-time. I have frequently gone in to him with some message, and found him occupied with this work."

produce, but of the great things it was calculated to accomplish. In his most confidential letters we find him often expatiating on the noble works he was engaged in designing or constructing, and the national good likely to flow from them, but never on the pecuniary advantages he himself was to reap. He doubtless prized, and prized highly, the reputation they would bring him; and, above all, there seemed to be uppermost in his mind, especially in the earlier part of his career, whilst many of his schoolfellows were still alive, the thought of "What will they say of this in Eskdale?" but as for the money results to himself, Telford seemed, to the close of his life, to regard them as of comparatively small moment.

During the twenty-one years that he acted as principal engineer for the Caledonian Canal, we find from the Parliamentary returns that the amount paid to him for his reports, detailed plans, and superintendence, was exactly 237*l*. a-year. When he conceived the works to be of great public importance, and promoted by public-spirited persons at their own expense, he refused to receive any payment for his labour, or even repayment of the expenses incurred by him. Thus, while employed by the Government in the improvement of the Highland roads, he persuaded himself that he ought at the same time to promote the similar patriotic objects of the British Fisheries Society, which were carried out by voluntary subscription; and for many years he acted as their engineer, refusing to accept any remuneration whatever for his trouble.[1]

Mr. Telford held the sordid money-grubber in perfect detestation. He was of opinion that the adulation paid

---

[1] "The British Fisheries Society," adds Mr. Rickman, "did not suffer themselves to be entirely outdone in liberality, and shortly before his death they pressed upon Mr. Telford a very handsome gift of plate, which, being inscribed with expressions of their thankfulness and gratitude towards him, he could not possibly refuse to accept."—'Life of Telford,' p. 283.

to mere money was one of the greatest dangers with which modern society was threatened. " I admire commercial enterprise," he would say; " it is the vigorous outgrowth of our industrial life : I admire everything that gives it free scope, as wherever it goes, activity, energy, intelligence—all that we call civilization—accompany it; but I hold that the aim and end of all ought not to be a mere bag of money, but something far higher and far better."

Writing once to his Langholm correspondent about an old schoolfellow, who had grown rich by scraping, Telford said : " Poor Bob L——! His industry and sagacity were more than counterbalanced by his childish vanity and silly avarice, which rendered his friendship dangerous and his conversation tiresome. He was like a man in London, whose lips, while walking by himself along the streets, were constantly ejaculating ' Money! Money!' But peace to Bob's memory : I need scarcely add, confusion to his thousands!" He himself was most careful in resisting the temptations to which men in his position are frequently exposed; but he was preserved by his honest pride, not less than by the purity of his character. He would not receive anything in the shape of presents or testimonials from persons employed under him as contractors; for he would not have even the shadow of an obligation stand in the way of performing his rigid duty to those who employed him to watch over and protect their interests.

Yet Telford was not without a proper regard for money, as a means of conferring benefits on others, and especially as a means of being independent. By the close of his life he had accumulated as much as, invested at interest, would bring him in about 800*l.* a-year, and enable him to occupy the house in Abingdon Street until he died. This was amply sufficient for his wants, and more than enough for perfect independence. It enabled him also to continue those secret acts of bene-

volence which constituted perhaps the most genuine
pleasure of his life.   It is one of the most delightful
traits in this excellent man's career to find him so constantly occupied in works of spontaneous charity, in
places so remote that it is impossible the slightest feeling
of ostentation could ever have sullied the purity of his
acts.   Among the large mass of Telford's private letters
which have been submitted to us, we find constant reference to sums of money transmitted for the support of
poor people in his native valley.   At new year's time he
regularly sent a remittance of from 30*l.* to 50*l.*, to be
distributed by Miss Malcolm of Burnfoot, and, after
her death, by Mr. Little, the postmaster at Langholm;
and the contributions thus so kindly made did much
to fend off the winter's cold, and surround with many
small comforts those who most needed help, but were
perhaps too modest to ask it.

Many of those in the valley of the Esk had known of
Telford in his younger years as a poor barefooted boy;
yet though become a man of distinction, he had too
much good sense to be ashamed of his humble origin;
perhaps he even felt proud that, by dint of his own
valorous and persevering efforts, he had been able to rise
so much above it.   Throughout his long life his heart
always warmed at the thought of Eskdale.   He rejoiced
at the honourable rise of Eskdale men as reflecting
credit upon his " beloved valley."   Thus, writing to his
Langholm correspondent, with reference to the honours
conferred on the different members of the family of
Malcolm, he said : " The distinctions so deservedly bestowed upon the Burnfoot family, establish a splendid
era in Eskdale ; and almost tempt your correspondent to
sport his Swedish honours, which that grateful country
has repeatedly, in spite of refusal, transmitted." [1]   It
might be said that there was narrowness and provin-

---

[1] Letter to Mr. William Little, Langholm, 24th January, 1815.

cialism in this; but when young men are thrown into the world, with all its temptations and snares, it is well that the recollections of home and kindred should survive to hold them in the path of rectitude, and cheer them in their onward and upright course in life. And there is no doubt that Telford was borne up on many occasions by the thought of what the folks in the valley would say about him and his progress in life, when they met together at market, or at the Westerkirk church porch on Sabbath mornings. In this light, provincialism or local patriotism is a prolific source of good; and may be regarded as among the most valuable and beautiful emanations of the parish life of our country. Although Telford was honoured with the titles and orders of merit conferred upon him by foreign monarchs, what he esteemed beyond them all was the respect and gratitude of his own countrymen; and, not least, the honour which his really noble and beneficent career was calculated to reflect upon "the folks of the nook," the remote inhabitants of his "beloved Eskdale."

When the engineer proceeded to dispose of his savings by will, which he did a few months before his death, the distribution was a comparatively easy matter. The total amount of his bequeathments was 16,600*l.* About one-fourth of the whole he set apart for educational purposes,—2,000*l.* to the Civil Engineers' Institute, and 1,000*l.* each to the ministers of Langholm and Westerkirk, in trust for the parish libraries. The rest was bequeathed, in sums of from 200*l.* to 500*l.*, to different persons who had acted as clerks, assistants, and surveyors, in his various public works; and to his intimate personal friends. Amongst these latter were Colonel Pasley, the nephew of his early benefactor; Mr. Rickman, Mr. Milne, and Mr. Hope, his three executors; and Robert Southey and Thomas Campbell, the poets. To both of these last the gift was most welcome. Southey said of his: "Mr. Telford has most kindly and unexpectedly left me 500*l*., with a share of his residuary property, which I am told will make it amount in all to 850*l*. This is truly a godsend, and I am most grateful for it. It gives me the comfortable knowledge that, if it should please God soon to take me from this world, my family would have resources fully sufficient for their support till such time as their affairs could be put in order, and the proceeds of my books, remains, &c., be rendered available. I have never been anxious overmuch, nor ever taken more thought for the morrow than it is the duty of every one to take who has to earn his livelihood; but to be thus provided for at this time I feel to be an especial blessing." [1]

Among the most valuable results of Telford's bequests in his own district, was the establishment of the libraries at Langholm and Westerkirk on a firmer foundation. That at Westerkirk had been originally instituted in the year 1792, by the miners employed to work an antimony mine (since abandoned) on the farm of Glendinning, within sight of the place where Telford was born. On the dissolution of the mining company, in 1800, the little collection of books was removed to Kirkton Hill; but on receipt of Telford's bequest, a special building was erected for their reception at New Burtpath, near the village of Westerkirk. The annual income derived from the Telford fund enabled additions of new volumes to be made to it from time to time; and its uses as a public institution were thus greatly increased. The books are exchanged once a month, on the day of the full moon; on which occasion readers of all ages and conditions,—farmers, shepherds, ploughmen, labourers, and their children,—resort to it from far and near, taking away with them as many volumes as they desire for the month's reading.

---

[1] 'Selections from the Letters of Robert Southey,' vol. iv., p. 391. We may here mention that the last article which Southey wrote for the 'Quarterly' was his review of the 'Life of Telford.'

Thus there is scarcely a cottage in the valley in which good books are not to be found under perusal; and we are told that it is a common thing for the Eskdale shepherd to take a book in his plaid to the hill-side—a volume of Shakespeare, Prescott, or Macaulay—and read it there, under the blue sky, with his sheep and the green hills before him. And thus, as long as the bequest lasts, the good, great engineer will not cease to be remembered with gratitude in his beloved Eskdale.

TELFORD'S NATIVE VALLEY, ESKDALE.

[By Percival Skelton.]

EDITOR'S NOTES

(For page 36)

Before mechanical engineering was recognized as a distinct form of civil engineering (around mid-nineteenth century), the expressions "millwright" and "factory architect" (in the early nineteenth century) denoted occupations which were early forms of mechanical engineering in many particulars. James Brindley and John Rennie both apprenticed as millwrights. It is less well known that Thomas Telford compiled a treatise "On Mills" in 1796–98 that gives a record of contemporary millwrighting practice ("Thomas Telford 'On Mills'," edited by E. Lancaster Burne, *Transactions of the Newcomen Society*, XVII (1936–37), pp. 205–214).

(For page 44)

Arthur Titley has examined notes written by Brindley about the Fenton Vivian engine. Titley believes that Smiles may have seen more of Brindley's diary and notebooks than is now extant, for Smiles in his preface refers to Brindley material in the possession of Joseph Mayer. On the basis of what he has seen, Titley does not believe that the engine had a wooden cylinder, but that it may have been lagged with wood. Titley finds that Smiles errs in giving Coalbrookdale as the place of origin for the boiler plates. Titley is not certain that the entries quoted by Smiles refer to the Fenton Vivian engine. Arthur Titley, "Notes Upon a Part of a Diary by James Brindley," *Transactions of the Newcomen Society*, XX (1939–40), pp. 67–74.

(For page 45)

A. W. Skempton writes that the first summit-level canal in the British Isles was built between 1737–45 from Newry in northern Ireland to Lough Neagh. Skempton surveys earlier canal engineering on the continent in "Canals and River Navigations Before 1750" in *A History of Technology*, edited by C. Singer, E. J. Holmyard, A. R. Hall, and T. I. Williams (New York: Oxford University Press, 1957), III, pp. 438–470. Smiles does not stress continental precedents, which may result, in part, from his enthusiasm for his engineers and their supporters. On the other hand, Brindley, Bridgewater, and contemporaries may have known surprisingly little of continental achievements. "Both in Parliament and out of doors these [the early canal works of Brindley] were treated as visionary and impossible; and no one seems to have known how much greater works had been executed in France before his [Brindley's] day." "Progress of Engineering Science", *Quarterly Review*, 114 (1863), p. 327.

(For page 47)

In a later edition Smiles added that "while passing through the south of France, the Duke was especially interested by his inspection of the Grand Canal of Languedoc or the Canal du Midi." *Lives* (New York, 1905), p. 189 in Brindley biography. Begun about one hundred years before Brindley began the Bridgewater Canal, the Languedoc was constructed on a grand scale. It extended 158 miles across the isthmus connecting France with Spain. There were more than one hundred locks, numerous aqueducts, and many bridges and tunnels. In the later edition Smiles included a twelve-page appendix on the Languedoc and its constructor, Pierre-Paul Riquet de Bonrepos. Smiles may well have reacted to criticism of his neglect of continental precedents.

(For page 58)

John Gilbert was a land agent who acted for the Duke as overseer, engineer, and general business manager. His contribution to the canal may have been greater than Smiles indicates.

Gilbert's importance is suggested in *Telford and Brindley* (London: W. & R. Chambers, 1895), pp. 100–101. He was an active promoter of the Trent and Mersey Canal; his brother, Thomas, was Earl Gower's land agent, a poor-law reformer, and member of Parliament.

(For page 73)

The fortuitous conjunction of the rise of steam power and canal transportation as noted here by Smiles is sometimes overlooked. The cotton-manufacturing industry developed rapidly in Lancashire, using available water-power sites before the introduction of the rotary steam engine.

(For page 93)

Granville Leveson-Gower, first Marquis of Stafford (1721–1803), because of his considerable wealth and political influence could help gain passage of parliamentary acts and raise capital necessary for engineering works. He was brother-in-law of the Duke of Bridgewater.

(For page 94)

Arthur Young visited the Bridgewater Canal and wrote in detail of the section from Manchester to Worsley. Generally he was favorably impressed by the engineering and economic aspects of it, but he noted some inadequacies of design and construction. *A Six Months Tour Through the North of England* (London, 1770), III, pp. 251–291.

(For page 108)

George Rennie, eldest son of John. Besides practicing civil engineering, he superintended the machinery manufacturing works established by his father. The biscuit-making machinery for Deptford Dockyard was made there.

(For page 121)

Robert Fulton (1765–1815) went abroad in 1786 and did not return to the U.S. for twenty years. In London he formed friendships with the Duke of Bridgewater and Lord Stanhope. Fulton studied canal engineering intensively.

(For page 122)

A tram-road was a road having tracks of timber beams, stone blocks, or slabs laid in parallel to form wheel tracks for easier transportation in "trams" or wagons used especially for hauling coal near mines.

(For page 141)

See Samuel Smiles, *Josiah Wedgwood*, London, 1894.

(For page 148)

Involved as he was in the management of a railway company, Smiles sophisticatedly analyzed and described the politics of technology. His abilities in this respect are displayed to advantage in his discussion of the Liverpool and Manchester Railway project and the opposition to it in the 1820's. See especially his "Parliamentary Contest on the Liverpool and Manchester Bill," *Life of George Stephenson*, London, 1857. The Duke of Bridgewater's trustees for his canal interests now numbered among the forces opposing innovation.

(For page 154)

When Brindley's tunnel was found too limited for the traffic, the Grand Trunk Canal Company consulted Thomas Telford. He recommended an entirely new tunnel of much larger dimensions. The work, begun in 1824, was completed in 1827.

(For page 160)

Smiles's account of the improvement in living conditions among the rural population should be contrasted with accounts of conditions in industrial cities. Smiles motivated reform by stressing objectives and ideals rather than the evils that beset society.

(For page 166)

The laudatory article in *Biographia Britannica,* 2nd ed. (London, 1780) states that, contrary to having been "taken by many for an idiot," as a writer to the *Morning Post* had observed (August, 1776), Brindley had an animated and sensible countenance.

(For page 168)

Hugh Henshall, engineer, carried on and completed works in which Brindley was engaged at the time of his death. "James Brindley," *Biographia Britannica,* 2nd ed. (London, 1780), p. 603. The article was based on materials from "Mr. Henshall, Mr. Wedgwood, and Mr. Bentley."

(For page 170)

"The crazy publication" of the last Earl of Bridgewater was Francis Henry Egerton, 8th Earl of Bridgewater, *A Letter to the Parisians, and the French Nation, upon inland Navigation, containing a Defence of the public Character of His Grace Francis Egerton, late Duke of Bridgewater . . . and including some Notices, and, Anecdotes, concerning Mr. James Brindley,* privately printed, 1820. Of Brindley's drunkenness, the 8th Earl wrote, "he [Brindley] used to drink, only a bason of milk in the morning . . . afterwards [he] left the house at Worsley, and removed to Stretford, where he became drunken: he carried in his pocket, a little Brandy-Bottle wattled, called a 'Pocket-Pistol,' " p. 66. Egerton also questions the importance of Brindley's role in some major engineering works along the Bridgewater Canal, pp. 62–65. The book of Egerton — if not crazy — is curious in style and content.

(For page 187)

Smiles included a short chapter in the **1861** edition on the millwright Andrew Meikle, who was especially inventive in improving agricultural machinery. His most notable invention was a thrashing-mill. In his patent of **1788** Meikle described himself as "engineer and machinist." During the twenty years from the date of the patent about three hundred and fifty thrashing-mills were erected in East Lothian alone.

(For page 193)

John Robison (1739–1805) was elected lecturer on chemistry in Glasgow University in 1766. In 1773 he became professor of natural philosophy at Edinburgh University. His interest in applications was great. He wrote on the steam engine, seamanship, the telescope, optics, waterworks, resistance of fluids, electricity, and magnetism for the *Encyclopaedia Britannica*. He published several volumes on "mechanical philosophy." Joseph Black (1728–1799) was named to the chair of anatomy and chemistry in the University of Glasgow in 1756; in 1766 he became professor of medicine and chemistry at Edinburgh. He wrote an influential paper on "Experiments upon Magnesia Alba, Quicklime, and some other Alkaline Substances" (see p. 10). He lectured on his discovery of latent heat, and among those greatly influenced by this was James Watt. Robison published Black's lecture notes in 1803: *Lectures on the Elements of Chemistry*.

(For page 194)

Colin Maclaurin (1698–1746) was a Scottish mathematician and natural philosopher. He taught at Edinburgh and was skilled in experimental physics and practical mechanics. He wrote, among other books, an elementary treatise on algebra which became a popular textbook. Bernard Forest de Belidor (1697–1761) was a French military engineer and mathematician. He published *Architecture hydraulique*, 4 vols, Paris, 1737–1754 and *Sommaire d'un cours d'architecture, militaire, civile, et hydraulique*, Paris, 1720. Smiles may refer to William Emerson

(1701–1782), an English mathematician, who wrote a series of mathematical manuals for students. William Jacob Gravesande (1688–1742) was a Dutch mathematician and philosopher. Among his published scientific works was *Institutes of the Newtonian Philosophy*. Switzer wrote on hydraulics.

(For page 207)

O. A. Westworth in "Albion Steam Flour Mill," *Economic History*, II (Jan. 1932), pp. 380–395, writes of differences between Rennie and Samuel Wyatt, a proprietor and the promoter of the enterprise. Westworth also quotes from correspondence showing the many engineering problems frustrating Rennie and Watt. William Murdoch came from Boulton and Watt to advise Rennie on at least one occasion. Writing about four years after publication of the Rennie biography, with the Boulton and Watt MSS. available, Smiles acknowledges that the mill was a commercial failure [*Lives of Boulton and Watt* (London, 1865), p. 326].

(For page 209)

Cyril Boucher notes that Rennie also designed water power mills throughout his life: *John Rennie* (Manchester, 1963), p. 80.

(For page 209)

Charles Stanhope, third Earl Stanhope (1753–1816). A close friend of the second William Pitt for a time, Stanhope was active in politics. He was elected member of the Royal Society in 1772 and was a member of the Philadelphia Philosophical Society. Throughout his life he experimented and invented. He contributed to the art of printing, perfecting among other things a process of stereotyping. In 1792 the *Gentleman's Magazine* announced that his steam-engine vessel of two hundred tons was being built.

(For page 215)

Throughout the *Lives* Smiles refers to members of the gentry and aristocracy who played the role of entrepreneur to his en-

gineers, but he does not stress the significance of their contributions, with the exception of those of the Duke of Bridgewater
and Sir William Pulteney. The Duke of Bedford, for example,
was a great patron and promoter of drainage of the Fens when
Telford was carrying out projects there.

(For page 228)

Sir Joseph Banks estimated his share of the expenses of the
drainage of the North Level of the Fens as £100,000 and his
final gain by improvement of his property as about £200,000.
Review of the *Life of Thomas Telford, Written by Himself,
Quarterly Review,* 63 (1839), pp. 403–457.

(For page 229)

Rennie was a university-trained engineer familiar with most
advanced structural science, and he incorporated this knowledge
in his structures. Because of this he was a rare type among
members of the profession in the eighteenth century; see Cyril
Boucher, *John Rennie* (Manchester, 1963), p. 37. Boucher has
a chapter on "structural theory and practice."

(For page 243)

Smiles describes the construction of the Conway and Britannia
bridges in the new edition of the life of George Stephenson which
was published as volume III of the *Lives* (London, 1862), pp.
420–440. This volume includes the biography of Robert Stephenson.

(For page 264)

Fulton later wrote to President James Madison and members
of Congress that he had unsuccessfully endeavored for many
years to have torpedoes introduced in France and England, and
that the numerous large-scale experiments in this connection
allowed him to discover and correct errors in design and application. Fulton describes his blowing up of the Danish brig
Dorothea, on October 15, 1805 as most satisfactory, proving

that an explosion of sufficient powder under a vessel would destroy it. Robert Fulton, *Torpedo War and Submarine Explosions* (New York, 1810), pp. 5–8.

(For page 264)

Other accounts bestow only nominal responsibility for the undertaking upon Rennie, crediting the original plans to Robert Stevenson. Robert Louis Stevenson, grandson of Robert, states that Rennie "did not design the Bell Rock, that he did not execute it, and that he was not paid for it." R. L. Stevenson, "Records of a Family of Engineers," *Letters and Miscellanies of Robert Louis Stevenson* (New York: Scribner, 1924), XVIII, p. 270. See also David Stevenson, *Life of Robert Stevenson, Civil Engineer* (London: E. & F. N. Spon, 1878), and Alan Stevenson, *Civil Engineers' and Architects' Journal* (1849). C. T. G. Boucher in *John Rennie* (Manchester, 1963) gives a balanced appraisal with Rennie as the influential consulting engineer (pp. 52–61).

(For page 273)

Sir John Rennie (1794–1874) civil engineer, and second son of John, stressed civil engineering while his brother George managed the machine manufactory in London. The father provided John with an excellent education and access to British society. He was knighted in 1831 after carrying out his father's designs for the London Bridge. Sir Hugh Myddelton was the first engineer to be knighted [Smiles, *Lives* (London, 1861) I, p. 125) ; Sir John Rennie, the second. See *Autobiography of Sir John Rennie, F. R. S.*, London: E. & F. N. Spon, 1875.

(For page 281)

Smiles has been criticized for his eulogistic character sketches of the engineers. L. T. C. Rolt observes that Smiles wrote of Telford and Rennie as paragons of virtue between whom a clash of personality would be inconceivable, but as a matter of fact we know that the two were at odds. *Thomas Telford* (London, 1958), pp. 148–149, 197.

(For page 282)

L. T. C. Rolt observes that Rennie's estimates were sometimes overly optimistic in the case of canals, *Telford* (London, 1958), p. 149.

(For page 284)

Catalogue of the splendid and valuable Library of the late John Rennie (sold at auction . . . 1829), London, 1829.

(For page 316)

Pulteney was reputed to be the richest commoner in England. He disposed of more Parliamentary pocket boroughs than anyone. Pulteney and John Wilkinson, the famous ironmaster, were on the committee recommending candidates for the position of Engineer of the Ellesmere Canal. This was one of Telford's major engineering works. Sir Alexander Gibb, *The Story of Telford* (London, 1935), pp. 27–28. Pulteney was also a commissioner for the Caledonian Canal for which Telford was principal engineer.

(For page 330)

See L. T. C. Rolt, *Telford* (London, 1958), pp. 34–58 for an informed account of the Ellesmere Canal. Smiles's account is in *Lives* (London, 1861), II, pp. 335–349.

(For page 331)

Contrary evidence leads to the conclusion that preliminary works "on the riverside" were not begun. L. T. C. Rolt, *Telford* (London, 1958), p. 142.

(For page 351)

William Jessop (1745–1814) was often relied upon by Telford for advice. See footnote on p. 256. Jessop, Matthew Davidson, and John Telford were engineers assisting him on the Caledonian

Canal. Gibb, *Story of Telford*, pp. 94–95. Smiles does not stress the importance of staff work to his engineers.

(For page 354)

Matthew Davidson (1756–1819), the former mason, also of Langholm, was brought to undertake bridgework when Telford became Surveyor of the County of Salop. Later Davidson had charge of construction for Telford on the Ellesmere Canal and then became chief resident engineer on the Caledonian Canal. Alexander Gibb, *The Story of Telford* (London, 1935), p. 21.

(For page 356)

Sir Alexander Gibb does not believe that Telford took the criticism and obloquy as seriously as suggested by Smiles, *The Story of Telford* (London, 1935), p. 225.

(For page 359)

Count Baltzar von Platen (1766–1829), a distinguished Swedish statesman of considerable political influence, was a promoter and executer of the Gotha Canal.

(For page 361)

Southey's trip taken with Telford to inspect the state of engineering works is described in Robert Southey, *Journal of a Tour in Scotland in 1819*, edited by C. H. Herford, London: John Murray, 1929.

(For page 382)

L. T. C. Rolt, looking at the drawings of the Runcorn design, believes it was as well that the bridge was not built, *Telford* (London, 1958), p. 119.

(For page 384)

W. A. Provis was one of a number of engineers who trained as young men under Telford. Provis had accompanied Telford on his Highland tours. His influence on these young men is discussed in, Sir Alexander Gibb, *The Story of Telford* (London, 1935), pp. 203–206.

(For page 403)

Smiles has written in detail of the Liverpool and Manchester Railway in *Life of George Stephenson*, London, 1857.

SELECTED BIBLIOGRAPHY

I. BOOKS BY SAMUEL SMILES

In the case of the first edition the place of publication is London and the publisher is John Murray unless otherwise shown. Dates only are given for later editions, and only British editions are noted. Smiles was published in America and in many languages.

Physical Education: or the Nurture and Management of Children. Edinburgh: Oliver and Boyd, 1838; 1868; 1905 (edited by Sir Hugh Beevor, London, Walter Scott).

In his preface, Smiles called this book popular, educational literature. He concluded — after writing of digestion, respiration, sleep, and other subjects — that physical education is the basis of moral and intellectual culture.

History of Ireland and the Irish People under the Government of England. London: William Strange, 1844.

Smiles thought the Irish worthy of study because of their patient endurance during centuries of oppression and wrong. He wrote this general survey not only because of his own sympathies but to remedy a deficiency in histories of Ireland.

Railway Property: Its Conditions and Prospects. London: Effingham Wilson, 1849.

This volume on the economics, politics, construction, operation, and maintenance of railway was published after Smiles became secretary of the Leeds and Thirsk Railway Company.

Life of George Stephenson: Comprising also a History of the Invention and Introduction of the Railway Locomotive. 1857 (four editions in first year), 1858, 1859, 1862, 1864, 1868, 1881.

Robert, son of George and a distinguished engineer, died in 1859, and subsequent editions of the life of George included a biography of the son. A revised edition of the *Life of George Stephenson* including the biography of Robert was published as volume III of the *Lives of the Engineers* in 1862. The revised edition of 1868 was enlarged to include material on the railway and the growth of London, and on Cugnot, Murdoch, and Trevithick.

Self-Help; with Illustrations of Conduct and Perseverance. 1859; 1860 (revised), 1866, 1879, 1891 (Lubbock's Hundred Books), 1905, 1912, 1958 (introduction by Asa Briggs).

Lives of the Engineers, with an Account of Their Principal Works; Comprising also a History of Inland Communication in Britain, 3 vols, 1861–62; 3 vols, 1862; 5 vols, 1874 (revised); 5 vols, 1904 (popular edition).

The first edition included eight parts in the first two volumes (1861). Part I was on early works of embanking and draining from ancient times to the eighteenth century, with a short biography of Cornelius Vermuyden, who undertook to drain the fens; and a memoir on Captain John Perry, who was engineer for Tsar Peter the Great and who repaired the breach in the Thames embankment at Dagenham. The second part is a biography of Sir Hugh Myddelton, London goldsmith, member of Parliament, who brought London fresh water from Hertfordshire in 1613. Part III on roads and modes of traveling from earliest times through the eighteenth century in Britain includes a memoir of John Metcalf, the blind Yorkshire road builder. William Edwards, the Welsh bridge builder, has a short memoir in Part IV on bridges, harbors, and ferries before the British Industrial Revolution. The biography of Brindley was the fifth part and concluded volume one. Part VI in volume two (1861) was the biography of John Smeaton; part VII the life of Rennie; and part VIII the life of Telford. Volume three (1862) was a revised edition of the life of George Stephenson, incorporating a life of his son Robert. The three-volume edition of 1862 was similar. The revised five-volume edition included the biography of Boulton and Watt published in 1865. The new division of the

Lives allocated volume one for early engineering (Vermuyden, Myddelton, Perry, and Brindley); volume two for harbors, lighthouses, and bridges (Smeaton and Rennie); volume three for roads (Metcalf and Telford); volume four for steam engineering (Boulton and Watt); and volume five for railway locomotion (George and Robert Stephenson). The lives of Brindley (and the early engineers), and of Telford were also published subsequently as separate volumes — the abridged Brindley in 1864, and that of Telford in 1867. The five-volume "popular" edition of 1904 was inexpensive.

Industrial Biography: Iron Workers and Tool Makers. 1863, 1879.

Lives of Boulton and Watt, Principally from the original Soho MSS: Comprising also a History of the Invention and Introduction of the Steam Engine. 1865.

This was included in the 1874 edition of the *Lives.*

The Huguenots: Their Settlements, Churches and Industries in England and Ireland. 1867, 1868, 1869, 1876, 1880, 1889, 1905.

Smiles included a study of the effects upon English industry of the Huguenots' emigration from France after the revocation of the Edict of Nantes. There is a section on Huguenot men of science and learning as well as one on men of industry.

Character: a Book of Noble Characteristics. 1871, 1879, 1910.

A Boy's Voyage Round the World: Including a Residence in Victoria, and a Journey by Rail across North America. Edited by Samuel Smiles. 1871.

This is a narrative of the two-year journey of the sixteen-year-old Samuel, Jr. The father edited the letters and notes of his son who made the trip because of inflammation of the lungs.

The Huguenots in France after the Revocation of the Edict of Nantes, with a Visit to the Country of Vaudois. London: Strahan, 1873, 1881, 1893.

Published after Smiles's paralytic stroke in 1871.

Thrift: a Book of Domestic Counsel. **1875.**

Life of a Scottish Naturalist: Thomas Edward. **1876, 1882, 1936** (edited and with wood engravings by Erich F. Daglish, London: J. M. Dent).

George Moore: Merchant and Philanthropist. London: George Routledge, 1878 (2nd edition).

Moore was a benevolent London warehouseman who was a native of Cumberland.

Robert Dick: Baker of Thurso, Geologist and Botanist. **1878.**

Duty; with Illustrations of Courage, Patience, and Endurance. **1880.**

James Nasmyth, Engineer, an Autobiography. Edited by Samuel Smiles. **1883, 1885, 1912.**

Smiles outlined the life of Nasmyth in *Industrial Biography.* Later, after Nasmyth prepared his reminiscences, Smiles agreed to edit the abundant notes which resulted in the *Autobiography.* Nasmyth, famous as the inventor of the steam hammer, developed many machine tools and used them in the engineering shops of his Bridgewater foundry. The *Autobiography* describes Nasmyth's association with many of the leading personalities and important events of the engineering world.

Men of Invention and Industry, **1884.**

Smiles has chapters on men of diverse interests, such as John Lombe, who helped introduce silk manufacture in England; John Harrison, the inventor of a marine chronometer; and Friedrich Koenig, an inventor of the steam printing machine. There is also a chapter on ship building in Belfast.

Life and Labour; or Characteristics of Men of Industry, Culture, and Genius. **1887, 1910.**

There are sections on "men and gentlemen," "great men," "great young men," "great old men," "health and hobbies," "overbrain-work," etc.

Jasmin: Barber, Poet, Philanthropist. 1891.

Longfellow translated the French poet's "The Blind Girl of Castel-Cuillé." Smiles has an appendix with translations of other Jasmin poems.

A Publisher and His Friends: Memoir and Correspondence of the Late John Murray; with an Account of the Origin and Progress of the House, 1768–1843. 2 vols., 1879–1891, 1911 (a one-volume edition edited by Thomas MacKay).

Josiah Wedgwood. 1894.

The Autobiography of Samuel Smiles, L.L.D., edited by Thomas MacKay. 1905.

A manuscript, "Conduct," was destroyed after Smiles's death at the suggestion of his publisher, Murray. Smiles also prepared guides to the colonies and America around 1843, and wrote many articles for periodicals, especially on social reform and personal conduct. He also wrote papers on engineering and engineers for the *Quarterly Review.*

II. SELECTED LIST OF BOOKS ON BRINDLEY, RENNIE, AND TELFORD

Cyril T. G. Boucher, *John Rennie, 1761–1821: the Life and Work of a Great Engineer.* Manchester: University Press, 1963.

Boucher has analyzed Rennie's mechanical work, considered his architecture, and studied his structural theory in writing a biography that deals especially with Rennie's engineering works. Bibliography and a list of Rennie's chief works.

Derbyshire Archaeological and Natural History Society, *James Brindley: Millwright and Civil Engineer.* 1958.

A biographical sketch and extracts from early comments on Brindley.

Sir Alexander Gibb, *The Story of Telford:' the Rise of Civil Engineering.* London: Alexander Maclehose, 1935.

Gibb, a distinguished engineer, characterizes the Smiles biography as elevating and highly moral, but inaccurate and incomplete. He notes several of the inaccuracies. Gibb gives a list of Telford's works and a full bibliography, including a list of reports by Telford. Gibb's definitive biography should prove especially helpful to students of engineering history; he does not attempt to provide understanding of economic and social change related to engineering.

Lionel T. C. Rolt, *Thomas Telford*. London: Longmans, Green, & Co., 1958.

L. T. C. Rolt has written extensively of Britain's great engineers and engineering. He had access to Telford materials used by Smiles, Sir Alexander Gibb, and some more recently discovered. Bibliography.

Telford and Brindley: The Story of Their Lives and Engineering Triumphs. London: W. & R. Chambers, 1895.

The name of the author of the two short biographies is not given. The style and emphasis are not unlike those of Smiles, but there is some new material from printed sources.

Thomas Telford, *Life of Thomas Telford, Civil Engineer: Written by Himself, Containing a Descriptive Narrative of His Professional Labours*. . . . Edited by John Rickman. London: Payne and Foss, 1838.

Concerned primarily with the works, there is little on the man and his times. Smiles had access to the *Life*. The editor, John Rickman, was a friend of Telford's and Secretary to the Commission for Highland Roads and Bridges and the Commission for the Caledonian Canal, for which Telford was Engineer. A handsome folio *Atlas* containing illustrations of Telford's works accompanies the text volume.

LIST OF ILLUSTRATIONS

INDEX